ICSU SHORT REPORTS

ICSU PRESS

VOLUME 7

ADVANCES IN GENE TECHNOLOGY: THE MOLECULAR BIOLOGY OF DEVELOPMENT

Proceedings of the Nineteenth
Miami Winter Symposium
Miami, Florida, U.S.A.
February 9–13, 1987

Edited by Richard W. Voellmy, Fazal Ahmad, Sandra Black,

David R. Burgess, Richard Rotundo, Walter A. Scott and William J. Whelan

The right of the
University of Cambridge
to print and sell
all manner of books
was granted by
Henry VIII in 1534
The University has printed
and published continuously
since 1584.

CAMBRIDGE UNIVERSITY PRESS
Cambridge
London   New York   New Rochelle
Melbourne   Sydney

Published by the Press Syndicate of the University of Cambridge, The Pitt
Building, Trumpington Street, Cambridge CB2 1RP, UK; 32 East 57th Street, New
York, NY 10022, USA; 10 Stamford Road, Oakleigh, Melbourne 3166, Australia

First published 1987

Printed in the United States of America

Library of Congress catalogue card number: 86-31713

Library of Congress Cataloguing in Publication Data

Miami Winter Symposium (19th : 1987)
  Advances in gene technology.
  (ICSU short reports : v. 7)
  Includes index.
1. Developmental genetics - Congresses.
2. Genetic engineering - Congresses. 3. Molecular biology - Congresses.
I. Voellmy, Richard W. II. Title. III. Series.
QH453.M53 1987   591.3
ISBN 0 521 34274 0

This is ICSU Short Reports, Volume 7. All have been published for the ICSU
Press by Cambridge University Press. Other volumes in the series are:

ICSU Short Reports, Volume 1
Advances in Gene Technology: Human Genetic Disorders
ISBN 0 521 26749 8

ICSU Short Reports, Volume 2
Advances in Gene Technology: Molecular Biology of the Immune System
ISBN 0 521 30486 5

ICSU Short Reports, Volume 3
Third European Bioenergetics Conference
ISBN 0 521 30813 5

ICSU Short Reports, Volume 4
Advances in Gene Technology: Molecular Biology of the Endocrine System
ISBN 0 521 32658 3

ICSU Short Reports, Volume 5
Fourth European Bioenergetics Conference
ISBN 0 521 33465 9
To be published in 1987

ICSU Short Reports, Volume 6
Fourth Congress of the Federation of Asian and Oceanian Biochemists
ISBN 0 521 33269 9

CONTENTS

CONTROL OF GENE EXPRESSION

EARLY DETERMINATION

DIFFERENTIATION

CELL SURFACE AND TISSUE INTERACTIONS

MORPHOGENESIS

GENE THERAPY

PREFACE

This is the record of the 19th Miami Winter Symposium, the joint venture of the Department of Biochemistry of the University of Miami School of Medicine and the Papanicolaou Cancer Research Center. The Symposium was held at the Hyatt Regency Hotel/University of Miami/James L. Knight International Center, during February 9-13, 1987.

This annual event is dedicated to exposing advances in gene technology, highlighting each year a fast-growing area of the new biology and honoring the leaders in the field by the Lynen Medal Lecture, given this year by Donald D. Brown (Carnegie Institute of Washington) and the Distinguished Service Award, presented to John C. Kendrew (St. John's College, Oxford) and Max F. Perutz (Medical Research Council, Cambridge, U.K.). We gratefully acknowledge the financial support for these awards, provided respectively by GIBCO/BRL Research Products Division of Life Technologies and the Biomedical Division of Elsevier Science Publishers.

The proceedings are published in the ICSU Short Reports format and are arranged in the order in which they were presented in the Symposium. In each section the 2-page Reports from invited speakers are followed by the relevant 1-page Reports that correspond to the 92 poster presentations.

Each presentation is intended to be a complete and original communication that may be cited in the literature as forming part of ICSU Short Reports, Volume 7, 1987.

The ICSU Short Reports represent a new method of reporting Symposia proceedings. This volume was part of the registration package for Symposium attendees, in a paperback version, and simultaneously placed on open sale in a hardback version by Cambridge University Press on behalf of the ICSU Press. We are grateful to GIBCO/BRL for meeting the cost of publishing the paperback edition, as noted on page xvii.

This reporting format began in 1984 with the 16th Miami Winter Symposium. It was intended to address the problem associated with the widespread use of posters to present short communications at scientific meetings. These posters are extremely informative. They are put together with a great deal of pains, but the poster is displayed for only a few hours, and, in a large meeting, is often not seen by many of the Symposium participants. All that remains for posterity is the abstract that was prepared many months before the meeting, which was already out-of-date at the time of the meeting, and which was contained in a book of abstracts which does not find its way into libraries and therefore the information cannot be retrieved.

The new format used here is an attempt to allow the presenter to commit the poster to paper. The page can be used to convey a great deal of information by way of introduction, methodology, results, conclusions, bibliography, diagrams, figures and tables. Produced to a deadline 3 months before the meeting, the Short Report acts as an up-to-date, permanent record of the poster. The simultaneous availability of the volume of Short Reports to libraries and persons who could not attend the meeting places the report in the body of retrievable literature.

Along with these advantages is the inclusion of two-page reports from the invited speakers, produced to the same deadline and available at the time of the Symposium, not, as is customary, many months later. What is presented here is a succinct but complete and documented statement. Via the bibliography, the reader is directed to explore the subject in more depth by being referred to the original seminal papers on which the speaker's contribution is based.

We have been much encouraged by the reception accorded to the Short Reports, of which this is the seventh volume. The series to date is listed elsewhere in the frontispiece. We regard this as the beginning of a new form of rapid communication which is certain to attract its imitators, a trend that we encourage. We are already modifying our format, based on experience, and have moved to a larger page size, halving the total number of pages, but still providing the same amount of space for each paper.

The ICSU Press, as a new publishing house, which began publishing in 1984, merits mention, and more details are to be found on page xv. It is the publishing house of ICSU, the International Council of Scientific Unions, an international non-governmental, scientific organization made up of 20 international scientific unions together with more than 80 national members, scientific and national associates. More familiar than ICSU itself are the individual members such as the International Union of Biochemistry, a union member, or the Royal Society of London, a national member. ICSU is the "United Nations" umbrella organization that allows these different unions and national members to come together to confront problems of common interest, and take advantage of opportunities for collaboration that are only possible on an international, interdisciplinary scale. Sir John Kendrew, this year's Distinguished Service Awardee, is the current ICSU President.

ICSU's headquarters are located at 51 Bd. de Montmorency, 75016 Paris, France.

THE ICSU PRESS

The ICSU Press, the publishing house of the International Council of Scientific Unions, was founded in 1983. The aim was to establish a publishing house to execute publication programs that will render service to the ICSU family and the cause of science. These programs could be sole initiatives of the ICSU Press or be in partnership with the Unions and/or Committees.

It was enjoined that all ICSU Press publications have the highest standard of content and reporting, should be consistent with the mission of ICSU, and bring an increased awareness of ICSU to scientists and those bodies that support and fund science.

The ICSU Press is intended to be of service to the ICSU Unions and Committees in respect of their publications, by way of advice on all technical and administrative aspects, and by way of negotiating contracts in which the Press may or may not be a partner with the ICSU family member. On request, the ICSU Press will act as the publisher of a Union or Committee publication.

The above objectives are achieved by the following means:

o   Undertaking publication projects on behalf of any ICSU family member or group of members.

o   Initiating interdisciplinary projects that are beyond the scope of any one ICSU member.

o   Acting as a source of advice to ICSU members engaged in publication activities, or acting as an expert negotiator with publishers on behalf of the ICSU member.

o   Developing recommendations concerning contracts for the publication of journals and books, and publication activities in general, so that these may be used as guides by ICSU family members.

o   By means of the ICSU Newsletter and brochures, advising the ICSU family of ICSU Press activities.

o   Producing service publications on behalf of ICSU and its members.

o   Studying new communications technologies and advising ICSU members thereon.

Publication began in 1984. Already 21 symposia and monographs have been published and two journals have been launched, BioEssays (1984-) and the Journal of Tropical Ecology (1985-). Another 30 works and journals are in production or planning, including our first semipopular volume.

Further information about the ICSU Press can be obtained from the Chairman, The ICSU Press, P.O. Box 016129, Miami, Florida 33101, USA or from the ICSU Secretariat, 51 Bd de Montmorency, 75016 Paris, France.

ACKNOWLEDGEMENT

The organizers of the Miami Winter Symposia are pleased to acknowledge the generosity of GIBCO/BRL, Life Technologies, Inc., in underwriting the cost of printing the edition of this volume that was given to the Symposium participants. The fact that a simultaneous print run was made of the Symposium edition and the library edition, lowered the unit cost of each version of the Symposium.

GIBCO/BRL

Life Technologies, Inc.

Gaithersburg, MD 20877

CONTROL OF GENE EXPRESSION

DIFFERENTIAL GENOMIC EXPRESSION IN THE SEA URCHIN EMBRYO

Eric H. Davidson, Barbara R. Hough-Evans, R. Andrew Cameron, Constantin N. Flytzanis*, Frank J. Calzone, and Roberta R. Franks

Division of Biology, California Institute of Technology, Pasadena, California 91125

One experimental approach to the basic problem of gene control in early development is to focus on the mechanism by which specific genes are activated in given regions of the early embryo. Presumably such genes encounter diverse sets of sequence-specific *trans*-activators as the initially equivalent nuclei of cleavage stage embryo are distributed. In some cases such regulatory information may be of maternal origin and disposition, and in others the presentation of spatially localized *trans*-activators may depend rather on inductive interactions between blastomeres. The sea urchin embryo offers special advantages for analysis of early development from this point of view. In recent years we have isolated and analyzed a number of genes that function in specific regions and cell lineages of this embryo. These include six actin genes, five of which are expressed in a lineage-specific manner in the embryo; the muscle myosin heavy chain gene; a gene coding for a major matrix protein of the embryonic spicule; the hyalin gene; and additional lineage-specific genes have been studied in other laboratories, such as the Spec gene family described by W. Klein and coworkers. The early cell lineage of the sea urchin embryo has been deduced from literature sources, and in our laboratory we have checked the fate assignments by microinjection of fluoresceinated dextrans into every blastomere of the 8-cell stage embryo. Our most successful recent molecular studies concern a cytoskeletal actin gene called CyIIIa, which is expressed exclusively in aboral ectoderm cells. Most of the aboral ectoderm descends from a single blastomere of the 8-cell stage. We have demonstrated that when injected into the egg a construct consisting of upstream CyIIIa sequences ligated to the bacterial CAT (chloramphenicol acetyl transferase) gene is activated ontogenically at the same time as is the endogenous CyIIIa actin gene, i.e., at 10-15 hours post-fertilization. *In situ* hybridizations display CAT mRNA in embryos developing from injected eggs only in aboral ectoderm cells. The key issue is now the mechanism causing the lineage specific expression of genes controlled by the CyIIIa regulatory system. We have obtained three forms of evidence indicating that this mechanism is in fact sequence-specific interaction between *cis* CyIIIa regulatory elements and *trans*-activators present in very limited availability. (1) Deletion experiments demonstrate that certain upstream regions of the CyIIIa sequence are required in order for the ontogenic regulation of the CyIIIa•CAT construct to occur. (2) *In vivo* competition experiments demonstrate that excess copies of CyIIIa regulatory sequences compete for some factor(s), so that the number of CAT enzyme molecules synthesized per CAT DNA molecule in injected embryos is inversely proportional to the number of CyIIIa DNA sequences. (3) Exonuclease III and DNase I "footprints" demonstrate that blastula-stage nuclei contain proteins that bind in a sequence-specific manner to at least four specific regions of the upstream CyIIIa sequence. The temporally and spatially regulated activation of this lineage-specific gene is thus probably accounted for by localized *cis-trans* interactions in the aboral ectoderm cell precursors, and the problem now becomes to discover the means by which these regulators are presented in the right cells at the right time.

*Present address: Department of Cell Biology, Baylor College of Medicine, Houston, TX 77030.

McMahon, A. P., Flytzanis, C. N., Hough-Evans, B. R., Katula, K. S., Britten, R. J., and Davidson, E. H. Introduction of cloned DNA into sea urchin egg cytoplasm: Replication and persistence during embryogenesis. *Dev. Biol.* **108**, 420-430, 1985.

Flytzanis, C. N., McMahon, A. P., Hough-Evans, B. R., Katula, K. S., Britten, R. J., and Davidson, E. H. Persistence and integration of cloned DNA in postembryonic sea urchins. *Dev. Biol.* **108**, 431-442, 1985.

Calzone, F. J., Jacobs, H. T., Flytzanis, C. N., Posakony, J. W., and Davidson, E. H. Interspersed maternal RNA of sea urchin and amphibian eggs. In: *Biology of Fertilization*, Vol. 3, The Fertilization Response of the Egg, C. B. Mertz and A. Monroy (eds.), Academic Press, Orlando, pp. 347-366, 1985.

Katula, K. S., Hough-Evans, B. R., McMahon, A. P., Flytzanis, C. N., Franks, R. R., Britten, R. J., and Davidson, E. H. A sea urchin gene transfer system. In: *Genetic Manipulation of the Early Mammalian Embryo, Banbury Report 20*, F. Costantini and R. Jaenisch (eds.), Cold Spring Harbor Laboratory, pp. 231-241, 1985.

Britten, R. J. and Davidson, E. H. Hybridization strategy. In: *Nucleic Acid Hybridisation - A Practical Approach*, B. D. Hames and S. J. Higgins (eds.), IRL Press, Oxford, pp. 3-15, 1985.

Roberts, J. W., Johnson, S. A., Kier, P., Hall, T. J., Davidson, E. H., and Britten, R. J. Evolutionary conservation of DNA sequences expressed in sea urchin eggs and early embryos. *J. Mol. Evol.* **22**, 99-107, 1985.

Davidson, E. H., Flytzanis, C. N., Lee, J. J., Robinson, J. J., Rose, S. J., III, and Sucov, H. M. Lineage-specific gene expression in the sea urchin embryo. *Cold Spring Harbor Symp. Quant. Biol.* **50**, 321-328, 1985.

Lee, J. J., Calzone, F. J., Britten, R. J., Angerer, R. C., and Davidson, E. H. Activation of sea urchin actin genes during embryogenesis. Measurement of transcript accumulation from five different genes in *Strongylocentrotus purpuratus*. *J. Mol. Biol.* **188**, 173-183, 1986.

Cox, K. H., Angerer, L. M., Lee, J. J., Britten, R. J., Davidson, E. H., and Angerer, R. C. Cell lineage-specific programs of expression of multiple actin genes during sea urchin embryogenesis. *J. Mol. Biol.* **188**, 159-172, 1986.

Davidson, E. H. Genomic function in sea urchin embryos: Fundamental insights of Th. Boveri reflected in recent molecular discoveries. In: *History of Embryology*, British Society for Developmental Biology Symposium, T. E. Horder, J. A. Witkowsky, and C. C. Wylie (eds.), Cambridge University Press, Cambridge, pp. 397-406, 1986.

Hwu, H. R., Roberts, J. W., Davidson, E. H., and Britten, R. J. Insertion and/or deletion of many repeated DNA sequences in human and higher ape evolution. *Proc. Natl. Acad. Sci. USA* **83**, 3875-3879, 1986.

Flytzanis, C. N., Britten, R. J., and Davidson, E. H. Expression of a cytoskeletal actin fusion gene following transfer into sea urchin embryos. In: *Gametogenesis and the Early Embryo*, J. G. Gall (ed.), A. R. Liss, New York, pp. 271-281, 1986.

Davidson, E. H. "Gene Activity in Early Development," third edition, Academic Press, Orlando, Florida, 1986.

INTERPLAY OF MULTIPLE SEQUENCE-SPECIFIC TRANSCRIPTION FACTORS WITHIN THE SV40 PROMOTER AND ENHANCER ELEMENTS.

R. Tjian, M. Briggs, P. Mitchell, W. Lee, J. Kadonaga, and K. Jones

Department of Biochemistry, University of California, Berkeley CA 94720.

Proteins that interact with specific DNA recognition elements are thought to contribute to the temporal and spatial restriction of important cellular processes such as replication and transcription. Studies of transcription have clearly established that eukaryotic protein coding genes contain a complex array of cis-regulatory elements that mediate induced, repressed, or basal transcription rates. (4,9,10,16). Although it is generally accepted that multiple classes of interactive DNA-binding proteins regulate gene expression in higher organisms, relatively few of the protein species that recognize RNA polymerase II promoter elements have been purified to homogeneity and analyzed biochemically. Nevertheless, the characterization of protein-DNA interactions at eukaryotic promoters provides one of the few means through which the mechanisms of transcription initiation can be investigated. Here, we summarize some of our recent efforts to dissect the protein-DNA transactions that take place within various viral and cellular promoters.

The SV40 control region contains multiple distinct transcriptional regulatory elements (5,6,7,14,15), including T antigen binding sites that autoregulate early mRNA synthesis, tandem GC box motifs that activate transcription, and the 72 bp repeats that function as transcriptional enhancers (Fig. 1).

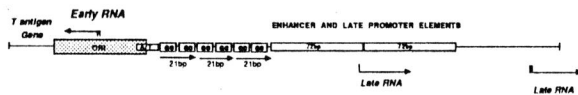

binding activity of polypeptides isolated from an SDS polyacrylamide gel (1). Purified Sp1 has subsequently been used to investigate the protein-DNA interactions at its tandem recognition sequences within the SV40 early promoter. In addition, we have characterized the transcriptional properties of Sp1 in reconstituted in vitro reactions.

Recent DNAse footprint protection analysis has revealed several additional cellular factors: activator protein 1, 2 , 3 and 4 (AP-1, AP-2, AP-3, AP-4) that bind to sequences in the 72 bp enhancers of SV40. In addition, AP-1 and AP-2 also bind to sequences in the basal level enhancer (BLE) of the human metallothionein II_A promoter (Fig. 2) and stimulate transcription of this cellular gene in vitro.

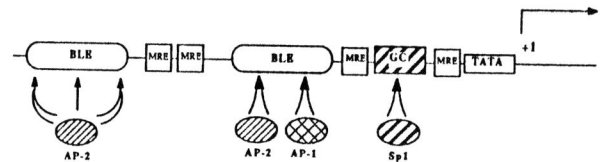

Fig. 2
We have developed DNA affinity columns to purify Sp1, AP-1, and AP-2 to > 95% homogeneity. The identity of Sp1 and AP-1 polypeptides were confirmed by excising the candidate polypeptides from a preparative SDS gel and renaturing sequence-specific DNA binding activity. In addition, monoclonal antibodies to the 105/95 species of Sp1 have been obtained by in vitro immunization and shown to immunoprecipitate a complex of Sp1 bound to the SV40 21 bp repeats. The relationships of the different transcription factors and DNA binding proteins are shown in Fig. 3.

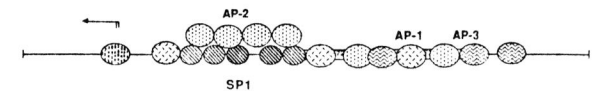

Fig. 1
Here, we report the identification and purification of multiple trans-acting factors that recognize and bind selectively to these different control sequences.

We had previously identified a cellular transcription factor, Sp1, that binds specifically to GC box elements in SV40 as well as other viral and cellular transcription units (2,3,6,7). In each case, binding of Sp1 to its recognition sites is correlated with activation of transcription in an in vitro reconstituted reaction. Although the DNA binding properties of auxiliary transcription factors such as Sp1 provide an important clue to their mode of action, the molecular mechanisms of transcriptional activation by Sp1 and other related proteins remain unknown. The initial studies that allowed detection of Sp1 and other cellular transcription factors had been carried out with either crude or partially purified preparations. However, continued progress and the success of future studies will depend largely on characterization of the biochemical properties of transcription factors. A major experimental barrier has been the purification and identification of transcription proteins because they are generally present in extremely low levels in the cell (1,11). Our primary objective was to identify the polypeptide(s) responsible for Sp1 DNA binding and transcriptional activation and thus define the polypeptide species comprising Sp1. To purify Sp1, we have developed a rapid and efficient sequence-specific DNA affinity chromatography procedure (11). We have identified Sp1 proteins by renaturing the DNA

Fig. 3
We have also used DNA affinity columns to purify the CAAT transcription factor, CTF, that binds selectively to the sequence, GCCAAT, found in a number of viral and cellular promoters, including HSV-TK, α-globin, β-globin, human heat shock and H-Ras. (8,9,12,13,16) In the case of the TK promoter, in vitro activation of transcription requires the combined action of CTF and Sp1. Binding sites for CTF in the human Ha-ras and alpha-globin promoters were highly homologous to sequences recognized by nuclear factor I (NF-I), a cellular DNA binding protein that potentiates the initiation of adenovirus DNA replication. To determine the relationship between CTF and NF-I, we compared the biochemical properties of these two independently-purified proteins. A comparison of DNA binding specificities, transcriptional activation properties, DNA replication activities, and polypeptide composition of purified CTF reveals that it is indistinguishable from nuclear factor I (NF1), a previously identified cellular protein involved in adenovirus DNA replication. This surprising finding suggests that the same set of cellular proteins are responsible for mediating both transcriptional activation and DNA replication.

REFERENCES

(1) Briggs, M.R., Kadonaga, J.T., Bell, S.P. and Tjian, R. (1986) Science 234, 47-52.

(2) Dynan, W.S. and Tjian, R. (1983a) Cell 32, 669-680.
(3) Dynan, W.S. and Tjian, R. (1983b) Cell 35, 79-87.
(4) Dynan, W.S. and Tjian, R. (1985) Nature 316, 774-778.
(5) Fromm, M. and Berg, P. (1982) J. Mol. Appl. Gen. 1, 457-481.
(6) Gidoni, D., Dynan, W.S. and Tjian, R. (1984) Nature 312, 409-413.
(7) Gidoni, D., Kadonaga, J.T., Barrera-Saldana, H., Takahashi, K., Chambon, P. and Tjian, R. (1985) Science 230, 511-517.
(8) Graves, B.J., Johnson, P.F. and McKnight, S.L. (1986) Cell 44, 565-576.
(9) Jones, K.A., Yamamoto, K.R. and Tjian, R. (1985) Cell 42, 559-572.
(10) Kadonaga, J.T., Jones, K.A. and Tjian, R. (1986) Trends in Biochem. 11, 20-23.
(11) Kadonaga, J.T. and Tjian, R. (1986) Proc. Nat. Acad. Sci. USA 83, 5889-5893.
(12) McKnight, S.L. (1982) Cell 31, 355-365.
(13) McKnight, S.L. and Kingsbury, R.C. (1982) Science 217, 316-324.
(14) Moreau, P., Hen, R., Wasylyk, B., Everett, R., Gaub, M.P. and Chambon, P. (1981) Nucl. Acids Res. 9, 6047-6067.
(15) Tjian, R. (1981) Cell 26, 1-2.
(16) McKnight, S. and Tjian, R. Cell 46, 795-805.

CONTROL OF GENE EXPRESSION IN STEM CELLS AND TERMINALLY DIFFERENTIATED CELLS

P. Gruss

Department of Molecular Cell Biology, Max Planck Institute of Biophysical Chemistry, 3400 Göttingen, Federal Republic of Germany

Several crucial developmental control processes in Drosophila are regulated through DNA-binding proteins, such as products deriving from homeotic genes or possibly members of the "gap" class of segmentation genes, e.g., Krüppel. Lacking the genetic tools available for Drosophila, one approach in the mammalian genome involves the isolation of related genes by identifying homologous sequences. Using the Drosophila homeo box as a probe we and others characterized a family of murine homeo-box containing genes located on chromosomes 6, 11, and 15 (see Fig. 1). We examined the developmental expression of several of these genes both in vitro in embryonal carcinoma cells and in vivo during mouse embryogenesis. In F9 cells Hox 1.1 is temporally regulated. Although no stable RNA can be detected in F9 stem cells, differentiation into parietal endoderm leads to transient expression of the Hox 1.1 gene. This expression must be controlled posttranscriptionally because no differences could be found using the nuclear run-on assay. The transient expression of Hox 1.1 could be confirmed by using antibodies prepared against the respective protein. This protein consists of 229 amino acids and migrates with an MW of 31 kD in SDS gels. Interestingly, the protein has been found to be located in the nucleus and was associated with chromatin. Presently, we are examining DNA sequences to which the Hox 1.1 protein can specifically bind and which might be involved in regulatory functions. Since homeo-box-containing genes are involved in the control of Drosophila development we were also interested in their expression during mouse development. Using in situ analyses we find their expression to be spatially restricted. Interestingly, this spatial restriction depends on the gene analyzed. Thus, specific mammalian homeo box transcripts are limited to discrete domains, which is highly reminiscent of Drosophila development. Using cDNA cloning techniques, we precisely defined the structure of Hox 1.1. With the help of cDNA expression vectors, we were able to express this cDNA in embryonal carcinoma stem cells. Polyclonal and monoclonal antibodies were used to monitor the protein. Taken together, our data are compatible with a control function of these genes during mammalian development.

In analogy to the experiments using the homeo box as a conserved DNA binding domain, by using Krüppel as a probe a second family defined by their potential "finger" structure was identified in mammalian mouse genomes. The sequence analysis of two independent isolates reveals the presence of seven possible "finger" structures harboring a putative metal-binding domain. Both genes were used as probes in order to monitor their expression during the differentiation of F9 cells. In contrast to homeo-box-containing genes such as Hox 1.1 we observe two "finger" containing genes to be active in F9 stem cells. Differentiation into parietal endoderm results in a switching off of both genes.

We are presently continuing our studies in order to define the role of these genes during mouse development.

Hox-1 (mouse chromosome 6)

Hox-2 (mouse chromosome 11)

Hox-3 (mouse chromosome 15)

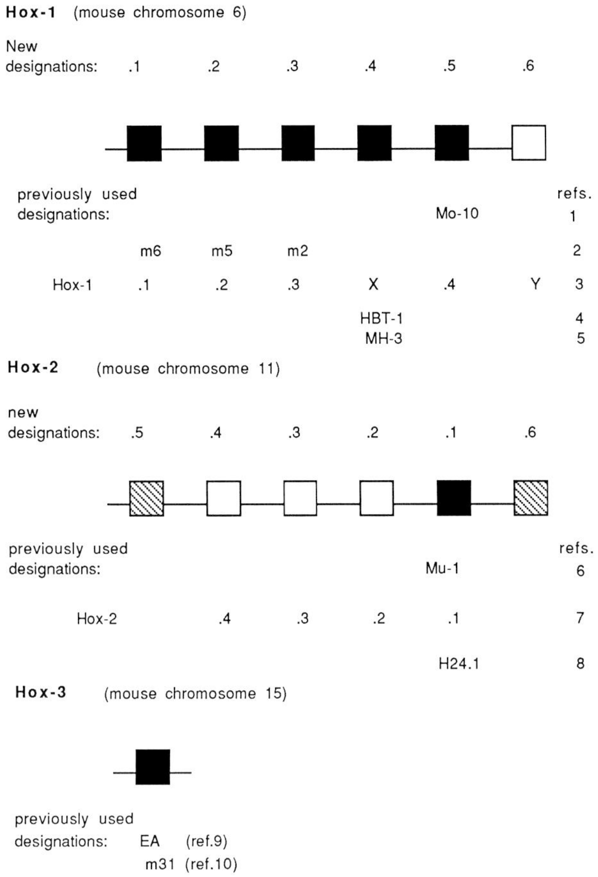

Fig. 1. The filled boxes represent homeo box sequences that have been published. The open boxes represent homeo box sequences that have been determined, but have not yet been published. The cross-hatched boxes represent homeo boxes that have not yet been sequenced; their existence is inferred from Southern blot hybridization analysis of cloned mouse genomic DNA fragments.

References

1. McGinnis, W., Hart, C.P., Gehring, W.J. and Ruddle, F.H. (1984) Cell 38, 675-680.
2. Colberg-Poley, A.M., Voss, S.D., Chowdhury, K., Stewart, C.L., Wagner, E.F. and Gruss, P. (1985). Cell 43, 39-45.
3. Duboule, D., Baron, A., Mahl, P. and Galliot, B. (1986) EMBO J. 5, 1973-1980; and unpublished data.
4. Wohlgemuth, D.J., Engelmyer, E., Duggal, R.N., Gizang-Ginsberg, E., Mutter, G.L., Ponzetto, C., Viviano, C. and Zakeri, Z.F. (1986). EMBO J. 5, 1229-1235.
5. Rubin, M.R., Toth, L.E., Patel, M.D., D'Eustachio, P. and Ngyuyen-Huu, M.C. (1986). Science 233, 663-667.
6. Hauser, C.A., Joyner, A.L., Klein, R.D., Learned, T.K., Martin, G.R. and Tjian, R. (1985). Cell 43, 19-28.
7. Hart, C.P., Awgulewitsch, A., Fainsod, A., McGinnis, W. and Ruddle, F.H. (1985). Cell 43, 9-18; and unpublished data.
8. Jackson, I.J., Schofield, P. and Hogan, J.B. (1985) Nature 317, 745-748.
9. Awgulewitsch, A., Utset, M.F., Hart, C.P., McGinnis, W. and Ruddle, F.H. (1986). Nature 320, 328-335.
10. Breier, G., Bucan, M., Francke, U., Colberg-Poley, A.M., and Gruss, P. (1986). EMBO J., in press.

REGULATION OF DROSOPHILA CHORION GENE AMPLIFICATION AND EXPRESSION

L. Cooley, R. Kelley, G. Leys, S. Parks, T. Orr-Weaver, and A. Spradling

Department of Embryology, The Carnegie Institute of Washington, 115 W. University Pkwy., Baltimore, MD 21210

During stages 10-14 of Drosophila oogenesis, the polyploid ovarian follicle cells surrounding the developing oocyte synthesize and secrete the multi-layered eggshell (chorion). Eggshell deposition entails the precise expression of ten genes whose structure and regulation have been studied in detail (1,2). In the germ line genome, single copies of each gene reside within one of two tandem, head-to-tail clusters located on the X and 3rd chromosomes. Prior to the onset of transcription, chorion genes undergo amplification in follicle cells. Multiple rounds of bidirectional replication initiate from a site(s) within each cluster, producing a 16- or 60-fold increase in the relative copy number of these chromosome regions. Subsequently, the transcription of each chorion gene is intricately controlled. Expression is tissue-specific since chorion transcripts are found only in follicle cells. In addition, in situ hybridization experiments (S. Parks, unpublished) revealed that each mRNA accumulates in a specific subset of follicle cells that changes rapidly with time. The spatial and temporal patterns of chorion gene expression have been related to the synthesis of specific chorion layers and of regional eggshell specializations such as the large anterior dorsal chorionic filaments. We have investigated mechanisms regulating chorion gene amplification and expression using two general approaches. First, sequences required in cis for normal regulation were defined by in vitro mutagenesis and transformation experiments. Second, an insertional mutagenesis strategy was developed to help indentify, clone and characterize genes whose products are required for eggshell production.

A small region within each cluster is essential for amplification

When reintegrated on P element vectors, specific genomic regions from within the chorion gene clusters induce developmentally-regulated amplification at their site of chromosomal insertion. This assay allowed a single "amplification control element" of 4-5 kb to be identified on both the X and 3rd chromosomes. Since these control elements function autonomously, they presumably contain replication origins used during amplification and sequences responsible for the tissue-specific and time-specific activation of disproportionate replication at the origins. To localize functionally important regions within the control elements, transposons containing small internal deletions were tested for their ability to amplify following germ line transformation. On the 3rd chromosome, a region essential for amplification was localized between positions -310 and -630 upstream from the s18 gene (3; T. Orr-Weaver, unpublished). No other portion of the 3.8 kb control element was required for amplification, although functionally redundant sequences would not have been detected. Similar experiments using sequences within the X chromosome gene cluster mapped an element essential for amplification between positions -9 and -488 upstream from the s38 gene.

Complex transcription control regions surround chorion gene initiation sites

Transformation experiments revealed that chorion genes are expressed with normal tissue- and stage-specificity when introduced into the genome surrounded by only about 1 kb of flanking DNA (1-4). We have studied the control region upstream from the s38 gene in greatest detail (B. Wakimoto et al., in preparation). Deleting the 5' flanking sequences between -2.5 kb and -488 had no detectable effect on s38 transcription. However removing 12 bp of additional DNA eliminated amplification, indicating the 5' boundary of the essential element lies between -488 and -476. In contrast, deletion up to position -476 did not alter s38 expression. A further series of 5' deletions identified several regulatory elements. A quantitative regulator lies between positions -476 and -266, since deletions to -266 reduced the level of s38 expression an average of about 5-fold but without altering developmental regulation. Deletion to -136 changed the normal temporal profile of s38 mRNA accumulation, resulting in the production of excess mRNA in stage 10, the earliest time it is normally transcribed. Tissue-specificity was not altered. An element required for correct initiation was localized between -106 and -55. A series of 3' deletions beginning within the transcription unit demonstrated that sequences up to position +74 could be eliminated without altering expression from the normal start site (L. Kalfayan et al., in preparation). Removing additional 3' sequences or substituting the corresponding sequences from the Drosophila alcohol dehydrogenase gene eliminated detectable expression.

Do transcriptional regulators control amplification?

Cis-regulatory elements essential for amplification and elements regulating the expression of a particular chorion gene, s18, are closely associated within the 3rd chromosome gene cluster (3). The results described above indicate that a similar close linkage exits within the X cluster between the region essential for amplification and s38 transcription control sequences. Transcription enhancers have been implicated in the regulation of replication from several eukaryotic viral origins, including those of polyoma and bovine papilloma virus. We have therefore studied the effect of linking amplification control sequences with a variety of other strong transcription control regions from Drosophila genes expressed in a wide variety of tissues.

Follicle cells contain factors that bind specifically to the s18 and s38 control regions

Recently G. Leys developed methods for isolating large numbers of nuclei from predominantly post-amplification follicle cells. Crude extracts of nuclear proteins from these preparations protected several sites within the s18 and 38 control regions from DNase I. Strong "footprints" were seen within the region essential for chromosome 3 amplification. The location of specific protected sites provides a guide for the further genetic analysis of these control regions.

Constructing a library of insertion mutations

Several unlinked recessive mutations have been identified that reduce or eliminate amplification (5; R. Kelley, unpublished). Amplification-defective mutations are expected to include genes directly controlling amplification, for example by encoding a protein that binds to the cis-regulatory seqeunces described previously. However a much larger number might represent products used for general functions such as genomic replication. Consistent with the latter interpretation, lethal alleles have been found for at least two amplification-defective mutants.

To facilitate the identification and cloning of genes disrupting amplification or other aspects of choriogenesis, L. Cooley has developed a method to produce large numbers of strains containing single, stable insertions of defective P transposable elements. A genetically marked, defective P element within a transformed strain is mobilized by crossing to a second strain containing a single integrated non-defective P element. The complete element is crossed out in the next generation, stabilizing new insertions of the defective element, which are recognized by the altered segregation of the marker gene. Ideally, the non-defective element should not be transposition-competent (for example due to the presence of defective terminal sequences) to prevent occassional contamination of the new insertion strains. Such a "genomic wings-clipped" element has yet to be sucessfully constructed. However starting with a strain provided by F. Spencer and G. Rubin, a derivative was obtained which transposes at a negligible rate but mobilizes a defective element in up to 30% of Fl males. This strain was used to generate 1500 new insertions on chromosome 2 or 3 of a defective P element encoding a G418 resistance marker. Female-sterile phenotypes associated with some of these insertions are currently under investigation.

REFERENCES
(1) Kalfayan, L., Levine, J., Orr-Weaver, T., Parks, S., Wakimoto, B., deCicco, D., and Spradling, A. (1986). Cold Spring Harbor Symp. Quant. Biol. 50, 527-535.
(2) Kafatos, F., Mitsialis, S., Spoerel, N., Mariani, B., Lingappa, J., and Delidakis, C. (1986). Cold Spring Harbor Symp. Quant. Biol. 50, 537-547.
(3) Orr-Weaver, T. and Spradling, A. (1986). Mol. Cell. Biol., in press (December)
(4) Wakimoto, B., Kalfayan, L., and Spradling, A. (1986). J. Mol. Biol. 187, 33-45.
(5) Snyder, P., Galanopoulos, V. and Kafatos, F. (1986). Proc. Nat. Acad. Sci. 83, 3341-3345.

ONCOGENESIS AND INSERTION MEDIATED MUTAGENESIS IN TRANSGENIC MICE

H. Westphal, K.A. Mahon, and J.S. Khillan
Laboratory of Molecular Genetics, National
Institute of Child Health and Human Development,
National Institutes of Health, Bethesda, MD 20892

P.A. Overbeek
Baylor College of Medicine, Texas Medical Center,
Houston, TX 77030

A.B. Chepelinsky and J. Piatigorsky
Laboratory of Molecular and Developmental
Biology, National Eye Institute, National
Institutes of Health, Bethesda, MD 20892

Transgenic mice allow us to study the expression of specific genes in the intact mammalian organism. The system has enormous potential, not only for molecular and developmental biology, but also for medical research. This report deals with malignancies induced in mice by oncogenes inserted in the germ line and with the mutagenic effect of gene insertion in general. Transgenic mice were generated by manual micro-injection of cloned DNA into pronuclei of one-cell embryos (1). Most of the resulting founder animals carry the integrated DNA at a single chromosomal locus in all tissues and transmit it to their progeny. Previous studies of a number of laboratories have shown that a tissue-specific gene is generally under correct spatial and temporal control in spite of its "out-of-place" chromosomal location (2). In the fused genes used in our study, these controls are conveyed by sequences located 5' to the coding region.

Oncogenesis was studied in mice carrying an alpha A crystallin-SV40 T antigen fused gene. The 5' control region of the mouse alpha A crystallin gene exerts a tight spatial and temporal control over coding regions placed 3' of it (3). Expression is confined to the epithelial and fiber cells of the eye lens and begins during mid-gestation, concomitant with lens development in the embryo. We, therefore, expected SV40 tumor antigen expression in the eye lens to begin at or shortly after lens differentiation. It was not self evident that this should lead to tumor formation in the lens, since tumors have never before been observed in the mammalian eye lens. Considering this fact, we were surprised to observe morphological transformation in stained sections of transgenic lenses as early as day 12 of gestation, coincident with the initial differentiation of lens fiber cells. Sections stained at a slightly later stage were positive for SV40 T antigen. All animals developed vascularized, invasive lens tumors within 2-3 months after birth. Some older animals also contained T antigen producing tumors in non-lens tissues. We currently study patterns of tumor progression in these animals.

Some of us (K.A. Mahon, P.A. Overbeek, J.S. Khillan, and H. Westphal) have noted mutant phenotypes in a number of transgenic mouse strains carrying a variety of gene inserts. These mutations cosegregate with the transgenic trait and are therefore thought to be a direct consequence of gene insertion. We presently investigate two such mutations. One is recessive and is characterized by a syndactyly in the fore and hind paws (4). The other mutation is dominant, results in early embryonic death of part of the litter and is characterized by a chromosome translocation. It remains to be seen whether these mutant traits are directly linked to the integration of foreign DNA in essential mouse genes involved in embryonic development. A molecular approach toward answering this question consists of analyzing the genomic mouse sequences interrupted by foreign gene inserts.

In summary, studies such as the ones presented here offer new and exciting molecular approaches to fundamental questions of mammalian oncogenesis and developmental biology.

REFERENCES

(1) Westphal, H. (1987) BioEssays, February issue.
(2) Palmiter, R.D. and Brinster, R.L. (1986) Ann. Rev. Genetics 20, 465-499.
(3) Overbeek, P.A., Chepelinsky, A., Khillan, J.S., Piatigorsky, J. and Westphal, H. (1985) Proc. Natl. Acad. Sci. USA 82, 7815-7819.
(4) Overbeek, P.A., Lai, S.-P., van Quill, K.P. and Westphal, H. (1986) Science 231, 1574-1577.

REGULATION OF THE HTLV III TRANSCRIPTION IN VITRO

Thomas Benter,Takashi Okamoto,Steven F.Josephs,
Reza M.Sadaie,Volker Heisig,Robert C.Gallo and
Flossie Wong-Staal

Laboratory of Tumor Cell Biology,National Cancer
Institut,National Institutes of Health,Bethesda,MD.

Introduction: The acquired immunodeficiency syndrome (AIDS) causing virus,human T-Lymphotropic virus type III (HTLV III),can infect certain Okt 4+ human T-cell lines in culture (Popovic et al.1984;. reviewed by Wong-Staal and Gallo,1985).A high level of virus expression can be observed after one week. This high gene expression results,at least in part, from transcriptional (Okamoto and Wong-Staal,1986) and post-transcriptional activation (Rosen et al. 1985;Feinberg et al.1986).In both mechanisms the virus encoded trans-activator protein TAT3 (Arya et al.1985) is involved (Okamoto et al.in preparation). The target site of this protein (TAR) is part of the long terminal repeat (LTR) downstream of the RNA initiation site (Rosen et al.1986;Josephs et al. submitted).Other regulatory elements within the LTR have been detected: a negative regulatory element (NRE) (Rosen et al.1985),an enhancer (Rosen et al. 1985;Josephs et al.submitted) and three binding sites for the Sp1 protein (GC box)(Jones et al.1986). In this study various up-stream deletion mutants of the viral LTR have been tested in a in vitro transcription assay.

Results: The mutants containing deletions in the up-stream region of the HTLV III LTR are shown in Figure 1.The plasmids were linearized with a restriction enzyme and processed for run-off transcription assay using a cell-free in vitro system.Since nuclear extracts of HTLV III infected H9 cells contain a trans-acting transcriptional activator protein (Okamoto and Wong-Staal,1986) the experiment was carried out either with or without H9/III nuclear extract. Deletions to position -117 showed an enhancement of transcription up to 1.5 times.A decrease was observed in the deletion to position -65.No substantial transscripional activity was detected in the deletion to

position -48.However,with transcription reactions supplemented with the H9/III extract,the activation was enhanced about ten fold.

Discussion: The cis-acting regulatory elements within the HTLV III LTR have been descripted in an in vitro transcripion system.The presence of the NRE on the transcriptional level has been demonstrated by removal of the 5' part of the U3 to position -117.A dramatic loss of transcriptional activity upon deletion of sequences between -117 to -65 suggested the presence of strong positive-regulatory element.The effect of this deletion was evident irrespective of preincubation with H9/III nuclear extract.However,trans-activation with this extract occured in all 5' deletion mutants.Recently,the HTLV III enhancer has been mapped between position -105 to -80 (Josephs et al.submitted).Three Sp1 binding sites (GC boxes) have been identified between -77 to -45 (Jones et al.1986).Complete loss of transcriptional activity could be attributed to the absence of the first Sp1 binding site (CD 54).However,reinsertion of the enhancer to CD 54 restored the accurate initiation (data not shown) suggesting that the presence of the Sp1 binding sites is not a prerequistite for the accurate initiation of transcription.

References:
1) Arya,S.K.,Guo,C.,Josephs,S.F.,and Wong-Staal,F. (1985).Trans -activator gene of human T-lymphotropic virus type III (HTLV III). Science 229 ,69-73
2) Feinberg,M.B.,Jarrett,R.F.,Aldovini,A.,Gallo, R.C.and Wong-Staal,F.(1986).HTLV III Expression and Production Involve Complex Regulation at the Levels of Splicing and Translation of Viral RNA. Cell 46 ,807-817
3) Jones,K.A.,Kadonaga,J.T.,Luciw,P.A.,and Tjian,R. (1986).Activation of the AIDS retrovirus promotor by the cellular transcription factor Sp1. Science 232 ,755-759
4) Okamoto,T. and Wong-Staal,F.(1986).Demonstration of virus-specific transcriptional activator(s) in cells infected with the Human T-cell Lymphotropic Virus Type III by in Vitro Cell-Free System. Cell,in press
5) Popovic,M.,Sarngadharan,M.G.,Read,E.,and Gallo,R.C.(1984).Detection,Isolation and contiuous production of cytopathic retroviruses (HTLV III) from patients with AIDS and pre AIDS. Science 224 ,497-500
6) Rosen,C.A.,Sodrowski,J.G.,and Haseltine,W.A. (1985).Location of cis-acting regulatory sequences in the human T-lymphotropic virus type III (HTLV III) long terminal repeat. Cell 41 ,813-823
7) Rosen,C.A.,Sodrowski,J.G.,Goh,W.C.,Dayton,A.I., Lippke,J. and Haseltine,W.A.(1986). Post-transcriptional regulation accounts for trans-activation of human T-lymphotropic virus type III. Nature 319 ,555-559
8) Wong-Staal,F. and Gallo,R.C.(1985). Human T-lymphotropic retroviruses Nature 317 ,395-403

This work was partly supported by a grant to T.B. from the Deutsche Forschungsgemeinschaft

Figure 1.

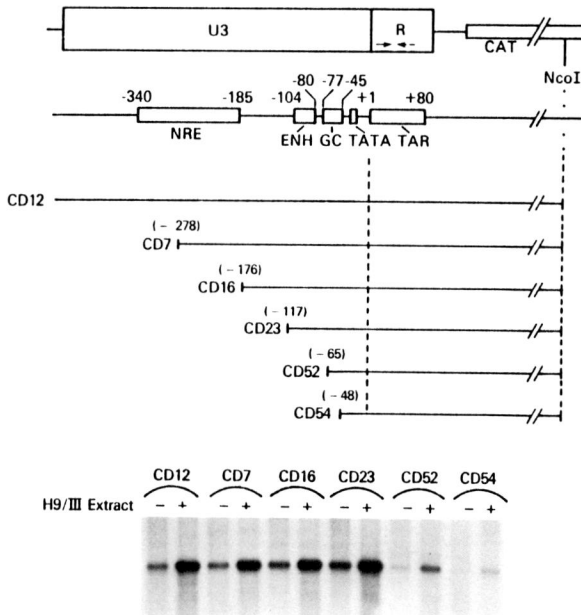

TRANSCRIPTIONAL ACTIVATION AND REPRESSION OF TRANSFECTED GENES STABLY INTEGRATED INTO THE GENOME OF RAT CELLS

L. Bouchard, F. Mathieu, C. Roberge,
J. Vass-Marengo, and M. Bastin

Department of Microbiology, University of
Sherbrooke, Sherbrooke, Que. J1H 5N4, Canada

In previous studies (1, 2) we described the isolation of phenotypically normal rat cell lines carrying transcriptionally inactive copies of the polyomavirus transforming gene, middle T (pmt). In this work we show that introduction of polyoma large T (plt), the second polyoma oncogene, into some of the flat cell lines can activate the resident pmt gene so as to increase the level of middle T antigen and convert the cells from a normal to a transformed state. Our results suggest that the large T antigen can induce conformational changes in the nucleosomal structure, thereby activating pmt. Changes in chromatin structure or configuration are also responsible for an occasional loss of transcriptional activity which inactivates the pmt locus as well as a cotransfected flanking marker.

The 8-2 cell line was selected for these experiments because of its particularly flat phenotype (1). Although 8-2 cells carried multiple copies of pmt in head-to-tail arrangements, they appeared untransformed in culture and did not produce any middle T antigen. The cells were retransfected with plt in the presence of neo and selected for growth in G418. Several of the colonies obtained by transfecting plt with the neo marker contained fully transformed cells before the appearance of foci in the control cultures (Table 1).

Table 1. Activity of recombinant plasmids

Plasmid transfected	Coding capacity		No of transformed cultures / no of cultures	
pSV2neo	neo		0/290	0
pneo-LT1	neo + plt		27/207	13.0
pneo-LT97	neo + plt (mutant)		16/152	10.5
pneo-LTtsa	neo + plt (tsa)	33C	20/196	10.2
		39C	1/182	0.5
pneo-ST3	neo + small T		0/219	0
pSVc-myc-1	neo + c-myc		0/144	0
pMSVp53G	neo + p53		0/179	0

Colonies of G418-resistant cells were picked and transferred into 15-mm Linbro microplates. Transformants were scored only when the change in morphology occurred before confluence.

The effect of large T is indirect. To determine whether the transfected cells expressed the large T protein, we examined several cell lines by immunofluorescence staining. The transformed cells showed positive nuclei but, upon passage in culture, some of the transformants tended to lose their ability to express plt. This suggested that the level of plt expression was not the determining factor in the conversion of 8-2 cells to the transformed state and that continuous expression of plt was not essential for maintaining the phenotype. We observed, however that all of the 8-2 cells that were transformed as a result of plt transfection expressed high levels of middle T antigen.

8-2 cells were also transfected with plt carrying the ts-a mutation. As expected, transfection at 33 C yielded dense foci of transformed cells (Table 1). However, when these cells were transferred to 39 C, the non-permissive temperature for ts-a mutation, they did not revert to a flat phenotype. On the other hand, transfection at 39 C did not produce transformation. Taken together, these results indicate that plt has the ability to transform 8-2 cells, but that continuous expression of the large T protein is not required for the maintenance of the transformed state.

The effect of plt is specific. To determine whether the effect of plt was specific, we analyzed several mutants with different abilities to immortalize primary cells. Table 1 shows that dl-97, which is very efficient in immortalization (3), had the ability to activate pmt expression. Polyoma small T was inactive in the process and so were the myc and p53 oncogenes.

Modulation of pmt transcriptional activity. We observed the spontaneous formation of foci on 8-2 monolayers 2 to 3 weeks after the cultures reached confluence. The cell lines derived from such foci expressed high levels of middle T antigen. Fluctuation analyses indicated that the 8-2 cell line mutated to the transformed state with a rate of 2×10^{-5} per cell per generation. We wished to see whether the transformants could also revert at high rates to a normal phenotype. To this end, FR3T3 cells were transfected with a hybrid plasmid encoding both the neo and pmt genes and selected either for G418 resistance or oncogenic transformation. Several cell lines expressing both neo and pmt were plated at very low densities and the morphology of each and every colony growing in the dish was recorded. Revertants were observed in populations even as small as 100 cells, indicating that the reversion from a transformed to a normal state occurred at a very high rate of the order of 10^{-2} per cell per generation. Surprisingly, all of the flat revertants also lost the ability to express neo, the gene adjacent to pmt. This showed that the changes in phenotype were not due to mutations in the pmt gene but, more likely, they were the results of epigenetic events modulating the level of gene expression in a given region of the chromosome. When we isolated retransformants from the revertants, they all recovered the resistance to G418, indicating that the adjacent gene was affected in the same manner. The revertants did not produce the middle T antigen nor the pmt and neo RNAs.

The effect of large T on 8-2 cells resembles that produced by epigenetic events, or genetic events operating at high rates, that induce the spontaneous conversion of cells carrying pmt to the transformed phenotype. We propose that large T can induce conformational changes in the nucleosomal structure, thereby indirectly activating transcriptionally inactive genes. By binding to specific DNA sequences large T could interact with the conformation of the chromatin and induce the phasing of the nucleosomes. Alternatively, large T could alter the chromatin structure by promoting excision, amplification or intramolecular recombination of integrated viral sequences.

REFERENCES

1. Gélinas, C. & Bastin, M. Virology 146: 233-245, 1985.
2. Bouchard, L., Vass-Marengo, J. & Bastin, M. Virology 154: in press, 1986.
3. Asselin, C., Vass-Marengo, J. & Bastin, M. J. Virol. 57: 165-172, 1986.

14

PROTO-ONCOGENE c-ets OF SEA URCHINS: TRANSCRIPTS AND SEQUENCE HOMOLOGY WITH v-ets AND HUMAN c-ets GENES

Z.Q. Chen, N.C. Kan, L. Pribyl, J. Lautenberger, E. Moudrianakis*, and T.S. Papas

Laboratory of Molecular Oncology, National Cancer Institute, Frederick, MD 21701, USA

*Department of Biology, Johns Hopkins University, Baltimore, MD 21218, USA

The proto-oncogene c-ets in normal vertebrate cells has been identified by its sequence homology with the v-ets gene present in the genome of avian erythroblastosis virus E26 (1). In addition to v-ets, E26 contains a second cell-derived oncogene, v-myb, that also is found in the genome of avian myeloblastosis virus (AMV) (2,3). Since unlike AMV, E26 causes erythroblastosis in chickens, v-ets may be responsible for the erythroblast tropism of E26. While mammalian c-ets sequences are divided into two domains, ets-1 and ets-2, located on separate chromosomes, the ets-1 and ets-2 sequences of chicken are contiguous (1,4). Since highly conserved genes may perform a common function in all animals in which they are found, the analysis of oncogenes in invertebrates may yield clues as to their function. For this reason, we have examined the c-ets gene in the sea urchin, an animal that has been intensely studied by developmental biologists.

Detection and molecular cloning of sea urchin c-ets sequences. Southern blot analysis of restricted Lytechinas variegatus DNA revealed a single strong band of hybridization with v-ets DNA for each of two enzymes (Fig. 1). Since the band seen with HindIII had a molecular weight of about 3.6 kb, a Charon 28 library was constructed from DNA fragments of this size prepared from a

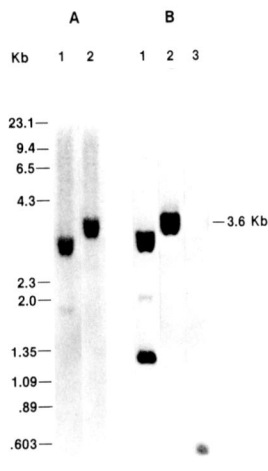

Fig. 1. Hybridization of molecularly cloned E26 v-ets and sea urchin proto-ets sequences to genomic sea urchin DNA. L. variegatus sperm DNA was cleaved with EcoRI (lanes 1), HindIII (lanes 2), HindIII and PstI (lane 3). The digested DNA was separated on agarose gels, transferred to nitrocellulose filters, and probed with ^{32}P-DNA. The hybridization probes were A: a 1.28 kb BglI-BglI fragment containing E26 v-ets sequences; B: a 376 bp HindIII-PstI fragment of sea urchin clone 12E3.

complete HindIII digest. A phage (12E3) containing sequences that hybridized to an E26 v-ets probe was isolated from this library.

DNA sequence analysis. The region of clone 12E3 homologous to v-ets was sequenced by the dideoxynucleotide chain termination method. A region of significant homology to E26 v-ets was found that corresponded to the region that is specific for Hu-ets-2 (Fig. 2). The homology with v-ets starts at a consensus splice acceptor sequence and ends at the point where v-ets and Hu-ets-2 are no longer homologous. A somewhat weaker homology with just Hu-ets-2 continues for 13 more codons ending at a common termination codon. 91 out of 97 (or 94%) specified amino acids are identical in sea urchin c-ets and E26 v-ets in the region of homology.

Analysis of transcription. Polyadenylated RNA was isolated from sea urchin embryos of the species Strongylocentrus purpuratus at different times after fertilization. In each case, a single 6.8 kb c-ets transcript was observed by Northern blot analysis.

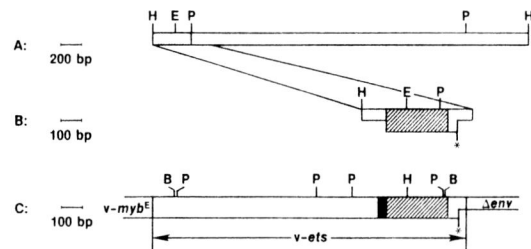

Fig. 2. Maps of SU-ets-2 and E26 v-ets. (A) Restriction map of 3.6 kb HindIII insert in clone 3E12. The 376 bp HindIII-PstI fragment shown to hybridize to E26 v-ets is stippled. (B) Details of sequenced region. The hatched area represents the region with sequence homology with E26 v-ets. This corresponds to the region homologous to Hu-ets-2 but not Hu-ets-1 in ref. 1. (C) Features of E26 v-ets. The boundaries of myb^E ets, and env sequences are from ref. 2. The hatched area represents the segment homologous only to Hu-ets-2 while the solid area represents the segment homologous to both Hu-ets1 and Hu-ets2 from ref. 1. The restriction sites indicated are: B (BglI), E (EcoRI), H (HindIII), and P (PstI).

REFERENCES

(1) Watson, D.K., McWilliams-Smith, M.J., Nunn, M.F., Duesberg, P.H., O'Brien, S.J., and Papas, T.S. (1985) Proc. Natl. Acad. Sci. U.S.A. 82, 7294-7298.

(2) Nunn, M.F., Seeburg, P.H., Moscovici, C., and Duesberg, P.H. (1983) Nature 306, 391-395.

(3) Leprince, D., Gegonne, A., Coll, J., de Taisne, C., Schneeberger, A., Lagrou, C., and Stehelin, D. (1983) Nature 306, 395-397.

(4) Watson, D.K., McWilliams-Smith, M.J., Kozak, C., Reeves, R., Gearhart, J., Nunn, M.F., Nash, W., Fowle III, J.R., Duesberg, P., Papas, T.S., and O'Brien, S.J. (1986) Proc. Natl. Acad. Sci. U.S.A. 83, 1792-1796.

REGULATION OF EXPRESSION OF PYRUVATE KINASE GENE OF NEUROSPORA CRASSA: A DNA-BINDING PROTEIN INTERACTS SPECIFICALLY WITH THE 5' END OF THE GENE.

M. Devchand and M. Kapoor

Cellular, Molecular and Microbial Biology Division, Department of Biology, University of Calgary, Calgary, Alberta, Canada T2N 1N4

A partial genomic library was prepared by inserting ∿ 2-4Kb EcoR1 fragments of N. crassa DNA in the plasmid pUC12. Screening with a [32P]-labelled fragment from the yeast pyruvate kinase (PK) gene (1) enabled the isolation of a positive clone that was characterized by restriction mapping and sequence analysis. The insert consists of a ∿ 1.85Kb fragment containing app. 75% of the coding region and ∿ 0.5Kb of the 5' noncoding sequence (Fig. 1).

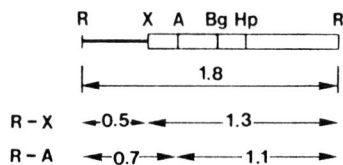

Fig. 1. Restriction map. Boxed area = coding region. R, EcoR1; X, Xba1; A, Acc1; Bg, BglII; Hp, HpaI.

The partial sequence of the 5' noncoding region (Fig. 2) shows the presence of two TATA boxes located at positions -100 and -130, respectively, relative to the translation start site; a CAAG 'cap' site consensus sequence at -26. Other notable features include stretches of T-residues--characteristic of constitutively expressed yeast genes--and CT-blocks, associated with low expression genes. The amino acid sequence, deduced from the nucleotide sequence of the coding region shows a high degree of homology to yeast PK, especially in the domains known to be conserved in PKs of other organisms.

Fig. 2. Partial sequence of PK gene 5' noncoding region.

```
        -140              -120             -100
         .        .        .        .       .
5'....GCCCTTTCTAGATATTAGGCCCCTTCTTAGTACCTCCTCTTCCTATAT-
           -80                    -60
            .        .             .           .
CATCNNTCACCTTTNGTTTTTTATTCCNNTCCTGTTCAAATCGCTCTTGTTTCN-
     -40                  -20
      .        .           .
ATTACCAAGTTCCCCNNCCATTATTTATAAAACCATCATCCTCAATG.......
```

PK GENE EXPRESSION IN VIVO. The expression of the PK gene was monitored by examining mRNA levels in mycelium by hybridization of Northern and dot blots of total cellular RNA with the 1.8Kb EcoR1 N. crassa PK gene fragment (see Fig. 1). Protein levels were followed by immunoprecipitation of crude cell-extracts using polyclonal antibody against purified PK (2). Two PK specific mRNA species were detected in cells of all ages. Carbon source- and age-dependent variation in PK protein and mRNA levels was evident: both increased up to 20 h and declined thereafter. Growth on acetate or sucrose resulted in an elevation in mRNA and protein levels in contrast with that on ethanol- and alanine-medium. The results suggested transcriptional controls.

A PK-SPECIFIC DNA-BINDING PROTEIN (DBP). A search for trans-acting factors, involved in the regulation of transcription, was conducted by using the protein blotting technique (3). Crude extracts of N. crassa cells--fractionated by SDS-PAGE on 7%-15% acrylamide gradient slab gels--were electrotransferred onto nitrocellulose and the blot probed with [32P]-labelled PK gene DNA. A 30Kd protein was found to bind strongly to the PK gene and weakly to plasmid pUC12 and total N. crassa DNA. Probing of Western blots with end-labelled individual restriction fragments demonstrated the specific interaction of the 30Kd DBP with the 0.5 Kb Xba1-EcoR1 and the 0.7 Kb Acc1-EcoR1 fragments (see Fig. 1), thus localizing the binding site in the 5' noncoding region of the gene.

Filter-binding assays (4), based on the retention of DNA:protein complexes on nitrocellulose, were used to study the kinetics of DNA:Protein binding. Interaction of native protein, from cell-extracts of the wild type strain with the recombinant plasmid pPK460 DNA (pUC12+PK DNA insert) and with pUC12 DNA alone, in the presence of increasing salt and calf thymus DNA conc. revealed stable binding to pPK460 but not to pUC12 DNA (Fig. 3A,B). In contrast, using extracts of a mutant (ace-8) lacking PK activity (5) stable binding to the 30 Kd protein could not be detected in filter-binding assays (Fig. 3C) or with Western blotting.

Fig. 3. Effect of NaCl and non-specific DNA on interaction of PK gene with proteins. A and B, end-labelled 1.8kb fragment, o, and pUC12 DNA o; A, 50mM NaCl; B,C, 1 µg CT DNA.

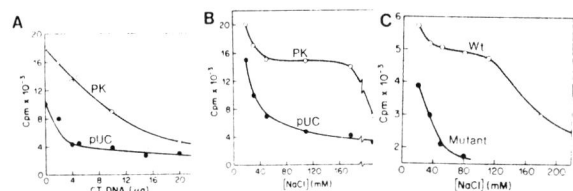

Formation of the DNA:protein complexes was also demonstrated by incubating crude extracts of the N. crassa wild type strain with end-labelled 0.7Kb Acc1-EcoR1 fragment, followed by electrophoresis on low-ionic strength, non-denaturing polyacrylamide gels (6). Judging by the mobility of the DNA:protein complexes--formed in the presence of increasing conc. of protein--as compared to that of free DNA, three complexes were discerned. These results, in conjunction with our Western blotting data suggest the occurrence of three binding sites for the 30 Kd DBP, localized primarily in the 5' noncoding region of the PK gene. The exact positions of the binding sites have not been identified yet. Multiple binding sites for regulatory proteins have been encountered in upstream activator sequences for several yeast genes (7).

REFERENCES

(1) Burke, R.L., Tekamp-Olson, P., and Najarian, K. (1983) J. Biol. Chem. 258, 2193-2201.

(2) Kapoor, M., and Bishop, M. (1982) Can. J. Biochem. 60, 771-776.

(3) Bowen, B., Steinberg, J., Laemmli, U.K., and Weintraub, H. (1980) Nucl. Acids Res. 8, 1-20.

(4) Riggs, A.D., Suzuki, H., and Bourgeois, S. (1970) J. Mol. Biol. 48, 67-83.

(5) Kuwana, H., and Kubota, M. (1982) Jpn. J. Genet. 58, 579-589.

(6) Garner, M., and Revzin, A. (1981) Nucl. Acids Res. 9, 3047-3060.

(7) Giniger, E., Varnum, S.M., and Ptashne, M. (1985) Cell 40, 767-774.

STIMULATION OF GROWTH BY EGF IN MDA-231 HUMAN MAMMARY CARCINOMA CELLS IS ACCOMPANIED BY THE RAPID AND PERSISTENT INDUCTION OF C-MYC ONCOGENE

J.A. Fernandez-Pol*#, D.J. Klos*, P.D. Hamilton* and V. Talkad*#

*V. A. Medical Center and #Deparment of Medicine, St. Louis University, St. Louis, Mo.63106

INTRODUCTION

Epidermal growth factor (EGF) is a potent mitogen for a number of human malignant mammary epithelial cells in culture (1). This peptide elicits a wide variety of rapid and delayed responses by binding to high affinity cell-surface receptors which are 170-kD glycoproteins (2). However, the molecular mechanism of the stimulatory action of EGF remains largely unexplained.

Recently, investigators have found that the expression of c-myc gene in cultured cells is enhanced by stimulation with EGF (3). Because of the role of EGF (1) and c-myc (4) in human mammary carcinomas we quantified EGF receptor and mRNA in MDA-231 human mammary carcinoma cells. This cell line was found to contain normal amounts of EGF receptors (1.7×10^5 binding sites/cell), to be strongly stimulated to grow by EGF, and to respond to EGF with a rapid and persistent induction of c-myc gene expression.

MATERIALS AND METHODS

Assay for growth promoting activity. Promotion of cell growth was assayed as previously described (1). Briefly, cells were cultured as monolayers in serum-free medium for 24-72 hrs. At the indicated times, cell number and ^{3}H-thymidine incorporation were determined.

Assay for oncogene expression. Dot blots were performed using the procedure of Thomas (5). The immobilized mRNAs from the cells were hybridized with ^{32}P-labeled nick-translated c-myc, c-fos, and c-ras^{Ha} probes.

RESULTS

The growth of MDA-231 cells in serum-free medium was promoted 3-fold by EGF (50 ng/ml) after 24-48 hrs.

In a time course study, induction of c-myc was observed 15 min. after addition of 50 ng/ml EGF to the cells. Maximal induction of approximately 7 to 8-fold over the basal level was reached at 4 to 5-hr after the addition of EGF. After 24 hr treatment with EGF (50 ng/ml) the level of c-myc was approximately 3-fold higher than the basal level.

A dose response study revealed that the maximal effect was obtained with 50 ng/ml EGF (Fig. 1). Increasing the concentrations of EGF above 50 ng/ml resulted in a pronounced decrease in c-myc mRNA accumulation (Fig. 1).

In contrast to the large fluctuations of c-myc mRNA synthesis, little change was detected in the amount of c-fos mRNA induced by EGF. MDA-231 cells expressed the 1.4 kilobase c-ras^{Ha} at very low levels in the presence of EGF. Thus, treatment of MDA-231 cells by EGF results in a very dramatic and specific stimulation of c-myc oncogene which appears to be associated with growth stimulation.

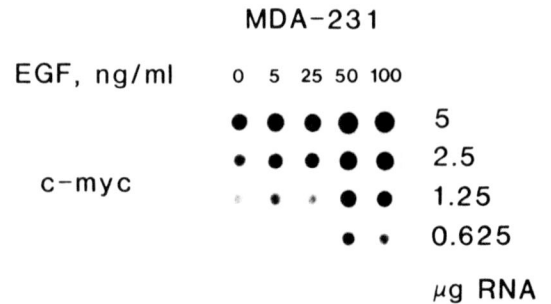

MDA-231

EGF, ng/ml   0 5 25 50 100

c-myc

Fig. 1. Dot blot analysis of c-myc mRNA from mammary cells MDA-231.

DISCUSSION

EGF promoted the growth of the human mammary cells MDA-231 and enhanced the c-myc gene expression. The expression of c-myc has been reported to be induced more than 20-fold after 1 hr in fibroblasts treated with serum, reaching the basal level of quiescent cells after approximately 18 hr (6). Our results show that the maximal enhancement of c-myc expression by EGF in defined serum-free conditions occurs later than by serum. Furthermore, the enhancement by EGF is not so pronounced as in the experiments with serum and it persists at high levels for at least 24 hr. These differences might be due to growth conditions, cell type, or to deregulation of c-myc expression in malignant cells (4).

REFERENCES

(1) Fernandez-Pol, J. A., Klos, D. J., and Grant, G. A. (October 1986) Cancer Res. 46, in press.

(2) Fernandez-Pol, J. A. (1985) J. Biol. Chem. 260, 5003-5011.

(3) Bravo, R., Burckhardt, J., Curran, T, and Muller, R. (1985) The EMBO Journal. 4, 1193-1197.

(4) Escot, C., et al (1986) Proc. Nat. Acad. Sci. 83, 4834-4838.

(5) Thomas, P.S. (1980) Proc. Nat. Acad. Sci.77, 5201-5205.

(6) Muller, R., Bravo, R., Burckhardt, J. and Curran, T. (1984) Nature, 321, 716-720.

DIFFERENTIAL INDUCTION OF HUMAN MT GENES BY HEAVY METALS

R. Foster, C. Sadhu and L. Gedamu

Department of Biology, University of Calgary, Calgary, Alberta, Canada T2N-1N4

INTRODUCTION- Metallothioneins (MTs) are ubiquitous, low molecular weight, cysteine-rich, heavy metal binding proteins(1). These proteins exist as two electrophoretically distinct isoforms,MT-I and MT-II, are transcriptionally regulated by glucorticoid hormones and heavy metals (Cd, Cu and Zn) and are encoded by a multigene family. To date one MT-II (MT-IIA) and five MT-I (MT-IA, MT-IB, MT-IE, MT-IF and MT-IG) functional genes have been isolated and characterized (2,3,4,5,6,7). MT-IIA and MT-IA exhibit differential induction phenotypes but are expressed in all cell types tested. The other MT-I genes (MT-IB, MT-IF and MT-IG) are induced by the same metals but are expressed in a tissue-specific manner. In this report we demonstrate that two divergently linked MT-I genes, MT-IF and MT-IG,are induced to different levels by various heavy metals.

RESULTS Expression of hMT-IG- In order to investigate the functionality of the hMT-IG promoter we linked the 5' flanking region of hMT-IG to the chloramphenicol acetyl transferase (CAT) reporter gene. Hepatoblastoma-derived cells, Hep G2, were transfected with this hMT-Icat construct and the expression of the fusion gene analysed by assaying for CAT activity. Fig.1 illustrates that a basal level of MT-IGcat expression can be induced to a higher level by metal ions: the relative order of induction being Cd>Cu>Zn.

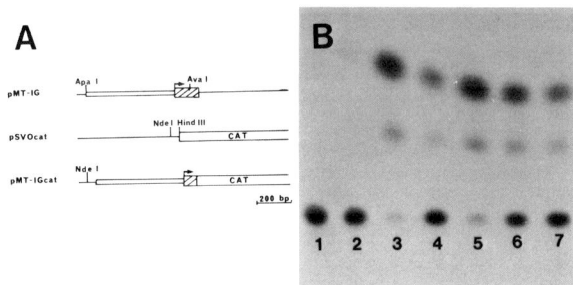

Fig. 1 Structure and expression of MT-IGcat. A) MT-IGcat was constructed by cloning a 600bp Apa I-Ava I fragment into the Hind III site of pSVO. The transcription initiation site is indicated by the arrow. Hatched block; first exon of MT-IG. B) Expression of MT-IG was assayed for in transfected Hep G2 cells. Choramphenicol(CM) and its acetylated forms (A,1-acetate chloramphenicol;B,3-acetate chloramphenicol) were detected by autoradiography. Lane 1, mock transfected cells. Lane 2+3,pSVO and pSV2 transfected cells. Lane 4-7, pMT-IGcat transfected cells induced with Cd(5),Cu(6)and Zn(7).

Differential Expression of MT Genes-Qualitative analysis of Fig. 1 suggests that the MT-IG promoter is differentially induced by heavy metals. To determine whether the MT-IIA and the MT-IF gene gave similar patterns of induction we constucted MT-IFcat and MT-IIAcat expression vectors. All three expression vectors were transfected into Hep G2 cells and their expression assayed for by CAT activity (Fig. 2). Comparative analysis of their transient expression suggests that:1) the three genes exhibit different levels of basal expression, the relative order being IIA>IG>IF, 2) of the three metal ions,cadmium is the best inducer, IIA,IG>IF, 3) copper induces MTIG to a higher level than either MT-IF or MT-IIA 4) zinc induces MTIIA to a higher level than either MT-IG or MT-IF.

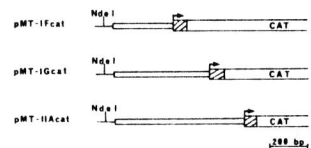

CATase ACTIVITY

PLASMID	CONTROL	CADMIUM	COPPER	ZINC
pMT-IFcat	15.02%	68.72%	32.38%	43.65%
pMT-IGcat	36.24%	96.63	61.13%	38.30%
pMT-IIAcat	68.71%	92.05%	40.08%	70.86%

Fig. 2 Expression of MTcat vectors. pMT-IFcat and pMT-IIAcat were constructed by cloning promotor containing fragments into the Hind III site of pSVO. Transcription initiation sites are indicated by arrows. Hatched box; exon 1 of respective MT genes. CAT activities were measured as a percentage of pSV2(contains SV40 early promotor) per 50 ug protein per hour.

CONCLUSION- Although six different functional MT genes have been characterized, the molecular mechanisms involved in the regulation of this multigene family remains unclear. At present certain regulatory elements, MREs,GC boxes and the TATA box, have been proposed to control transcription of the MT-I genes. In addition to these, MT-IIA also has a GRE and an enhancer. Our results have demonstrated that three MT genes are differentially induced by Cu and Zn. Thus differential transcriptional control of MT genes may be effected by the differential affinity of trans-acting factors for the various MT promotors.
    Funded by MRC(Canada)and AHFMR.

REFERENCES
1.Kagi, J.H.R. and M. Nordberg. (1979). Metallothionein. Basel:Birkhauser.
2.Karin,M. and R.I. Richards. (1982). Nature 299:797-802.
3.Richards,R.et.al.(1984).Cell37:263-272.
4.Heguy,A. et.al. (1986). Mol.Cell.Biol. 6:2149-2157.
5.Schmidt,C.J.et.al.(1985). J.Biol.Chem. 260:7731-7737.
6.Varshney,U.et.al.(1986). Mol.Cell.Biol. 6:26-37.
7.Foster, R. et.al. in press.

PRIMARY STRUCTURE AND DEVELOPMENTAL EXPRESSION OF THE D. MELANOGASTER src FAMILY HOMOLOG, Dsrc4

R.J. Gregory, K.L. Kammermeyer, W.S. Vincent and S.W. Wadsworth. Worcester Foundation for Experimental Biology, Shrewsbury, MA 01545

INTRODUCTION

The genome of the fruit fly, *Drosophila melanogaster*, has been shown to contain three distinct genes with homology to the vertebrate oncogene v-*src* (1,2,3). Hybridization and DNA sequence analyses have revealed that two of these genes, designated D*src* and D*ash* appear to be homologous to the vertebrate oncogenes *src* and *abl* respectively (1,4). In this report we show that the third gene, designated D*src4*, also exhibits a high degree of homology to both the *src* and *abl* oncogenes. Transcripts of all three *D. Melanogaster src* homologs are maternally inherited and are present on polysomes in developing embryos, indicating that they encode gene products that participate in early embryonic development (5). In addition, we have examined the expression of the D*src4* gene throughout fly development by RNA blot hybridization.

RESULTS

Isolation of Dsrc4 complementary DNA clones

Drosophila D*src4* complementary DNA (cDNA) clones were isolated from λ phage libraries by hybridization with a previously described genomic D*src4* sequence (5). Two separate phage libraries (provided by Drs. B. Yedvobnick and L. Kauvar) were screened for D*src4* cDNA. The first was prepared from RNA isolated from 3 to 12 hour old embryos, the second with poly(A)$^+$ RNA from adult females. Two independent clones comprising a total of 3.0 kb of DNA were isolated from these libraries. Analysis of the nucleotide sequence of these clones revealed the presence of a 601 amino acid open reading frame. This open reading frame is initiated by an ATG codon whose flanking sequences conform well to the *D. melanogaster* initiator consensus (6). Although this is the first in-frame initiator methionine codon within our cDNA sequence, no in frame stop codon is present 5' to the potential initiator. Thus, we cannot presently rule out the possibility that the open reading frame may initiate upstream of the 5' end of the cDNA clone.

Homology of Dsrc4 with other src family genes

In Figure 1, we present a conceptual translation of the D*src4* open reading frame and compare this protein with the prototypical member of this gene family, chicken c-*src*. Several gaps have been introduced into the sequences to maximize homologies we judge to be significant. This comparison reveals that D*src4* bears considerable homology to c-*src* throughout 70% of its length; only the N-terminal 140 amino acids of D*src4* appear completely unrelated. Within the kinase domain of these two proteins (defined here as extending from codon 270 to 470 of c-*src* gene) 54% of the positions contains identical residues. The kinase domain segment of the D*src4* open reading frame was also compared with other members of the *src* kinase family (Table 1).

Developmental expression of Dsrc4 mRNA

The expression of D*src4* RNA at various developmental timepoints was estimated by RNA blot hybridization utilizing a complementary RNA probe transcribed *in vitro* from a 0.6 kb D*src4* genomic DNA fragment. 5μg of total RNA from each developmental stage were electrophoresed in denaturing formaldehyde gels, transferred to nitrocellulose, and hybridized with the ^{32}P-labeled probe. This analysis revealed that D*src4* is expressed primarily during embryogenesis and, with the exception of expression in imaginal discs, expression drops off abruptly prior to and during the larval stages. There also appears to be some

A = identical amino acid
Δ = conserved amino acid

Figure 1: Amino acid comparison of the conceptual translation of the D*src4* gene product with the amino acid sequence of the c-*src* gene product. Conserved amino acids have been assigned according to Chambon et al., (Proc. 18th Miami Winter Symp., 1986, 246-249). ♦ indicates the conserved lysine residue, equivalent to c-*src* lysine 295, present in all known tyrosine protein kinases.

Table 1: Homology of *src* Family Oncogenes with the Dsrc4 Kinase domain.				
c*src*=54%	D*ash*=54%	D*src*=51%	v*fgr*=51%	v*yes*=51%
v*abl*=50%	v*fps*=44%	v*fes*=43%	v*fms*=39%	c*raf*=29%

transient expression of D*src4* RNA during early metamorphosis. The RNA is present only at low levels in male adults and most of the expression observed in adult females has previously been shown to be confined to ovarian tissue (5). Data pertaining to the *in situ* localization of D*src4* mRNA during development, will also be presented.

DISCUSSION

The protein encoded within the D*src4* open reading frame has several interesting properties. Among these is a stretch of 27 amino which is 74% glycine acids in the N-terminal third of the protein. Also, the protein encoded by the open reading frame would have a pI of 10.4, higher than any other *src* family protein. The conceptual translation also reveals no N-terminal hydrophobic domain capable of anchoring the protein to a cell membrane. In addition, the glycine at position two of the avian *src* protein, which serves as the myristylation site responsible for anchoring p60^{src} to the cell membrane (7), is not present adjacent to the potential initiator methionine of D*src4*.

Interestingly, the degree of homology of D*src4* with chicken c-*src* is as high or higher than its homology with the D*src* or D*ash* (see Table 1). The divergence of D*src4* from the other *D. melanogaster src* family genes is therefore probably an evolutionarily distant event and suggests that these kinases have evolved to fulfill different roles in *Drosophila* development. The homology of D*src4* with the *src* and *abl* members of the kinase family extends into the N-terminal regions of these proteins. Only *src*, *yes* and *abl* have previously been shown to have homology in this region (8). In addition, the point at which the D*src4* and c-*src* sequences appear to diverge in the N-terminal region is also the point at which the sequences of *src* and *abl* appear to diverge (8).

REFERENCES

1) Hoffman-Falk, H. et al., (1983) Cell 32, 589-598.
2) Simon, M. et al., (1983) Nature 302, 837-838.
3) Simon, M. et al, (1985) Cell 42, 831-840.
4) Lev, Z. et al., (1984) Mol. Cell. Biol. 4, 982-984.
5) Wadsworth, S.W. et al., (1984) Nucl. Acids Res. 13, 2153-2170.
6) Cavener, D. (1986) personal communication.
7) Cross, F.R. et al., (1984) Mol. Cell. Biol. 4, 1834-1842.
8) Ben-Neriah, Y. et al., (1986) Cell 44, 577-586.

ANALYSIS OF THE DIHYDROFOLATE REDUCTASE REPLICON FROM METHOTREXATE-RESISTANT CHINESE HAMSTER OVARY CELLS

J. L. Hamlin, J. E. Looney, J. P. Vaughn, and S. D. Creacy.

Department of Biochemistry, University of Virginia School of Medicine, Charlottesville, VA 22908.

Our laboratory is interested in the structure and function of mammalian chromosomal replicons. As a model system, we have developed a methotrexate-resistant CHO cell line (CHOC 400) that has amplified the dihydrofolate reductase (DHFR) gene and flanking sequences approximately 1,000 times. The amplified DHFR domains (amplicons) are arrayed end-to-end in homogenously-staining chromosome regions. We have previously reported that DNA replication appears to initiate in the early S period preferentially at one location within each amplicon (1), suggesting that the unit of amplification may be equivalent to a parental replicon. This model makes at least three predictions: 1) all of the amplicons in a single cell should have approximately the same boundaries (defined by the termini of the parental replicon), provided that no sequence rearrangements occur subsequent to amplification per se; 2) the DHFR amplicons in independently-isolated MTX-resistant cell lines derived from the same parental cell should be similar in size and sequence arrangement; and 3) each amplicon should have a single, fixed origin of replication. We report the results of studies designed to test these predictions.

(1) **ISOLATION OF DHFR AMPLICONS.** We have isolated a series of overlapping recombinant cosmids that represent the equivalent of two different amplicon types from the CHOC 400 cell line. The larger type I amplicons are 260 kb in length and appear to be arranged in head-to-tail configurations. The smaller type II amplicons represent a truncated version of the type I amplicon, are 220 kb in length, and are arrayed head-to-head and tail-to-tail. Both amplicon types appear to have precisely-defined endpoints. We have utilized the amplicon cosmids as probes on restriction digests of genomic DNA from less resistant cell lines, and have shown that the larger type I amplicon was the original unit of amplification. However, coincident with a chromosomal breakage event that occurs in a progenitor cell line with a lower copy number of the DHFR gene, the truncated type II version appears. Therefore, there appears to be a defined unit of DHFR gene amplification in the CHO genome, but this unit can be modified by subsequent chromosomal rearrangements.

2) **ANALYSIS OF THE DHFR AMPLICONS IN OTHER MTX-RESISTANT CHINESE HAMSTER CELL LINES.** When cosmids spanning the amplicon are used to probe digests of genomic DNA from two other independently-isolated MTX-resistant cell lines, all of the sequences in the CHOC 400 amplicon are detected in the other three genomes (2). This result suggests that the amplicons in these cell lines are at least as large as that found in CHOC 400, but does not rule out the possibility that they might be larger. However, by an in-gel renaturation procedure (3), we show here that the pattern of amplified restriction fragments is nearly identical in all three cell lines. We conclude that cells are constrained to amplify a unit of DNA whose boundaries are defined, possibly by the termini of the parental DHFR replicon.

3) **EVIDENCE FOR A FIXED ORIGIN OF DNA SYNTHESIS IN THE DHFR AMPLICON.** By an in vivo labelling method, we showed previously that DNA replication initiates somewhere within a region defined by three contiguous EcoRl fragments in the DHFR amplicon (1). In order to determine whether the origin is a fixed genetic element, we have characterized this initiation locus by a variety of criteria. Nuclease hypersensitivity studies on isolated nuclei detect an altered chromatin structure within a 6 kb EcoRl fragment that is the first fragment labelled at the beginning of S phase. We have also detected a topoisomerase II cleavage site within this fragment. In addition, a HindIII subclone from the 6 kb EcoRI fragment has been demonstrated to replicate autonomously in yeast. However, we have not yet observed autonomous replication of this sequence in mammalian cells. Thus, this locus has properties that might be expected of a fixed origin of replication, although the biological activity of the cloned origin has not yet been demonstrated in mammalian cells. We will present preliminary results of electron microscopic studies in which we have used an R-looping procedure to mark the position of the DHFR gene in genomic DNA enriched for amplicon sequences. By this protocol, we will be able to measure the distance between the gene and the center of replication bubbles (origin) in the DHFR amplicon, and thus define the precise location of the origin of replication.

REFERENCES

(1) Heintz, N. H., and Hamlin, J. L. (1982) Proc. Natl. Acad. Sci. USA 79, 4083-4087.
(2) Montoya-Zavala, M., and Hamlin, J. L. (1985) Mol. Cell. Biol. 5, 619-627.
(3) Roninson, I., Abelson, H. T., Housman, D. E., Hoell, N., and Varshavsky, A. (1984) Nature 309, 626-628.

CHARACTERIZATION OF A U1RNA GENE FROM THE SILKMOTH, BOMBYX MORI. R.J. Herrera and J. Wang

Department of Biology and Biotechnology, Worcester Polytechnic Institute, Worcester, MA 01609

In this report, the cloning and sequencing of a Bombyx mori DNA fragment containing a U1 snRNA gene is detailed.

Materials and Methods

1.Genomic Library Screening. Several genomic library equivalents of B. mori were constructed in EMBL-4. The genomic bank was screened by in situ hybridization with purified B. mori, 32P-labeled U1 snRNA. Twenty clones positive for the U1 gene were isolated and purified. Dot-blot hybridization experiments followed by progressive melt-outs were used to select "real" U1 genes from pseudogenes. Six clones exhibiting the stronger signals after a melting temperature minus 2°C wash and autoradiography were selected for further analysis.

2.Restriction Enzyme Analysis. Prior to restriction enzyme analysis, a 3.75 Kb fragment was recovered from one of the clones displaying a strong signal. The fragment was then subcloned into pUC-13. Several restriction enzymes were selected and the physical map of the insert was determined.

3.DNA Sequencing. A 513 bp SalI-EcoRI fragment displaying a positive signal when probed with [32P] ATP-labeled U1RNA was subcloned into M13mp11 and M13mp18 and sequenced using the dideoxy-chain termination procedure.

Results and Discussion

Through the sequencing of the 513 bp, U1 fragment, it was found that the U1RNA gene began at the 130th base from the EcoRI site. The initial 98 bp of the gene has 83% homology (see Fig.1) with the D. melanogaster U1RNA gene (1). However, no homology is observed after this point.

Fig.1: Nucleotide sequence of the B. mori U1 fragment and its comparison to a D. melanogaster U1 gene.

```
       X      10        20        30        40        50
       TTTATTTATGTGGCAATGGTTAATATTGAATGATTTATCTTTGTCAAGTACGTAGTCTAC

              70        80        90        100       110
       ATTTTTTATTATATATAGTATTTTAAATTGAATGATGAAATCCTAAATAATATTTTTGAGTC

              130       140       150       160       170
B. mori TATTTTTATATACTTACCTGGCGTAGGGGAT-ACCGTGATCATGCAGGCGGTTCCCCCAG
       :::::::::: :::::: :: :::::::::: : :::::::::: :
D. melanogaster ATACTTACCTGGCGTAGAGGTTAACCGTGATCACGCAAGGCGGTTCCTCCGG
              X      10        20        30        40        50

       180       190       200       210       220       230
       GGCGAGGCTGTTCCATTGCACTGCGGAGTGGTTGACCT-TGCCGATTATTTAGATTTTTAT
       : ::::::  ::::::::: ::: :::::::::::::::::::  : :::::
       AGTGAGGCTTGGCCCATTGCACCTCGGCTGAGTTGACCTCTGCGGATTATTCCTAATGTGAA
              60        70        80        90        100

       240       250       260       270       280       290
       CAATAAAAAATCAATAAATAAAAATGTATACTCGGAATAATTATTTTCTTTATACCTCGAATA
       ::         ::::          :       :       :
       TAACTCGTCGGTGTAATTTTTGGTAGCCGGGAATGGCGTTCGCGCCGTCCCGA
                      120       130       140       150       160 X

       300       310       320       330       340       350
       GAACTTAAAATTAATATTTTTATAATAAAATAAGTAACTTCATTCCTATCCGTGTCCCACG

       360       370       380       390       400       410
       ACACCACACTTCTTTTTATTTATTTATTTACTTAGCGAGATATGTGTATGACTCCTCCTGAG

       420       430       440       450       460       470
       TTAGTTAATCGGTTTCACAGAGATCACATTGCAAGAGATAGGCAGGATGGTATCCTT

       480       490       500       510
       TATTTATTTATTTATTGCTTAGTTGGGTCGACCTG
```

Sequence comparison with Human (2), D. melanogaster (1) and Sea Urchin (3) U1RNA gene flanking sequences show no homology in the 5' and 3' flanking regions. The "ACUUACCUGG" sequence which occurs at the 5' end of all U1RNAs and is thought to be involved in the base-pairing interaction between U1RNA and the splice junction consensus sequence (4) is found in the correct position and orientation in this B. mori U1. Sequences similar to the regulatory signal TATAAA (2) were found at

positions -20 to -28, -45 to -52, -54 to -64 and -105 to -109. Also the sequence TTTA is present in several tandem direct repeats at the 5' (-122 to -129) and 3' (+242 to +256 and +347 to +362) flanking regions. The polyadenylation signal "CAATAAA" (1) is found in the middle of the gene (+109 to +115) instead of 3' to the gene.

All U1RNAs sequenced to date have a similar secondary structure (5) and high GC content at the 3' end ranging from 60-75% (1-3,6-8). High GC content may represent a requirement for strong base pairing of hairpin loop structures and could be indicative of functional U1RNA genes as opposed to pseudogenes. The U1RNA gene we have sequenced is AT rich (77%) in the last 30 nucleotides. A similar AT rich area is seen at the 3' end of a human U1RNA pseudogene (9). This data together with the fact that the polyadenylation signal is located inside the gene just after an abrupt lack of homology with known U1RNA genes may be indicative of a deletion mutation involving the last 65 nucleotides of the U1RNA gene. Since there is more than one type of U1RNA found in some organisms (10-13), an additional possible explanation is that the gene sequenced codes for a different species of U1RNA not yet characterized in B. mori. Indeed, several bands have been observed after denaturing polyacrylamide electrophoresis of anti-cap and Lupus immunoprecipitated RNA in the U1RNA area (6). It is interesting that the secondary structure of U1RNA is conserved in the hypothetical transcript of the sequenced U1RNA gene, in spite of divergent nucleotide sequence at the 3' end (see Fig. 2).

Fig.2: Secondary structure of hypothetical U1 transcript.

Based on the above discussion, we can conclude that, most likely, the gene sequenced is a U1RNA pseudogene. However, before we exclude the possibility that this gene codes for a U1 variant, functional studies must be performed.

References
1. A. Alonso, J.L. Jorcano, B. Hovemann, T. Schmidt. J. Mol. Biol. 184:825.
2. E. Lund, J.E. Dahlberg (1984) J. Biol. Chem. 259:20131.
3. D.T. Brown, G.F. Morris, N. Chodchoy, C. Sprecher, W.F. Marzluff (1985). Nuc. Acids. Res. 13:537.
4. M.R. Lerner, J.A. Boyle (1980). Nature 283:220.
5. S. Mount, J.A. Steitz (1981). Nuc. Acids. Res. 9:6351.
6. D.S. Adams, R.J. Herrera, R. Luhrmann, P.M. Lizardi (1984). Biochemistry 24:117.
7. D.J. Forbes, M.W. Kirschner, D. Caput, J.E. Dahlberg, E. Lund (1984). Cell 38:681.
8. H. Nojima, R.D. Kornberg (1983). J. Biol. Chem. 258:8151.
9. R.A. Denison, S.W. VanArsdell, L.B. Bernstein, A.M. Weines, PNAS 78:810.
10. L. Elsebet, K. Brenda (1985). Science 29:1271.
11. N. Kato, F. Harada (1985). J. Biol. Chem. 260:7775.
12. R. Reddy, D. Henning (1979). J. Biol. Chem. 254:11097.
13. E. Lund, B. Kahan, J.E. Dahlberg (1985). Science 229:1271.

Acknowledgement: We would like to thank Mr. Mark A. Osborne for his editorial assistance.

METALLOTHIONEIN GENE EXPRESSION AND ITS CORRELATION TO HYPOMETHYLATION.

N. Jahroudi, C. Sadhu and L. Gedamu

Department of Biology, University of Calgary, Calgary, Alberta, Canada T2N-1N4

The genes coding for human Metallothioneins (MTs) are members of a multigene family (1,2) We have studied the induction of two members of MT-I (MT-IF and MT-IG) and the MT-IIA genes in response to the heavy metals Cd, Zn, Cu and the glucocorticoid analogue dexamethasone (Dex) (3,4). The results of these studies in two human hepatoma (Hep G2 and Hep 3B2) and a human lymphoblastoid (WI-L2) cell lines have indicated that the expression of these genes is cell-type specific and differentially regulated. Furthermore, our preliminary results have indicated that methylation may have a role in cell-type specific expression of these MT genes.

1. TIME COURSE OF TOTAL MT-mRNA INDUCTION

The synthesis of total MT-mRNA was studied in hepatoma (Hep G2, Hep 3B2) and lymphoblastoid (WI-L2) cell lines exposed to Cd, Zn, Cu and Dex. In Hep G2, maximal induction was obtained when cells were exposed to heavy metals or Dex between 5-9 hrs. Similar studies indicated that the times of maximal MT-mRNA induction, in response to heavy metals, in the WI-L2 cell line is approximately 12-15 hrs. (data not shown). However, WI-L2 cells do not respond to Dex up to a concentration of 10 uM.

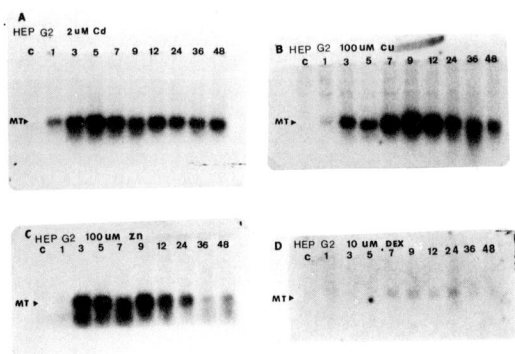

Figure 1. Time course of total MT-mRNA induction in Hep G2 cells. Total nucleic acids were isolated from Hep G2 cells induced for various time intervals (1-48 hrs) with 2 uM Cd (A), 100 uM Cu (B), 100 uM Zn (C) and 10 uM Dex (D). Total (10 ug) nucleic acid was electrophoresed on a methyl-mercury hydroxide agarose gel. The RNA was transferred to diazobenzyoxy methyl (DBM) paper and the filters were hybridized to the MT-II processed gene coding region to detect total MT-mRNA.

2. EXPRESSION OF MT-IIA GENE.

Using a specific MT-IIA probe, we studied the expression of the MT-IIA gene in Hep G2 and WI-L2 cells in response to heavy metals and Dex induction. Time course experiments have shown that the maximum MT-IIA mRNA accumulation is the same as total MT-mRNA accumulation as shown in Figure 1. Furthermore, upon Dex induction, MT-IIA mRNA accumulation is observed only in Hep G2 cells and there is no response in WI-L2 cells. Figure 2 shows that azacytidine (Aza) treatment of WI-L2 cells results in an increase in basal level of MT-IIA mRNA and also a significant induction in response to Dex.

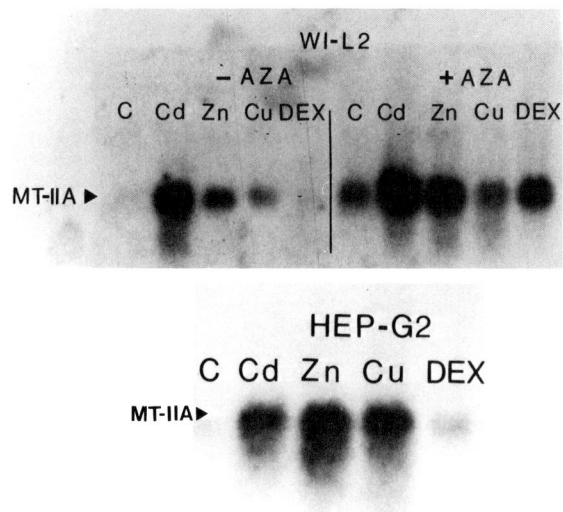

Figure 2. Concentration of heavy metals and Dex used were 2 uM Cd, 100 uM Zn, 150 uM Cu and 10 uM Dex. WI-L2 cells were treated with 8 uM Aza for 72 hours, then allowed to recover in Aza free medium before induction. Total nucleic acids were analyzed by Northern Blot using a 3'-noncoding specific MT-IIA probe.

CONCLUSION

The observation of MT-IIA induction in response to Dex upon Aza treatment suggests that methylation may have a role not only in cell-type specific expression of this gene, but also, in differential regulation of this gene in response to various inducers. To further investigate whether methylation plays a direct role on activation of MT-IIA in response to Dex, we are currently investigating the methylation pattern of MT-IIA gene in Hep G2, WI-L2 and Aza treated WI-L2 cells with specific interest at the Glucocorticoid Regulatory Element (GRE) of this gene.

This work was supported by MRC Canada to L.G. N.J. and C.S. are supported by AHFMR.

REFERENCES

1. Karin, M and R.I.Richards (1982). Nature 299: 797-802.
2. Varshney, U. et. al. (1984) Mol. Biol. Med. 2: 193-206.
3. Varshney, U. et. al. (1986) Mol. Cell. Biol. 6: 26-37.
4. Foster, R. et. al. In press.

22

CHICKEN U4 RNA GENES: STRUCTURE AND EXPRESSION

G.M. Korf, M.L. Hoffman, K.J. McNamara, and W.E. Stumph

Department of Chemistry and Molecular Biology Institute, San Diego State University, San Diego, California 92182

U4 small nuclear RNA (snRNA) is one of the six major snRNAs found in eucaryotic cells. Recent evidence indicates that U4 RNA, like the U1 and U2 RNAs, is required for messenger RNA splicing. To further analyze U4 RNA biosynthesis and function, we have cloned a region of the chicken genome that contains two genes homologous to U4 snRNA. One of these genes, if expressed, codes for a variant U4 RNA not previously identified in chicken cells.

(I) TWO CLOSELY LINKED GENES HOMOLOGOUS TO U4 RNA. Three λ-phage clones were isolated from a chicken DNA library using a human U4 pseudogene sequence as a hybridization probe. These were subsequently found to be overlapping clones derived from the same region of the chicken genome. A 2.1 kb BamHI restriction fragment common to each clone was found to contain the U4-hybridizing sequences. A restriction map depicting the genomic organization of the U4 homologies in the 2.1 kb fragment is shown in Fig. 1.

(II) SEQUENCE OF THE U4 HOMOLOGIES. The 1665 bp of DNA between the sites marked by asterisks in Fig. 1 were sequenced. The two U4 RNA genes have the same transcriptional orientation and are separated by 365 bp of non-coding DNA. The downstream gene, denoted U4B, agrees perfectly in sequence with a major chicken U4 RNA species known as U4B RNA (1). Unexpectedly, the upstream gene has seven base changes relative to the U4B gene and does not code for any previously known chicken U4 RNA sequence. These sequence comparisons are illustrated in Fig. 2.

(III) IS THE U4X GENE A REAL GENE OR A PSEUDOGENE? We have recently shown that the chicken U4B gene is accurately expressed in frog oocytes when injected as part of a cloned fragment with 280 bp of 5'-flanking DNA and 283 bp of 3'-flanking DNA (2). In contrast, we were unable to detect expression of the U4X gene in oocytes when injected with 324 bp of 5'-flanking DNA and 116 bp of 3'-flanking DNA. (From present knowledge of snRNA gene expression, this amount of flanking DNA would be expected to contain all of the transcriptional signals required for efficient expression.) Nevertheless, several features of the U4X gene sequence suggest that it may be an expressible gene. For example, four of the seven base substitutions in the U4X gene shown in Fig. 2 are identical to the bases at the corresponding positions in Drosophila U4 RNA (3). Furthermore, none of the seven base substitutions in the U4X gene would have a significant effect on the proposed secondary structure of U4 RNA nor on the

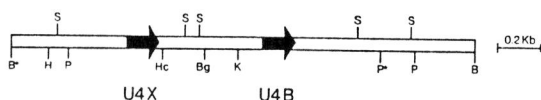

Fig. 1. Restriction Map and Genomic Organization of Chicken U4 RNA Genes. The bold arrows represent the locations and transcriptional orientations of the two U4 RNA genes. Enzyme abbreviations: B, BamHI; H, HindIII; S, SstI; P, PstI; Hc, HincII; Bg, BglI; K, KpnI.

Fig. 2. Sequence Comparison of the Chicken U4B and U4X Gene Sequences with the Chicken U4B RNA Sequence. Dashes indicate agreement with the U4B RNA sequence, whereas base changes relative to U4B RNA are shown explicitly.

postulated base pairing interactions between U4 and U6 RNAs in the U4/U6 snRNP particle. Therefore, the pattern of base substitutions does not appear to result from random drift, but rather is more consistent with a selection for changes that maintain U4 RNA function. Moreover, we have recently shown that the U4X gene, when placed downstream of the U4B gene promoter, can code for a stable U4X RNA transcript in oocytes (unpublished data). Therefore, the defect for efficient expression of the chicken gene in oocytes lies in the U4X gene promoter, and not in the coding region.

There is present in the literature a considerable amount of data indicating that two separate regions of 5'-flanking DNA are required for the efficient transcription of vertebrate snRNA genes by RNA polymerase II. The proximal region is located approximately 55 bp upstream of the cap site and is responsible for fixing the site of transcriptional initiation. The distal region, located approximately 200 bp upstream of the cap site, enhances the level of transcription. Interestingly, the U4X gene contains the sequence CTGTG in the proximal region, whereas the sequence CCGTG is perfectly conserved in six other cloned chicken snRNA genes (2). In the distal regulatory region, the six chicken snRNA genes contain the functional sequence octamer ATGCAAAT or a closely related variant, whereas the U4X gene contains the severely altered sequence ATGGTAAT. Therefore, the U4X gene has potentially inactivating mutations in both its proximal and distal regulatory regions.

The promoter alterations described above are probably responsible for our inability to detect efficient expression of the U4X gene in frog oocytes. Nevertheless, that does not mean that the U4X gene is never expressed in chicken cells. If expressed, it may require a transcription factor present in chicken cells but not in frog oocytes. An intriguing DNA sequence, which could be a recognition site for such a gene-specific factor, exists just upstream of and partially overlaps the proximal regulatory region of the U4X gene. Here an unusual palindromic sequence, unique to U4X, is found: GCGCGCCGGCGCGC. Further experiments are in progress to study the expression of the U4X gene in chicken cells.

REFERENCES
(1) Reddy, R. (1985) Nucleic Acids Res. 13, r155-r163.
(2) Hoffman, M.L., Korf, G.M., McNamara, K.J., and Stumph, W.E. (1986) Mol. Cell. Biol., in press.
(3) Myslinski, E., Branlant, C. Wieben, E.D., and Pederson, T. (1984) J. Mol. Biol. 180, 927-945.

EVOLUTIONARY CONSERVATION AND MULTIPLE PRESENCE OF DUAL-SITE 18S rRNA·mRNA COMPLIMENTARITY IN EUKARYOTIC VIRUSES.

L.E. Maroun and R.B. Adams

Southern Illinois University School of Medicine, P.O. Box 3926, Springfield, IL 62708

The evolution of the small sub-unit rRNA from the prokaryotic 16S to the 18S eukaryotic form involves the deletion of the Shine and Dalgarno sequence and a pair of coupled transversions that create the 18S rRNA sequence GAAGG....UUUGG. We have presented observations that eukaryotic cellular mRNA 5'-leader sequences frequently have the ability to form a uniformly structured dual-region hybrid with this 18S rRNA sequence (21). Our previous study specifically excluded viral mRNA because of the potential role of this site in interferon action (22). We present here examples of dual-region ribosome binding sites as observed in eukaryotic viral mRNA and some not previously presented examples of oncogene and cellular mRNA potential 18S rRNA annealing sites (FIG. 1).

To allow space for presentation of a larger number of examples the structures do not show the unpaired bases of the central loop or the 2,6 dimethyladenosine rRNA loop. These can quickly be deduced from the 36 base 3'-hydroxyl end sequence of 18S rRNA up to a putative colicin-E3-like endonuclease cleavage site 3'-AUUACUAGGAAGGCGUCCAAGUGGAU GCCUUUGGAA* (23). The number of bases in the single mRNA interior loop is given above the sequence gap (AVG=3.5, range:0-10).

Note that G:U pairs constitute 23% of the total bonds in this compilation, a figure which is comparable to the 21% G:U pairs seen for the first 100 (24) bonds at the 3' end of 16S rRNA. In the new group of mRNA structures presented here, the average distance to the AUG start codon is 6.8 bases (range:0-14) and the average interstrand ΔG is -15.1 kcal (range:-8.2 to -26.6).

The list includes two examples of eukaryotic viruses that transcribe more than one of their small number of mRNAs with potential rRNA annealing sites (Vaccinia and Rabies), and a demonstration of the extensive evolutionary conservation of rRNA annealing ability in the major transcript of a family of related cancer viruses (5,6,7 and 8, FIG.1). An interesting example of a conserved site is found in RSV (and FSV). This site provides an example of a hairpin loop juxtaposing the two mRNA regions of complimentarity and, as seen in the Fc receptor mRNA (#20), the dual site bond is absent from the sequences near upstream unused AUG codons.

REFERENCES

1. Broyles, S.S., et al. (1986) PNAS 83:3141
2. Earl, P., et al. (1986) PNAS 83:3659
3&4.Tordo, N., et al. (1986) PNAS 83:3914
5. Schwartz, D.E., et al. (1986) Cell 32:853
6. Ratner, L., et al. (1985) Nature 313:277
7. Sagata, N., et al. (1985) PNAS 82:677
8. Sanchez-Pescador, R., et al. (1985) Sci. 227:484
9. Marth, J.D., et al. (1985) Cell 43:393
10. Coussens, L., et al. (1986) Nature 320:277
11. Tsujimoto, Y., et al. (1986) PNAS 83:5214
12. Durkin, J. et al. (1986) Mol. Cel. Bio. 6:1386
13. Mercken, L., et al. Nature 316:647
14. Dahlback, B., et al. (1986) PNAS 83:4199
15. Rotwein, P., (1986) PNAS 83:77
16. Tsou, A.P., et al. (1986) Mol. Cel. Bio. 6:786
17. Lobe, C.G., et al. (1986) Sci. 232:858
18. Bakhshi, A., et al. (1986) PNAS 83:2689
19. Coussens, L., et al. (1986) Sci. 233:859
20. Hibbs, M.L., et al. (1986) PNAS 83:6980
21. Maroun, L.E., et al. (1986) J. Th. Bio. 120:85
22. Maroun, L.E., (1979) Biochem. J. 179:221
23. Alberty, H., et al. (1978) Nuc. Aci. Res. 5:425
24. Brimacombe, R., et al. (1985) Biochem. J. 229:1

FIGURE 1

INFLUENCE OF A DNA SEQUENCE ON THE AMPLIFICATION AND EXPRESSION OF ADJACENT GENES.

Sally G. Pasion, Jennifer A. Hartigan, Vipin Kumar, David T.W. Wong and Debajit K. Biswas.

Laboratory of Pharmacology, Harvard School of Dental Medicine and Department of Pharmacology, Harvard Medical School, Boston, MASS 02115.

Several multihormone-producing clonal strains of rat pituitary tumor cells in culture (GH cells) synthesize and secrete in the medium different amounts of prolactin (PRL) and growth hormone. Prolactin synthesis in the GH cell strains GH_12C_1 and $F_1BGH_12C_1$ cannot be detected (1). However, PRL-synthesis can be induced following treatment of these cells with 5-bromodeoxyuridine (BrdU) (1). The BrdU-induced PRL synthesis in $F_1BGH_12C_1$ is apparently mediated via amplification of the rPRL gene.

Role of the 5' End Flanking Sequences of rPRL Gene on the Amplification of Other Adjacent Genes.

Gene transfer studies reveal that 10.3kb DNA segment, flanking the 5' end of the rat prolactin (rPRL) gene of 5-bromodeoxyuridine-responsive GH ($F_1BGH_12C_1$) cells induces amplification of the adjacent Herpes simplex virus type 1, thymidine kinase gene (HSV1TK), following transfection and integration into the chromosome of the recipient mouse L cells (2). In this report this characteristic property of the DNA sequence ("Amplicon") was further verified by studying the amplification of other linked or unlinked genes by DNA-mediated transfer studies. Our results demonstrate that both the unselected human growth hormone (hGH) gene and the selected HSV1TK gene are amplified in the recipient cells in response to 5-bromodeoxyuridine (BrdU) when transferred with the 10.3-5'rPRL gene sequence of BrdU-responsive cells. This observation is further substantiated by the BrdU-induced amplification of the bacterial "Neo" (Neomycin phosphoribosyl transferase) gene cotransferred with the "Amplicon" sequence. Similar cotransfection studies reveal that, the amplification capability is limited to a 4kb subfragment of the 10.3-5'rPRL DNA sequence of the ("Amplicon") BrdU-responsive cells. These results demonstrate that genes of heterologous origin, linked or unlinked and selected or unselected, are coamplified when located within the amplification boundary of the "Amplicon". The "Amplicon" region is identified within a 4kb DNA segment at the 5' end of 11.2-5'rPRL sequence.

Defective Distal Regulatory Element at the 5' Upstream of rPRL Gene of Steroid-Nonresponsive GH-Subclones.

The prolactin nonproducing (PRL⁻) GH cells GH_12C_1 (3, 4) and $F_1BGH_12C_1$ (1) do not respond to steroid hormones estradiol (E_2) or hydrocortisone (HC). However, the stimulatory effect of E_2 and inhibitory effect of HC on prolactin synthesis can be demonstrated in the prolactin-producing GH cell strain, GH_4C_1 (3, 4). In this investigation we have analyzed the cloned 11.2 kilobase (kb) DNA sequence upstream of rat prolactin gene (11.2-5'rPRL) of steroid-responsive GH_4C_1 and that of steroid nonresponsive $F_1GH_12C_1$ cells to i) identify the steroid responsive positive and negative regulatory elements and ii) to verify the status of these regulatory elements in steroid-nonresponsive cells. Results presented in this report demonstrate the following: i) the basal level expression of the cotransferred "Neo" (Neomycin Phosphoribosyl transferase) gene is inhibited by dexamethasone and is stimulated by E_2 treatment of transfectants carrying the 11.2-5'rPRL or the 4kb subfragment of this upstream sequence of steroid-responsive cells; and ii) these effects of the steroids on the adjacent "Neo" gene expression are not observed in transfectants with either of these rPRL upstream sequences derived from steroid-nonresponsive cells. We report here the identification of the distal regulatory elements located in a DNA segment 3.8-7.8kb upstream of the transcription initiation site of rPRL gene. Both the positive and the negative effects of steroid hormones can be identified with this distal regulatory element. This distal regulatory element appears to be nonfunctional in steroid-nonresponsive cells, presumably due to a structural defect in this region.

REFERENCES:

1. Biswas, D.K., Lyons, J. and Tashjian, A.H. Jr. (1977). Induction of Prolactin Synthesis in Rat Pituitary Tumor Cells by 5-Bromoeoxyuridine. Cell. 11, 431-439.

2. Biswas, D.K., Hartigan, J.A. and Pichler, M.H. (1984). Identification of DNA Sequence Responsible for 5-Bromodeoxyuridine-Induced Gene Amplification. Science 225, 941-943.

3. Bancroft, F.C. (1981). GH Cells Functional Clonal Lines of Rat-Pituitary Tumor Cells. In: Functionally Differentiated Cell Lines. Gordon Sato (Ed.), Alan R. Liss, New York, NY. pp 47-59.

4. Martin, T.F.J. and Tashjian, A.H. Jr. (1977). Cell Culture Studies of Thyrotropin-Releasing Hormone Action. In: Biochemiscal Action of Hormone. G. Litwick (Eds.), Vol IV, Academic Press, New York, NY. pp 269-312.

ORNITHINE DECARBOXYLASE mRNA EXPRESSION AND TRANSLATION

Lo Persson[1], Ingvar Holm[2] and Olle Heby[2]

Departments of [1]Physiology and [2]Zoophysiology, University of Lund, S-223 62 Lund, Sweden.

INTRODUCTION. The polyamines putrescine, spermidine and spermine play important roles in growth and development (1). ODC, the first enzyme in the biosynthetic pathway, appears to be regulated at the transcriptional, translational as well as post--translational level. Inhibition of the ODC activity by treatment with α-difluoromethylornithine (DFMO), an enzyme-activated irreversible inhibitor, causes arrest of growth and development. In Ehrlich ascites tumor (ELD) cells, treated with DFMO and thus lacking ODC activity, we were surprised to find an increased ODC content (2). This was found to be mainly a result of increased synthesis of ODC. Since there was no change in ODC mRNA content our data indicated that the DFMO-mediated depletion of putrescine and spermidine had caused the increase in ODC content by specifically affecting the rate of ODC mRNA translation (2). The present study addresses the obvious question whether an increase in cellular polyamine content causes a decrease in translation of ODC mRNA (without affecting transcription).

MATERIALS AND METHODS. ELD cells, grown in suspension culture (3), were seeded in the absence or presence of DFMO (5 mM) or spermidine (1 μM-10 mM). After 24 h the cells were analyzed for ODC activity, ODC synthesis and ODC mRNA content. ODC activity was determined by measuring the release of $^{14}CO_2$ from $1-^{14}C$-ornithine (3). ODC synthesis was determined by measuring the incorporation of ^{35}S-methionine (25-min pulse) into ODC protein, which was then immunoprecipitated (using a monospecific ODC antibody) and analyzed by electrophoresis and autoradiography (3). ODC mRNA content was estimated by Northern blot analysis (3), using a ^{32}P-labeled ODC cDNA probe (pODC 934) (4).

RESULTS AND DISCUSSION. We recently reported (2), that DFMO-mediated depletion of the cellular polyamine content results in a compensatory increase in ODC content, without markedly affecting the ODC mRNA content or ODC turnover rate. This suggested that the polyamines act as negative regulators of ODC translation. We now provide direct evidence for this means of regulation, by demonstrating an elevated incorporation of ^{35}S-methionine into ODC protein in polyamine-depleted cells (Fig. 1).

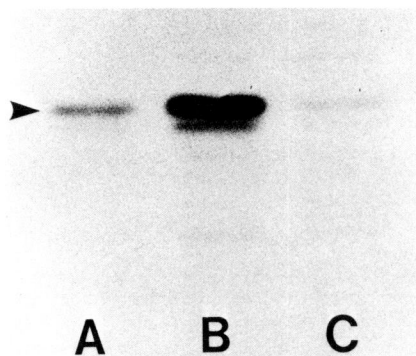

Fig. 1. Autoradiogram of a polyacrylamide gel, showing the incorporation of ^{35}S-methionine into ODC protein in ELD cells grown for 24 h in the absence (A) or presence of 5 mM DFMO (B) or 10 μM spermidine (C). Arrow, migration of ^3H-DFMO--labeled purified mouse kidney ODC (MW $\sim$ 53 kD).

Table 1. Effects of DFMO and spermidine (SPD) on the ODC activity and ODC content in ELD cells.

	ODC activity (U*/10^6 cells)	ODC content (ng/10^6 cells)
Control (0 h)	< 0.01	n.d.
Control (24 h)	2.05 ± 0.08	1.19 ± 0.06
5 mM DFMO (24 h)	0.03 ± 0.003	4.30 ± 0.46
10 μM SPD (24 h)	0.11 ± 0.07	n.d.

*1 Unit (U) = 1 nmol CO_2/h; n.d. = not determined.

Table 2. Effect of spermidine on the ODC mRNA content in ELD cells.

	ODC mRNA content (relative to 0 h control)
Control (0 h)	1.0
Control (24 h)	10.2
Spermidine, 1 mM (24 h)	11.8

Provision of putrescine or spermidine at the time of seeding has previously been found to abolish the induction of ODC activity (3) (Table 1). Spermidine was consistently found to be more effective than putrescine, causing greater than 50% inhibition of the ODC activity at a 1 μM concentration. To determine whether spermidine acts directly on ODC protein synthesis, ELD cells were pulse-labeled with ^{35}S--methionine after spermidine treatment. Analysis of immunoprecipitated ODC by electrophoresis and autoradiography showed that 10 μM spermidine had effectively suppressed the incorporation of ^{35}S-methionine into the enzyme (Fig. 1). The suppression of ODC synthesis caused by spermidine was not a result of inhibition of the increase in ODC mRNA content (Table 2). Therefore, we conclude that polyamines exert a negative feedback control of ODC expression by affecting translation, but not transcription. The translational control is particularly interesting in view of the fact that ODC mRNA has a long 5´ non--coding leader. It is conceivable that the polyamines affect translation by interfering with the leader, e.g. by changing its secondary structure.

SUMMARY. The present study shows that the polyamines act as feedback inhibitors of ODC mRNA translation. Thus, in exponentially growing ELD cells ODC synthesis is inhibited by an increase and is stimulated by a decrease in cellular polyamine content. The effects of polyamines on ODC synthesis are not attributable to interference with transcription.

REFERENCES
(1) Heby, O. (1981) Differentiation 20, 1-19.
(2) Persson, L., Oredsson, S.M., Anehus, S. and Heby, O. (1985) Biochem. Biophys. Res. Commun. 131, 239-245.
(3) Persson, L., Holm, I. and Heby, O. (1986) FEBS Lett., in press.
(4) Berger, F.G., Szymanski, P., Read, E. and Watson, G. (1984) J. Biol. Chem. 259, 7941-7946.

ACKNOWLEDGEMENTS. This study was supported by grants from the Swedish Medical (04X-02212) and Natural Science (B-BU/CF 4086-112) Research Councils and the Medical Faculty (University of Lund). DFMO was a generous gift from the Merrell Dow Research Institute, Strasbourg, France. pODC 934, encoding mouse kidney ODC, was a kind gift of Dr. Franklin G. Berger (4).

A DEVELOPMENTALLY REGULATED PROMOTER-SPECIFIC FACTOR IS ALSO REGULATED BY THE ADENOVIRUS E1A GENE

R. Reichel, I. Kovesdi, and J.R. Nevins

Howard Hughes Medical Institute
The Rockefeller University
1230 York Avenue
New York, N.Y. 10021

INTRODUCTION

One particularly attractive system for the study of gene control is the differentiation of teratocarcinoma cells in culture (1,2). It is clear from numerous studies that genes are both activated and inactivated during this process and that at least part of this control is transcriptional (3-5). Thus, an identification of the factors mediating this process, both promoter-specific binding factors and activities that might regulate these factors, is crucial to a final understanding of the basis for differentiation. It was previously shown that F9 teratocarcinoma cells contain a transcriptional regulatory activity similar in nature to the adenovirus E1A gene and upon differentiation the activity disappears (6). We have recently identified a cellular transcription factor, utilized by the adenovirus E2 promoter, in adenovirus infected HeLa cells (7). The level of this factor is markedly increased by the action of the adenovirus E1A gene and thus appears to mediate E1A control of the E2 gene. We have now asked whether this factor could be detected in F9 cells and if so, will it be affected by the differentiation process?

RESULTS

Using a gel retardation assay, we have detected an E2-specific factor in mouse F9 cells. Competition experiments with E2 promoter mutants suggested that this F9 protein occupies the same E2 promoter sequences as does the factor from adenovirus infected HeLa cells that is regulated by E1A. Exonuclease III footprinting assays revealed that the F9 factor does protect the same E2 promoter region as does the HeLa protein, it binds between -32 and -71 (+1 refers to the cap site of the E2 promoter). We therefore conclude that most likely a murine equivalent of the factor induced by the adenovirus E1A gene in a productive infection of HeLa cells is present at significant levels in F9 cells. To investigate the influence of differentiation on the E2 promoter-specific mouse protein, we have prepared nuclear extracts from F9 cells which had been differentiated with retinoic acid and dibutyryl cyclic AMP. By means of gel shift assay or exonuclease III mapping, we were not able to detect the factor in extracts from differentiated F9 cells. In addition, the same result was obtained with a whole cell extract suggesting that the change in the level of the factor was not due to extraction differences of the cells. Therefore, we suggest that the level or binding activity of the E2 promoter binding protein is specifically regulated as a function of F9 cell differentiation.

It seems very likely that the E1A-like activity in F9 cells that disappears upon differentiation is responsible for the control of the E2-specific factor. In order to test this possibility, differentiated F9 cells were infected with wildtype adenovirus. Nuclear extracts were prepared from the infected cells and assayed for the E2 binding protein. In sharp contrast to differentiated F9 cells, which contain virtually no E2 factor, nuclear extracts from adenovirus infected F9 cells have a high level of E2 binding activity. Again, the same result was obtained when a whole cell extract was used to detect the E2 promoter binding protein, indicating that the induction of the factor by an adenovirus infection is not an extraction artefact. Application of an exonuclease III mapping experiment revealed that the binding sites on the E2 promoter are identical for the factor in F9 cells and the factor induced by the infection of differentiated F9 cells. The E2 promoter-specific protein could not be detected in differentiated F9 cells which had been infected with the E1A mutant d1312, a result which is consistent with the conclusion that an E1A-like activity in F9 cells controls the E2 factor.

CONCLUSION

We have detected a cellular transcription factor in mouse F9 cells that recognizes the E1A-inducible E2 promoter. This factor was previously found in HeLa cells and is induced by the action of the adenovirus E1A gene product. We have demonstrated that upon differentiation of F9 cells with retinoic acid and dibutyryl cyclic AMP, this E2-specific protein declines to near undetectable levels, a result consistent with the control of this factor by E1A and the presence of an E1A-like activity in F9 cells but not differentiated F9 cells. Indeed, if the adenovirus E1A gene is introduced into differentiated F9 cells by means of viral infection, the E2 factor increases in abundance. The described experiments have thus identified a cellular transcription factor, the concentration or binding activity of which is regulated during cellular differentiation and in response to the E1A oncogene.

REFERENCES

(1) Martin, G.R. (1980) Science 209, 768-776

(2) Strickland, S. (1981) Cell 24, 277-278

(3) Croce, C.M., Linnenback, A., Huebner, K., Parnes, J.R., Margulies, D.H., Appella, I., and Seidman, J.G. (1981) Proc. Natl. Acad. Sci. USA 78, 5754-5758

(4) Wang, S.Y. and Gudas, L.J. (1983) Proc. Natl. Acad. Sci. USA 80, 5880-5884

(5) Colberg-Poley, A.M., Voss, S.D., Chowdhury, K., and Gruss, P. (1985) Nature 314, 713-718

(6) Imperiale, M.J., Kao, H.T., Feldman, L.T., Nevins, J.R., and Srickland, S. (1984) Mol. Cell. Biol. 4, 867-874

(7) Kovesdi, I., Reichel, R., and Nevins, J.R. (1986) Cell 45, 219-228

REGULATED EXPRESSION OF A CLONED HUMAN RIBOSOMAL PROTEIN GENE IN CULTURED CHINESE HAMSTER CELLS.

D.D. Rhoads and D.J. Roufa

Center for Basic Cancer Research. Division of Biology. Kansas State Univ. Manhattan, KS. 66506

Ribosomal protein (rprotein) genes are regulated stringently by mammalian cells growing in tissue culture. That approximately eighty single copy rprotein genes and moderately reiterated rRNA genes are co-regulated and that their products are transported to the nucleolus and assembled efficiently into ribosomes promises an unique experimental system for study of a tightly regulated mammalian multigene family. Toward this end we have isolated and characterized somatic cell mutants expressing a recessive drug-resistance mutation affecting ribosomes -- emetine resistance (1) -- and have cloned the rprotein gene responsible for this phenotype (RPS14) from wild type and mutant Chinese hamster and human cells (2-4). Although Chinese hamster and human rprotein S14 possess identical primary amino acid sequences, genes that encode them display several single-base differences that result in discernible species-specific restriction site and fragment length markers.

RPS14 is located on the long arms of human chromosome 5 (5) and Chinese hamster chromosome 3 (6); and emetine resistant RPS14 alleles differ from wild type by single base changes affecting adjacent 3'-Arg codons (3). Human RPS14 spans 5.5 kbp of DNA and is composed of five exons and four introns. The entire nucleic acid sequence of the human gene is known (4); it encodes a 650 base mRNA that constitutes approximately 0.01% of the cell's poly(A)$^+$-RNA (2-4).

RESULTS

Construction of an expressible human RPS14 clone. A 10.1 kbp fragment of human RPS14 DNA was purified from the lambda genomic clone HGS14-1 (4) and inserted into the Bam HI site of the selectible mammalian shuttle vector pSV2Neo (7). Two subclones (pSVNS14-1 & -4) contain an entire human RPS14 gene as well as SV40 and bacterial plasmid replication origins and prokaryotic drug resistance genes (amp^r and neo^r). The human DNA fragment includes 4 kbp of chromosome sequence upstream of RPS14 and 600 bp of downstream sequence in both molecular orientations relative to the transfection vector.

Expression of human RPS14 genes in transformed Chinese hamster cells. An emetine-resistant Chinese hamster cell clone, Em^r-2 (1), was transformed with pSVNS14-1 and -4 (8); and 66 neo^r clones were isolated in medium containing the neomycin analog G418 (500 ug/ml). These were recloned in tissue culture and now are being tested for emetine resistance, expression of human S14 mRNA, and the number and organization of integrated pSVNS14-1 and -4 DNA sequences.

Initial studies indicate that approximately 40% (23/53) of the G418 resistant clones are sensitive to emetine. This suggests that they have acquired an intact, functional human RPS14 gene and that they express it. All 14 of the emetine sensitive transformants analyzed by an S1 nuclease protection assay involving species-specific RPS14 DNA probes elaborate both the Chinese hamster and human S14 mRNAs. Total RPS14 transcription in transformants is quantitatively about the same as in untreated Chinese hamster and HeLa cells despite the fact that they appear to contain human S14 genes in a variety of copy numbers and chromosomal arrangements. Two experimental controls, an emetine resistant transformant and a clone transformed by the pSV2Neo vector (without human RPS14), express only the Chinese hamster RPS14 gene. Data obtained so far are summarized in Table I.

Table I. S14 mRNA Content of CHO Cells Transformed with Human RPS14 DNA Sequences.

Clone	Tot. S14[a] mRNA	CHO S14[a] mRNA (%)	Hum S14[a] mRNA (%)
Em^r-2	1.00	100	0
HeLa	0.99	0	100
Emetine-Sensitive Transformants			
2a	2.92	63	37
10a	3.37	68	32
10b	1.96	65	35
16b	2.50	54	46
18a	1.93	53	47
Emetine-Resistant Transformants			
Neo-1[b]	2.27	100	0
12a	3.20	100	0

[a] densitometry of S1 nuclease protection assays
[b] transformed by the pSV2Neo vector alone

DISCUSSION

That emetine resistant Chinese hamster cells transfected by pSVNS14 plasmids elaborate exogenous human RPS14 transcripts which complement the mutant phenotype demonstrates: i) the human RPS14 DNA clone used is indeed a functional ribosomal protein gene ii) the human clone used contains all of the cis-acting DNA sequences required to promote accurate RPS14 transcription iii) expression of exogenous RPS14 sequences is regulated coordinately with resident Chinese hamster rprotein genes and iv) the protein product of exogenous rprotein gene transcripts is transported efficiently to recipient cell nucleoli for assembly into functional, emetine sensitive ribosomes.

Acknowledgments. The authors wish to thank Ms. Barbara Van Slyke for capable technical aid. This research is supported by grant GM-23013 from The National Institutes of Health and is contribution No. 87-94-A from the Kansas State Agricultural Experiment Station.

REFERENCES

(1) Boersma, D., McGill, S., Mollenkamp, J. and Roufa, D.J. (1979) J. Biol. Chem. 254, 559-567.
(2) Nakamichi, N., Rhoads, D.D. and Roufa, D.J. (1983) J. Biol. Chem. 258, 13236-13242.
(3) Rhoads, D.D. and Roufa, D.J. (1985) Mol. Cell. Biol. 5, 1655-1659.
(4) Rhoads, D.D., Dixit, A. and Roufa, D.J. (1986) Mol. Cell. Biol. 6, 2774-2783.
(5) Nakamichi, N., Kao, F.-T., Wasmuth, J. and Roufa, D.J. (1986) Somat. Cell and Molec. Genet. 12, 225-236.
(6) Campbell, C.E. and Worton, R.G. (1980) Somat. Cell Genet. 6, 215-224.
(7) Southern, P.J. and Berg, P. (1982) J. Molec. Appl. Genet. 1, 327-341.
(8) Kawai, S. and Nishizawa, M. (1984) Mol. Cell. Biol. 4, 1172-1174.

TUMORIGENESIS IN TRANSGENIC MICE CONTAINING MTV-SV40 EARLY REGION HYBRID CONSTRUCTS

S.R. Ross, Y. Choi, I. Lee and D. Henrard

University of Illinois Medical Center, Chicago, IL 60612

Mouse mammary tumor virus (MTV) is a type B retrovirus whose sequences are found in the murine germ line and which is responsible for the transmission of mammary carcinomas in mice (1). The transcription of MTV RNA occurs primarily in the lactating mammary gland (2,3). In order to map the tissue-specific regulatory elements coded for by the provirus, we constructed a plasmid containing the SV40 early region coding sequences (large T and small t tumor antigens) under the control of the MTV long terminal repeat (LTR) (pLTag, Fig. 1); the level of SV40 RNA isolated from cells transiently transfected with this plasmid is induced 5-10 fold with dexamethasone. This molecule was introduced into the germ line of mice by microinjection.

Fig. 1. Restriction map of pLTag. ⬚ is the MTV LTR with a GRE, ▨ represents the t antigen coding region, ▬ the T antigen coding region. Abbreviations: Bs, BstXI; P, PstI; B, BamHI; C, ClaI; Sa, Sau3a; R, Rsa I; H, HaeIII; Hp, HpaII; G, BglII; Bc, BclI.

We have produced 7 independent lines of mice containing this plasmid in 3 different mouse strains, ICR outbred and C3H/HEN, two C3H MTV-susceptible strains and C57BL/6, a resistant strain (Table I). The number of copies of the injected DNA found in each animal ranges from 1.5 - >50. All of the mice contain intact LTR and SV40 sequences, based on restriction digestion with ClaI and BclI and Southern blot analysis (Fig. 1).

Table I. pLTag transgenic mice.

Mouse	Sex	Copy	Strain	Age≠	Tissue*
LTag1	M	4	ICR	3.5 m	TESTES, PROSTATE, LUNG, KIDNEY, ADRENAL, spleen, salivary gland
LTag2	F	2	ICR	3.5 m	KIDNEY, LUNG, spleen, salivary gland
LTag 3	F	1.5	C3H	5 m / 1d**	lung, kidney
LTag 4	nd	>50	C3H	1d	LUNG, SKIN LYMPHOMA
LTag 5	F	2	ICR	2m	nd
LTag 6	nd	6	C57	1d	nd
LTag 7	M	2	ICR	2m	nd

Abbreviations: nd, not determined; m, month; d, day
≠ age at death or present age
*UPPER CASE, tumor; lower case, RNA &/or protein analyzed
**RNA was analyzed from the tissues of offspring found dead on day of birth

All of the mice, except LTag 4 and 6, which were dead on the day of birth, were mated with the same strain animals of the opposite sex. (LTag 6 has not yet been analyzed.) LTag 4 was found dead in utero after a C section delivery and was considerably degraded; however, it appeared to have adenocarcinoma of the lung and a skin lymphoma. LTag 5 has produced one litter of 11, of which 4 are alive; the dead offspring are presently being analyzed. LTag 3 has produced 2 litters, one of 8; the two transgenic animals in the G2 generation of this litter were found dead. One of these animals showed expression of the MTV-SV40 hybrid transcript in kidney and lung, but not brain, liver, heart and thymus; the other was too degraded to analyze. All of the G2 offspring in the 2nd litter (11) are dead and are presently being analyzed. LTag 3 is showing signs of illness.

Neither LTag 1 or LTag 2 ever produced any offspring. At approximately 6 weeks of age, LTag 2 began to show signs of illness (scruffy coat, lethargy) and was sacrificed. LTag 1 developed visible abdominal masses at about 3.5 months; this male never mated with any females placed in the cage with him. Both mice were sacrificed at 3.5 months. LTag 1 had multifocal in situ adenocarcinomas of the prostate gland, lung, kidney and seminal vesicles, a neuroblastoma of the adrenal gland, a testicular lymphoma and possibly a Leydig cell carcinoma. LTag 2 had multifocal in situ carcinomas of the lung and kidney and mildly dysplastic mammary tissue. All tissue was examined by immunofluorescence with anti-SV40 antibody, and by Northern analysis and S1 protection assays. The tumor tissue contained both the hybrid transcript and SV40 early proteins t and T. Non-tumor tissue did not contain these transcripts or proteins, with the exception of the salivary gland and the spleen; in addition, SV40-positive lymphoid cells were frequently seen in negative tissue (i.e. brain, liver, etc.).

These results were unexpected in light of our results with a pLTR-HSV1 thymidine kinase plasmid, with which only mammary gland and testes expression was seen in transgenic mice (4) and those reported for transgenic mice containing an MTV 3' LTR-myc construct, in which predominantly mammary carcinomas were detected (5). The MTV-SV40 hybrid construct is consistently expressed in lung, kidney, salivary gland, lymphoid cells, and possibly testes. We do no know why no mammary carcinomas have developed; however, only LTag 3 has lactated more than once. We are presently constructing transgenic mice with greater portions of the MTV genome, in order to determine whether regions of the genome in addition to the LTR contribute to tissue specificity.

REFERENCES

(1) Nandi, S. and McGrath, C.M. (1973) Adv. Canc. Res. 17, 353.
(2) Varmus, H.E., Quintrell, N., Medeiros, E., Bishop, J.M., Nowinski, R.C., Sarkar, N.H. (1973) J. Mol. Biol. 79, 663.
(3) Hu, W.-S., Fanning, T.G., Cardiff, R.D. (1984) J. Virol. 49, 66.
(4) Ross, S.R. and Solter, D. (1985) Proc. Natl. Acad. Sci. USA 82, 5880.
(5) Stewart, T.A., Pattengale, P.K., Leder, P. (1984) Cell 38, 627.

NUCLEOSIDE TRIPHOSPHATE DEPENDENT DNA BINDING PROPERTIES OF mos

Arun Seth, Friedrich Propst and George Vande Woude

BRI-Basic Research Program, NCI-Frederick Cancer Research Facility, Frederick, Maryland

The viral mos (v-mos) gene of Moloney murine sarcoma virus encodes a 37 kilodalton env-mos fusion protein. The mos protein expressed by strains MSV ts110 and MSV124 have been shown to possess serine/ threonine autophosphorylation activity (1,2), however similar autophosphorylation activity has not been observed with the HT1 MSV v-mos product. The HT1 mos gene is identical in amino acid sequence to c-mos (mouse) (3).

Expression of c-mos RNA in Mouse Tissues

We have shown that c-mos mRNA is expressed in several mouse tissues such as testes, ovary, epididymis, kidney, brain and embryo (4). Our results indicate that RNA expression is developmently regulated in ovary and testes. It is highest in ovaries from day 14 to day 25 after birth, but thereafter decreases and reaches adult level after 35 weeks. In testes there is little or no mos expression before day 25. However, by day 33 the maximum level is reached and is equivalent to the adult level.

mos-nucleic acid binding is ATP dependent

The v-mos product has been localized to the cytoplasm and is expressed at very low levels (5). The low level of mos product in transformed cells makes the characterization of its biochemical properties difficult therefore, we over-produced the HT1 v-mos (p40^{mos}), N-terminal (p19^{mos}) and c-terminal (p25^{mos}) mos deletion mutant products in E. coli (Fig. 1) using the high level expression vectors pJL6 and pANH-1 (3,6). Previously we showed that p40^{mos} binds ATP and possesses ATPase activity (3). In the present study we investigated whether mos, like other ATPases possess DNA binding properties. Protein-DNA interactions were monitored by nitrocellulose blot protein-DNA binding assay. We show that p40^{mos} binds DNA and RNA in the presence of Mg^{2+} ATP and certain other nucleoside triphosphates whereas no binding is observed in the absence of ATP (Fig. 2).

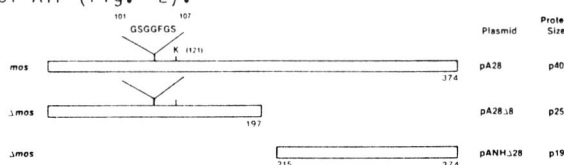

Fig. 1. Map of the mos and deletion mutant products expressed in E. coli. Plasmid names and the proteins they encode are indicated.

While the C-terminal deletion mutant protein (p25^{mos}) shows DNA binding activity, no DNA binding is observed with the N-terminal deletion mutant protein (p19^{mos}) (Fig. 2). These results indicate that all of the mos protein nucleic acid binding activity is localized to the N terminus of the protein. DNA binding to p40^{mos} was not observed in the presence of AMP or the nonhydrolysable ATP analog β,γ methylene adenosine 5'-triphosphate (AMPPCP), and in the presence of ADP the binding was only 20% of that observed with ATP. In contrast, the C-terminal deletion mutant protein p25^{mos} containing the ATP binding domain showed essentially no difference in

DNA binding in the presence of either ADP or ATP and also showed 4% and 45% binding in the presence of AMP and AMPPCP, respectively. Approximately 90% of the bound DNA was dissociated when any of the mos-DNA complexes were washed in the presence of 1M NaCl or in the absence of ATP.

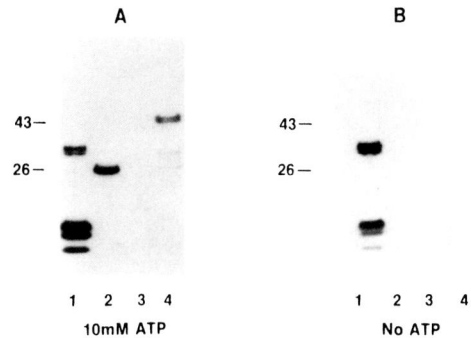

Fig. 2. mos DNA binding is ATP dependent. The mos DNA binding was assayed on nitrocellulose blots as described (Seth et al. submitted). The DNA binding was performed in the presence (A) or absence (B) of ATP. 1 μg of protein was loaded per lane. Lane: 1, (histones); Lane 2, (p25^{mos}) Lane 3, (p19^{mos}); lane 4, p40^{mos}.

These results suggest that DNA binding to mos protein occurs via a conformational change effected by the nucleoside triphosphate and since the p25^{mos} mutant protein lacks ATPase activity, the ATP hydrolysis is not required. The significance of mos protein-DNA binding we observe is not clear but our results raise the possibility that mos-nucleic acids interaction may play an important role in the transformation process.

REFERENCES

(1) Kloetzer, W.S., S.A. Maxwell, and R.B. Arlinghaus. 1983. Proc. Natl. Acad. Sci. USA 80:412-416.

(2) Maxwell, S.A., and R. B. Arlinghaus. 1985. Virology 143:321-333.

(3) Seth, A., and G.F. Vande Woude. 1985. J. Virol. 56:144-152.

(4) Propst, F. and G.F. Vande Woude. 1985. Nature 315:516-518.

(5) Papkoff, J., E.A. Nigg, and T. Hunter. 1983. Cell 33:161-172.

(6) Seth, A., P. Lapis, G.F. Vande Woude and T. Papas. 1986. Gene 42:49-57.

Research sponsored by the National Cancer Institute, DHHS, under Contract NO. N01-CO-23909 with Bionetics Research, Inc.

EXPRESSION OF A mRNA SPECIES OF THE MOUSE HOMEOTIC GENE LOCUS (HOX-3) IN YEAST

Martin J. Walsh, Mount Sinai Medical Center, Department of Pediatrics, New York, New York 10029

Using a galactose-inducible promoter (GAL 10), we have been able to express and induce a mRNA species which corresponds to 1.8kb in length of the mouse Hox-3 locus. Polyadenylated RNA isolated from yeast grown in galactose has been identified by Northern Blot analysis to be a Hox-3 transcript. Further studies of hybrid selected poly A$^+$ RNAs from both control and galactose-induced yeast using a cell-free translation assay indicates the appearance of a band of M$_r$ 27,000. This model presents a system which will help enable us to identify the structure and function of mammalian early determination gene products which otherwise would prove difficult to study.

RESULTS: Using cloning vector YEp-51 (gift from J. Broach, Princeton University), we have inserted the 2.2kb Bam H1 - Eco R1 fragment from pMoEA (gift from W. McGinnis, Yale University) adjacent to the GAL 10 promoter ligated between Bcl 1 - Sal I sites (see fig. 1). In addition to the Gal 10 promoter, YEp51 contains a transcription termination signal which lies distal to the promoter region. Orientation of the clone was determined by Xho I digestion of the plasmid DNA. The vector clone YEp51 (Hox-3) was then transfected and propagated in E. coli using HB101 to obtain sufficient quantities of plasmid DNA. Yeast strain of S. cerevisiae (KPPK-1) which requires leucine for growth was transformed with YEp51 (Hox-3) and grown in media supplemented with glucose or galactose and appropriate amino acids.

Figure 1 Schematic diagram of the construction of YEp51 (Hox-3).

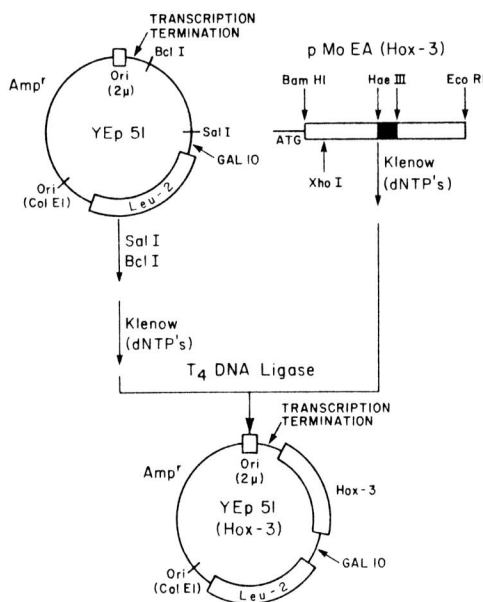

To examine the presence of RNA: RNA was isolated by procedure of D. Beach (unpub.) and was selected for poly A$^+$ on oligo-dT columns from yeast grown in glucose and galactose with the vector, with and without the Hox-3 insert. Aliquot of 3.0ug of Poly A$^+$ was applied to a 1.4% formaldehyde gel and transferred to nitrocellulose and was hybridized using the Hae III fragment of pMoEA DNA (Data shown, Figure 2). A 1.8kb fragment is shown to hyridize specifically to the Hox-3 probe.

Figure 2a. Northern blot analysis. 2b. In vitro translation using rabbit reticulocyte lysate.

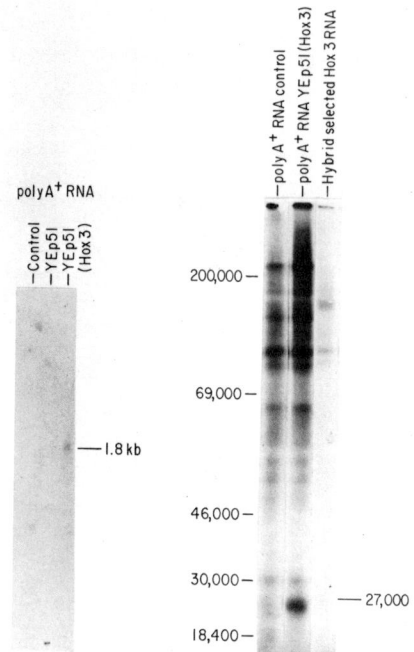

We hybrid selected 50ug Poly A$^+$ RNA to the Hox-3 genomic plasmid to obtain approximately 1ug of messenger RNA. Both polyadenylated RNA and hybrid selected Hox-3 transcript was translated by rabbit reticulocyte lysate (Promega Biotec). The appearance of a band at 27,000 M$_r$ indicates the possible translation of the gene transcript.

DISCUSSION: To this date, the roles and patterns of homeotic gene expression are still conjective, and still in question are the relative structure and function of homeotic gene products. An alternate means of current study is to introduce and express this gene and other homeotic genes in both prokaryotic and simple eukaryotic systems for isolation of large quantities of both RNA and protein. In the Northern Blot Analysis (fig.2) we see the presence of the Hox-3 transcript and although there has been reported homology with MAT alpha 2 of yeast S. pombe, there seems to be no cross hybridization with S. cerevisiae (3). In more recent experiments, we have identified a much less intense intermediate band of approximately 2.1kb (data not shown) which indicates that the yeast has processed further than the actual transcript and/or may be another species of mRNA, this is still currently being investigated. In vitro translation data, although inconclusive as to the final gene product, does demonstrate that this protein product seems to be stable under the growth conditions of S. cerevisiae. Future experiments will determine both the half-life of the RNA and to develop antibodies to help identify the structure and function of the protein.

References
(1) Tamanoi, F., Walsh, M.J., Kataoka, T., and Wigler, M. (1984) PNAS 81:6924.
(2) Awgulewitsch, A., Utset, M.F., Hart, C.P., McGinnis, W. and Ruddle, F.H. (1986) Nature 320:328.
(3) Porter, S.D. and Smith, M. (1986) Nature 320:766.

Acknowledgements:
I thank Drs. J. Broach and W. McGinnis for clones of YEp51 and pMoEA, respectively. This work was funded by the N.I.H. & the Mt. Sinai School of Medicine.

THE VIRAL ETS GENE CONTAINS DOMAINS FROM NONCODING AND CODING REGIONS OF THE CHICKEN CELLULAR ETS GENE

D. K. Watson[1], M. J. McWilliams[2] and T. S. Papas[1]. [1]Laboratory of Molecular Oncology, National Cancer Institute, Frederick, MD 21701; [2]Program Resources, Incorporated, Frederick Cancer Research Facility, Frederick, MD 21701.

E26 is a replication-defective avian retrovirus that induces erythroblastosis and myeloblastosis in chickens and transforms erythroblasts and myeloid cells in culture. The E26 provirus has been cloned and sequence analyses has identified two cell derived sequences, myb and ets, which are fused with gag in the p135 transforming protein (1). Comparison of the oncogenic properties of E26 and AMV suggests that the ets domain of E26 may be responsible for the erythroblast tropism of this virus.

Mammalian ets has unique domains on different chromosomes.

To define the origin of E26 and the function of the ets gene, we have initiated the characterization of the mammalian and avian ets genes. Using somatic cell hybrid panels, the mammalian homologs of the 5' v-ets domain, ets-1, were mapped to chromosome 11 in man, to chromosome 9 in mouse and to chromosome D1 in cat. The homologs of the 3' v-ets domain, ets-2, were similarly mapped to human chromosome 21, to mouse chromosome 16 and to feline chromosome C2 (2,3). Both human ets genes are transcriptionally active. The human ets-1 locus encodes a single mRNA of 6.8 kb; the second locus, ets-2 encodes three mRNAs of 4.7, 3.2, and 2.7 kb. At the predicted amino acid level, the human and viral gene products are over 95% conserved.

Chicken ets has contiguous ets-1 and ets-2 sequences.

We have determined the nucleotide sequence of the v-ets homologous exons of overlapping chicken genomic clones. Whereas the mammalian ets genes are present on two chromosomes, the chicken ets gene we have isolated has contiguous ets-1 and ets-2 sequences (Figure 1). The chicken c-ets gene appears to be a typical eukaryotic gene with nine viral homologous regions separated by introns of varying size, defined by consensus splice signals.

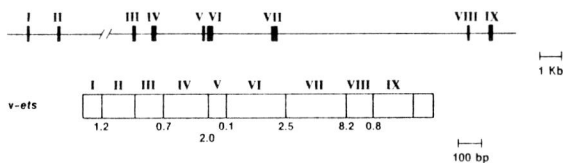

Figure 1. Genetic organization of the chicken c-ets gene. The previously sequenced v-ets portion of E26 is diagrammed below the chicken c-ets locus to display relative exon size.

Because ets-1 and ets-2 sequences can be found either on different chromosomes (mammals) or as contiguous sequences (chicken), we conclude that the v-ets is likely to contain at least two disso-ciable functional domains. In conclusion, E26 may actually include a total of four domains (Δgag, myb, ets-1 and ets-2).

Viral ets contains noncoding regions from the cellular ets gene.

The chicken locus is primarily expressed in normal cells as a 7.5 kb mRNA transcript (2). Since the v-ets domain of E26 is only 1.5 kb, the chicken gene product has much greater complexity, being five times larger. To determine whether all of the regions in c-ets homologous to v-ets are found in c-ets mRNA, probes from specific regions of exons I through exon IX were prepared and hybridized to polyadenylated RNA isolated from chicken thymus.

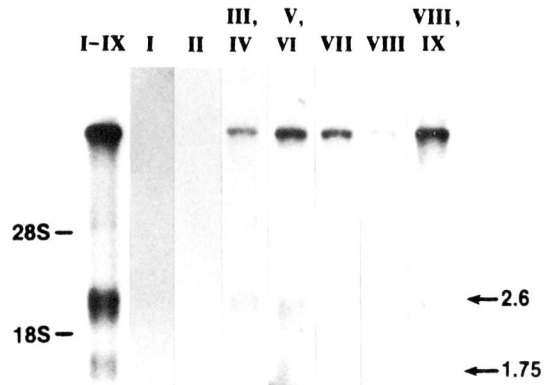

Figure 2. Chicken ets expression. Chicken thymus polyadenylated RNA samples (1 μg) were resolved on a 1.2% formaldehyde/agarose gel. Lanes were hybridized to v-ets (I-IX) or to chicken exon probes (as indicated) under stringent conditions. Mobility of the 28S and 18S ribosomal RNAs are indicated.

As can be seen (Figure 2), probes "I" and "II" fail to detect any RNA, while all other probes detect the 7.5 kb message. Thus, not only has E26 captured only a small portion of the normal chicken gene, but also has incorporated over 200 nucleotides not transcribed in the normal gene. In the virus, these noncoding nucleotides encode a significant proportion (14%) of the amino-acids of p135.

REFERENCES

1. Nunn, M.F., Seeburg, P.H., Moscovici, C., and Duesberg, P.H. (1983) Nature 306, 391-395.
2. Watson, D.K., McWilliams-Smith, M.J., Nunn, M.F., Duesberg, P.H., O'Brien, S.J., and Papas, T.S. (1985) Proc. Natl. Acad. Sci. U.S.A. 82, 7294-7298.
3. Watson, D.K., McWilliams-Smith, M.J., Kozak, C., Reeves, R., Gearhart, J., Nunn, M.F., Nash, W., Fowle III, J.R., Duesberg, P.H., Papas, T.S., and O'Brien, S.J. (1986) Proc. Natl. Acad. Sci. 83, 1792-1796.

EARLY DETERMINATION

Molecular and Developmental Analysis of the Homeobox and the Antennapedia Gene in Drosophila

W.J. Gehring, S.Schneuwly and J. Wirz
Biozentrum, University of Basel, Klingelbergstr. 70
CH-4056 Basel, Switzerland

The **homeobox** is a short DNA segment of ~180 bp characteristic of homeotic genes and other genes controlling the body plan in Drosophila (reviewed in Gehring, 1986). DNA sequences of 11 Drosophila homeoboxes and 19 homeoboxes of other insects and of vertebrates have been compared and analyzed. These sequences encode a structural domain of the respective proteins, the homeodomain, which is highly conserved during evolution: 13 out of 60 amino acids are invariant. Their consensus sequence is almost identical to the sequence of the homeodomain encoded by Antennapedia (Antp). It is proposed that Antp represents the **ancestral** or **prototype sequence** and specifies the developmental and evolutionary **ground state**. Homologies have also been found between the homeodomain and the mating type proteins of yeasts, which in turn are partially homologous to prokaryotic gene regulatory proteins containing the characteristic helix-turn-helix motif. The degree of conservation of the homeodomains is highest in the putative recognition helix thought to make the contacts to the nucleotides in the major groove of the DNA. This suggests that the homeodomains of various genes bind to similar or identical DNA sequences. When DNA sequences from homologous genes of different species are compared, e.g. the **engrailed** sequences from Drosophila, honeybee and mouse, the homology extends beyond the homeobox, suggesting that the homeobox-containing genes of vertebrates may serve a similar function as the homologous genes in Drosophila. Therefore, the genetic regulatory mechanisms controlling development may be more universal than anticipated.

On the basis of his extensive studies on the Bithorax-complex, Lewis (1978) has proposed that homeotic genes control developmental pathways and that the combination of homeotic genes that are active in a particular segment specify the kind of segment that is formed, i.e. segmental identity. Two types of homeotic mutations have been described which essentially have opposite effects; (i) **recessive loss of function mutants**, which in extreme cases are due to the deletion of the homeotic gene, lead to a transformation of a given segment into a more anterior one; (ii) a considerable number of **dominant gain of function mutants** are known that transform in the opposite (posterior) direction. The nature of these dominant mutations has been an enigma for molecular biologists. In Antp the recessive loss of function leads to an anterior transformation of T2 and T3 towards T1 and head segments (Wakimoto and Kaufman, 1981; Struhl, 1981; Schneuwly and Gehring, 1985), whereas dominant gain of function mutants lead to a transformation of the antennae into second legs, i.e. in the posterior direction. Most of these dominant mutations at the Antp locus are due to chromosomal inversions which separate the 5´ end of the gene from the protein coding region. Molecular analysis of one of these mutants (In(3R)Antp[73b]) indicates that the inversion leads to the fusion of the promoter and leader sequences of a foreign gene to the protein coding region of Antp (Schneuwly, Kuroiwa and Gehring, submitted). Since several inversions which result in similar homeotic phenotypes are known to have different chromosomal breakpoints, it seems unlikely that in all of those cases Antp is fused to an antenna-specific promoter. Therefore, we assume that any promoter that is active in the antennal imaginal disc at the right stage of development can generate the antennal legs. The reason why legs are only formed in place of the antennae has to be sought in the control circuits. Previous studies on transdetermination have indicated that the antennal disc cells can transdetermine frequently into leg cells without any detectable

mutational change (Gehring 1966 and 1967) suggesting that the antenna is a weak point in the circuits controlling developmental pathways. If this were the case, it should be possible to induce the formation of antennal legs by a promoter that can be induced in all cells, including the antennal disc.

In order to test this prediction we chose the heatshock promoter of the hsp70 gene which is active in all cells and, since the Antp gene is too large for transformation experiments, we had to use a cDNA clone. The Antp cDNA was inserted into a newly constructed P-vector under the control of the heatshock promoter hsp70 (Schneuwly, Klemenz and Gehring, submitted). The hsp70-Antp fusion gene was transfered into the germ line of normal flies. The resulting transformants possess two normal Antp[+] genes and an additional fusion gene under heatshock control. At normal temperatures these flies do not show any change in phenotype. However, if the transformed larvae are heatshocked during the early third larval stage, they develop into flies with antennal legs. Thus, the dominant gain of function phenotype appears to be due to overexpression of the Antp protein in the antennal disc, an ectopic site where the protein is normally not expressed at detectable levels. Therefore, it is possible to **alter the body plan** of Drosophila by altering the expression of the normal gene (or the normal homeotic protein) in a predictable way, and to "redesign the fruit fly".

References:
Gehring, W. (1966). J. Embryol. exp. Morph. **15**, 77-111.

Gehring, W. (1967). Dev. Biol. **16**, 438-456.

Gehring, W.J. (1986). Cold Spring Harb. Symp. Quant. Biol. **5**, 243-251.

Gehring, W.J. and Hiromi, Y. (1986). Ann. Rev. Genet. **20**, 147-173.

Lewis, E. (1978). Nature **276**, 565-570.

Schneuwly, S. and Gehring, W.J. (1985). Dev. Biol. **108**, 377-386.

Schneuwly, S., Klemenz, R. and Gehring, W.J. (1986). Redesigning the body plan of Drosophila by ectopic expression of the homeotic gene Antennapedia. (submitted to Nature).

Schneuwly, S. Kuroiwa, A. and Gehring, W.J. (1986). Molecular analysis of the dominant hoemotic Antennapedia phenotype. (submitted to The EMBO Journal).

Struhl, G. (1981). Nature **292**, 635-638.

Wakimoto, B.T. and Kaufman, T.C. (1981). Dev. Biol. **81**, 51-64.

Wirz, J. and Gehring, W.J. (1986). Localization of the Antennapedia protein in Drosophila embryos and imaginal discs. (submitted to The EMBO Journal).

MATERNAL GENES ORGANIZING THE ANTERO-POSTERIOR PATTERN IN THE DROSOPHILA EMBRYO

R. Lehmann, H.G. Frohnhöfer, T. Berleth, and Christiane Nüsslein-Volhard

Max-Planck-Institut für Entwicklungs-biologie, Abteilung Genetik, Spemannstr. 35/III, 7400 Tübingen, FRG

In insect embryos, experimental manipulations have accumulated evidence that the antero-posterior pattern is organized by two centers of activity, located at the two egg poles (1). While Drosophila is the system of choice for the genetic analysis of pattern formation, it has not been studied much with methods of classical embryology. However, recent experiments involving removal and/or transplantation of cytoplasm from the two egg poles provide evidence also in Drosophila for the localization of factors organizing the pattern (2). The experimental animals strongly resemble the phenotype of several maternal effect mutants, suggesting that the genes code for factors required for the establishment and function of the terminal organizing centers.

THE POSTERIOR CENTER. The development of the abdomen in the Drosophila embryo requires the activity of seven maternal effect genes. Lack of function of these genes lead to a deletion of the abdominal region while head, thorax and the somatic hind end, the telson, develop normally. Five of these loci affect at the same time pole cell formation and mutant embryos lack the specialized pole plasm including polar granules (oskar, vasa, valois, tudor, staufen) (3,4). Transplantation of cytoplasm from the posterior tip, but not from more anterior egg regions of wild type embryos into the abdominal region of mutant embryos can completely restore abdomen formation in most instances (4). Interestingly, rescue activity for the two mutants which affect only abdomen but not pole cell formation (pumilio and nanos) is also strictly localized in the posterior pole plasm. The mutants of six of these genes (with the exception of pumilio) do not complement in the cytoplasmic transplantation assay. Cytoplasm from pumilio embryos, when transplanted into the abdominal region, can restore abdominal segmentation in all of the other mutants and also in pumilio embryos. Our data suggest that pumilio is not affecting the posterior activity as such but the transmission of a signal from the posterior pole plasm to the abdominal region. This signal might be produced by nanos while the activity of the other five genes is required for the establishment and function of its source, the pole plasm.

Transplantation of the source to more anterior regions, while restoring abdominal segmentation, in general does not lead to a reorganization of the embryonic pattern and polarity. Only when transplanted to the anterior tip of recipient embryos, a suppression of head formation, which often is accompanied by the induction of posterior telson structures at the anterior is observed (2, 4). This activity is dependent on the wild type function of oskar, suggesting that this group of genes is involved in organization of posterior pattern and polarity (4).

THE ANTERIOR CENTER. Anterior development is affected by the maternal effect gene bicoid. In embryos from bcd females, head and thorax are lacking entirely and are replaced by a duplication of the somatic hind end, the telson. In weaker alleles, the reduction of anterior structures is incomplete and no telson duplications are formed. The gastrulation pattern and fate mapping studies indicate that the reduction in size of anterior anlagen is accompanied by an expansion of posterior anlagen in a coordinated manner. Temperature shift and pulse experiments using a ts allele indicate a phenocritical stage during syncytial blastoderm. Thus the antero-posterior pattern is labile until briefly before blastoderm cellularization, up to the time of spatially restricted gene expression of zygotic segmentation genes.

The mutant phenotype can be normalized and modified by the transplantation of cytoplasm from wild type embryos. bcd$^+$ activity is localized at the anterior tip of unfertilized eggs and early embryos until the cellularization of the blastoderm. Mutant embryos injected with bcd$^+$ cytoplasm anteriorly can develop a completely normal set of head and thorax, with concommitant suppression of the anterior telson duplication. The extent of anteriorness reached is dependent on the amount of transplanted cytoplasm as well as the bcd$^+$ gene dosage in the donor females suggesting that it is the quantity of one substance that determines different qualities. When transplanted to the middle or the posterior of mutant embryos, bcd$^+$ activity can ectopically induce head and thorax (or only thorax) formation while suppressing abdominal development. The efficiency of induction reached in such heterotopic transplantations decreases with distance from the anterior tip. The transplanted activity has long range organizing influence on polarity and pattern (5).

INTERACTIONS OF ANTERIOR AND POSTERIOR CENTER. Cytoplasmic transplantation experiments show a strong mutual inhibition of the activities localized in the polar centers. The inhibitory effects are dependent on gene activities which are also responsible for the inductions. Thus transplantation of posterior pole plasm into the anterior of bcd embryos can induce an abdomen with reversed polarity. Near-reciprocal effects are observed upon transplantation of anterior plasm to the posterior in osk embryos. Double mutants of osk and bcd lack polarity entirely and head can be induced at the posterior, abdomen at the anterior using the appropriate cytoplasm without encountering inhibitory effects.

MOLECULAR ANALYSIS. Cloning of the gene oskar is underway in our laboratory. The bcd gene has been tentatively identified in the genomic walk of Scott et al. (6) of the Antp region in a region around -40 by a 2,5kb maternal transcript. SP6 polymerase in vitro transcripts of a genomic clone from this region, when injected into the anterior of early embryos, induces bcd phenocopies with high frequency. In situ hybridization on early embryos shows that bcd mRNA is localized at the anterior tip of the egg, in agreement with the localization of the head-inducing activity (7).

REFERENCES

(1) Sander, K. (1976) Adv. Insect Physiol. 12, 125-238.

(2) Frohnhöfer, H.G., Lehmann, R., and Nüsslein-Volhard, C. (1986) J. Embryol. Exp. Morph. 97, Supplement.

(3) Schüpbach, T. and Wieschaus, E. (1986) Wilh. Roux Archives 195, 302-317.

(4) Lehmann, R. and Nüsslein-Volhard, C. (1986) Cell 47, in press.

(5) Frohnhöfer, H.G. and Nüsslein-Volhard, C. (1986) Nature, in press.

(6) Scott, M.P., Weiner, A.J., Hazelrigg, T.I., Polisky, B.A., Pirrotta, V., Scalenghe, F. and Kaufman, T.C. (1983) Cell 35, 763-776.

(7) Frigerio, G., Burri, M., Bopp, D., Baumgartner, S. and Noll, M. (1986), submitted for publication.

THE DEVELOPMENTAL ROLE OF MAMMALIAN HOMEO BOX GENES

F. H. Ruddle

Department of Biology, Yale University,
New Haven, Connecticut 06511

Homeo box sequences are 180 base pair coding domains in homeotic and other developmentally relevant genes. First described in Drosophila, these sequences are highly conserved in a broad range of species -- some distantly related. The homeo box function has not been definitively determined, but it is known that it shares the features of a DNA binding domain of the lambda repressor type. Experimental studies in the engrailed gene in Drosophila have provided direct evidence for a DNA binding function of the engrailed homeo box protein.

In our studies on the mammalian homeo boxes which show a high level of conservation with respect to the homeo box of the Drosophila Antennapedia gene, we have concentrated on linkage patterns, gene organization of the molecular level, gene expression, and mechanisms of gene regulation. A minimum of 12 different homeo boxes have been mapped to three loci, namely Hox-1, -2, and -3. In the mouse, these map respectively to chromosomes 6, 11, and 15, whereas in man the cognate sites map to chromosomes 7, 17, and 12. Hox-1 contains six homeo boxes clustered over approximately 80 kb. These have been designated in serial order from left to right: Hox-1.1, -1.2, -1.3, -1.4, -1.5, and -1.6. Hox-2 is also a cluster of at least five homeo boxes spanning a distance of approximately 60 kb. These are designated in serial order from left to right: Hox-1.5, -1.4, -1.3, -1.2, and -1.1. A sixth homeo box (Hox-1.6) may exist to the right of Hox-1.1. Hox-3 contains at least one homeo box, designated Hox-3.1.

Northern analysis has shown that each homeo box can be detected in individual transcripts. These patterns of expression suggest that each homeo box resides within an individual gene which we call homeo box genes, and which are given the same specific designations as the homeo boxes themselves. The individual homeo boxes appear to code for a number of different RNA transcripts determined most probably by differential promotion start sites, splicing, and processing. The specifics of this variation remain to be determined. One also can discern specificity of transcript expression dependent on tissue type and temporal patterns during development. Major organ systems that show expression are the central nervous system (CNS), testis, and ovary.

We have focused our attention on the specificity of expression of transcripts Hox-1.5, -2.1, and -3.1 in the CNS. In 13.5 day old fetuses, we see distinctive patterns of expression for these transcripts. Hox-1.5 shows high expression with an anterior limit at the anterior limit of the hind brain. The anterior limit of Hox-2.1 is in the mid-hind brain and Hox-3.1 has its anterior limit at the level of the third cervical vertebra. Each transcript shows a pattern of high expression at its anterior limit and a gradient of reduced expression in an anterior-posterior direction. These patterns appear to be established during organogenesis and extend into the newborn and young adult. Data on additional transcripts and expression on earlier stages will be presented.

We have attempted in the case of Hox-1.5 to demonstrate a DNA binding function of the protein product in the 5' region of the Hox-1.5 gene. These experiments have successfully demonstrated such a function and we shall describe these data in detail.

Altogether, the homeo box system in mammals appears to function in a developmental regulatory manner. We shall focus on the possibility that the mammalian homeo box clusters may indeed be controller genes that regulate developmental events.

GENES THAT CONTROL ASPECTS OF NEMATODE DEVELOPMENT

R. Horvitz

Department of Biology, MIT, Cambridge, MA 02139

Genetics provides one approach toward a molecular understanding of development. For example, the characterization of the phenotype of a mutant animal can indicate where and when the product of a particular gene is utilized during development. In addition, the isolation and characterization of the products of genes found to be important for development may lead to an elucidation of the molecular mechanisms that control development.

The nematode *Caenorhabditis elegans* is an appropriate organism for the genetic analysis of development. This animal is well suited for genetical research: it is only one mm in length and has a three-day life cycle, so that both large numbers of individuals and many generations are easily examined (1). Furthermore, the development of *C. elegans* is known in great detail: the complete pattern of cell divisions, migrations, deaths and differentiations has been elucidated (2,3,4). Thus, mutant animals defective in specific aspects of development can be identified and readily analyzed. We have focussed our efforts on the isolation and characterization of mutants abnormal in cell lineage, cell death, and neuronal differentiation and functioning.

(I) ISOLATION OF CELL LINEAGE MUTANTS. Cell lineage mutants have been isolated by identifying mutants abnormal in specific morphological structures or behaviors and examining such mutants for defects in the cell lineages that generate the cells of those structures or that control those behaviors. In addition, mutants abnormal in cell number have been identified directly. Some cell lineage mutants have been obtained by isolating mutations that suppress or enhance the defects of other cell lineage mutants.

(II) MANY CELL LINEAGE MUTANTS ARE HOMEOTIC. Most of the cell lineage mutants that have been characterized in detail display transformations in the fates of particular cells: certain cells express lineages normally associated with other cells instead of their own lineages. These mutants can be considered to be single-cell homeotic mutants (5). Some cause spatial transformations in cell fates (so that cells behave like cells located elsewhere in the animal), whereas others cause temporal or sexual transformations (altering the time of expression or the sexual-specificity of particular cell fates). Of the approximately 40 homeotic genes that have been identified, detailed developmental and genetic studies have established two of these genes as particularly good candidates for directly controlling development: *lin-12* (6) and *lin-14* (7). Both of these genes have been cloned and characterized molecularly (8; G. Ruvkun, V. Ambros and R. Horvitz, unpublished results).

(III) GENETIC PATHWAYS THAT CONTROL CELL LINEAGE. Multiple genes involved in particular cell lineages have been identified. The interactions among such genes have been examined to define genetic pathways. The most extensively characterized genetic pathway for *C. elegans* cell lineage is that of vulval development. More than 200 mutations defining 31 genes involved in vulval development have been identified (9; E. Ferguson, S. Kim and R. Horvitz, unpublished results). Most of these genes have been placed within a branched pathway for the specification of the vulval cell lineages (E. Ferguson, P. Sternberg and R. Horvitz, unpublished results). Some of these genes are required for the generation of the cells involved in vulval development, some are involved in the determination of the fates of these cells, and others are necessary for the expression of particular fates once those fates have been determined.

(IV) PROGRAMMED CELL DEATH. Of the 1090 somatic nuclei generated by the *C. elegans* cell lineage, 131 undergo programmed cell death. Mutants in which these deaths fail to occur have defined two genes, *ced-3* and *ced-4* (10). Surviving cells in these mutants differentiate and, in at least some cases, are capable of functioning as specific neurons (10; L. Avery and R. Horvitz, unpublished results). The construction and analysis of genetic mosaics for *ced-3* and *ced-4* have indicated that these genes act within the cells that die (J. Yuan and R. Horvitz, unpublished results).

(V) DIFFERENTIATION AND FUNCTIONING OF THE SEROTONINERGIC HSN MOTORNEURONS. The genetic analysis of the *C. elegans* behavior of egg laying has led to the identification of a set of 47 mutants that lack the functioning of the HSN motorneurons that innervate the vulval muscles and drive egg laying (C. Desai and R. Horvitz, unpublished results). These mutants define 16 genes that are involved in distinct steps of HSN neurogenesis: the establishment of sexual identity, long-range cell migration, neuronal process elaboration, expression of the neurotransmitter serotonin, and the development of functional synaptic interactions.

(VI) PROSPECTS. We hope that the further genetic and molecular characterization of genes involved in these various aspects of *C. elegans* development should help elucidate the biochemical bases of animal development and differentiation.

REFERENCES

(1) Brenner, S. (1974) Genetics 77, 71-94.
(2) Sulston, J. and Horvitz, R. (1977) Devel. Biol. 56, 110-156.
(3) Kimble, J. and Hirsh, D. (1979) Devel. Biol. 70, 396-417.
(4) Sulston, J., Schierenberg, E., White, J., Thomson, N. and Brenner, S. (1983). Devel. Biol. 100, 64-119.
(5) Sternberg, P. and Horvitz, R. (1984) Ann. Rev. Genetics 18, 489-524.
(6) Greenwald, I., Sternberg, P. and Horvitz, R. (1983) Cell 34, 435-444.
(7) Ambros, V. and Horvitz, R. (1984) Science 226, 409-416.
(8) Greenwald, I. (1985) Cell 43, 583-590.
(9) Ferguson, E. and Horvitz, R. (1985) Genetics 110, 17-72.
(10) Ellis, H. and Horvitz, R. (1986) Cell 44, 817-829.

AXIS DETERMINATION IN THE AMPHIBIAN EMBRYO

J. Gerhart

Department of Molecular Biology, Univ. of California, Berkeley CA 94720

We are studying the series of developmental processes by which the intracellular organization of the amphibian egg is changed into the multicellular organization of the embryo. We are particularly interested in the determination of the dorsal anatomical elements of the embryonic body axis, the notochord, somites, and central nervous system. These structures, and gill slits, define vertebrates. The main processes are outlined as follows:

Establishment of primary polarity and the animal-vegetal axis: The oocyte originates by mitosis from a stem cell, the oogonium. At this earliest stage the oocyte already has a polarity manifested by an axis of organelles and regional specializations: 1) the unclosed division bridge to neighboring oocytes, 2) the centrosome surrounded by a sphere of mitochondria and Golgi vesicles, and 3) the eccentric nucleus. The nuclear end of the cell eventually becomes the animal hemisphere of the egg, whereas the centrosome and division-bridge end becomes the vegetal hemisphere (Coggins, 1973; Wiley et al 1985). The organellar axis, which constitutes the primary polarity of the young oocyte, differentiates into the animal-vegetal axis of the full grown oocyte (and egg) over a period of months. Differentiation is especially evident with the build-up of close-packed large yolk platelets in the vegetal hemisphere. Although vitellogenin enters the oocyte uniformly by receptor-mediated endocytosis, the yolk platelets (membrane-bounded endosomal intermediates) newly formed from it are directionally transported to the vegetal end of the cell and accumulate there (Danilchik and Gerhart, 1986). Thus, the oocyte has means to localize materials within itself. Other materials such as maternal mRNA's may be localized by this same transport process. The interface of the animal and vegetal cytoplasmic regions is important as the future site of the blastopore at gastrulation. The different materials of the oocyte's two hemispheres represent true cytoplasmic localizations.

Secondary polarity by cortical rotation of the egg: The unicellular egg preserves the organization of the oocyte, with an important modification following fertilization. The egg's thin (2-5 μm) surface layer rotates 30 degrees relative to the egg's large cytoplasmic core (1300 μm), in an animal-vegetal direction. The linear distance of displacement is large, 300 μm. By rotation, the animal-vegetal contacts between the two cytoplasmic layers of the egg are systematically offset, and the primary polarity of the egg is altered. Ancel and Vintemberger discovered this rotation process (in Rana) and Elinson has added to our knowledge of it. Vincent et al (1986) have recently analyzed it in Xenopus. It is an entirely maternal process. The direction of rotation invariably predicts the position of the future dorsal midline of the embryonic body axis of the embryo, and the extent of rotation predicts quite well the anterior limit of the dorsal structures of the body axis. Rotation is needed for dorsal but not for ventral development. Without rotation the egg develops as a radially ventralized "invertebrate" embryo (Scharf and Gerhart, 1980, 1983; Elinson and Manes, 1980). We think rotation affects one quadrant of the cytoplasm of the vegetal hemisphere (but not the animal hemisphere), locally activating it for promoting later dorsal development, as discussed below. Other regions of the vegetal hemisphere can be also activated, causing twins or even radially dorsalized embryos to form (Black and Gerhart, 1986). We can force artificial rotation to occur in a chosen direction to a chosen extent, using gravity or centrifugation; this initiates dorsal development normally. We can also alter the magnitude of the egg's response to cortical rotation, and thereby alter the extent of dorsalization.

Vegetal induction of the place and time of gastrulation: At the midblastula stage, 8 hrs after fertilization, vegetal cells start inducing nearby animal hemisphere cells to prepare for gastrulation (Nieuwkoop, 1977; Gimlich and Gerhart, 1984). The affected cells will constitute the gastrula's marginal zone, a wide band between the blastopore (40 degrees latitude from the vegetal pole) and the equatorial pigment line. Cells of this zone will take the major part in gastrulation: their motility and repacking will transform the egg axis into the embryonic body axis (for a review, see Gerhart and Keller, 1986). Any animal hemisphere cell, if transplanted into the marginal zone 2 hrs before gastrulation will respond to the vegetal cells and engage in active motility and rearrangement, while uninduced animal hemisphere cells will just flatten, as part of epibolic expansion of the blastocoel roof.

Cortical rotation by the egg is not needed in order for vegetal cells to be active in inducing gastrulation, since embryos completely blocked for rotation can still gastrulate. However, their gastrulation is radially symmetric and of a ventral type, that is, limited in extent and late in time, beginning at 12 hrs. Thus, vegetal cells, because of the cytoplasm they receive from the oocyte, are inherently capable of inducing a ventral-type gastrulation in marginal zone cells. Rotation is needed for a subset of vegetal cells to exceed this level of inductive activity, to induce a dorsal-type gastrulation which is extensive and early, starting at 10 hrs. Thus, we think rotation in the egg spatially patterns the vegetal cytoplasm so that when cellularized, one subregion will be more inductive than the others and will cause its marginal zone neighbors to start early, extensive gastrulation. Rotation seems simply to cause a local exaggeration of an inductive process that is inherent in vegetal cells anyway. Induction and the response to induction seem to require new gene expression (see Gerhart et al, 1986) as well as the scheduled translation of maternal RNA's, although this is not yet well analyzed.

Formation of the body axis: In gastrulation, the embryonic body axis comes into existence as the result of three types of integrated movement by cells of the marginal zone: 1) the formation of 3 embryonic germ layers, 2) the convergence of cells toward the dorsal midline, building up that region of the embryo, and 3) the elongation of cell populations into rows on the dorsal midline, forming the anteroposterior dimension of the body axis (Keller et al, 1985). Normally, the earliest gastrulating cells eventually organize the head, becoming the most anterior dorsal parts of the body axis, and inducing the formation of the anterior neural plate. Hama, Kaneda, Suzuki and others (see Suzuki et al, 1984) have shown in explants that early gastrulating cells of the marginal zone are not determined as anterior dorsal cells at the start of gastrulation but, paradoxically, as posterior ones. These early cells acquire anterior dorsal character <u>during</u> gastrulation, in connection with their movements. We have come to similar conclusions based on preliminary experiments in which we inhibit gastrulation at various times after it has begun. We find indeed that early gastrulating cells develop as part of the tail if blocked early (and the tail is the only completed part of the body axis); as part of the trunk if blocked intermediately (and the tail and trunk comprise the entire body axis), and head parts if allowed to persist longer in gastrulation (and a complete axis is formed). Thus, gastrulating cells may sequentially gain the capacity for ever more anterior dorsal development based on the <u>amount</u> of gastrulation they accomplish. If they start late or stop early in gastrulation, they qualify only for ventral posterior fates. Their final fates are not determined before gastrulation, whereas their types of movement, interactions, and responses are determined.

Thus, we propose that cortical rotation of the egg is needed to activate a region of vegetal cytoplasm such that cells cleaved from it will be sufficiently inductive to cause their neighboring marginal zone cells to start gastrulation early, thereby to have a long duration to proceed though the determinative steps leading to evermore anterior and dorsal fates. Late gastrulating cells of the less induced regions never progress beyond ventral fates. We assume that the final anteroposterior and dorsoventral determination of marginal zone cells somehow goes hand in hand with their explicit success in forming parts of the body axis. Developmental processes must succeed at all stages for dorsal anterior development to occur. Failure in any one leads to loss of the anterior end of the axis. We can produce cyclopic (one central eye) or headless (acephalic) embryos by slightly blocking rotation in the egg, or by removing some of the most inductive vegetal cells of the blastula, or by stopping gastrulation early. Radially ventralized embryos result if <u>any</u> of these three processes is blocked entirely.

In summary, we now have a tentative idea of the relationship of each process to the next, that is, how each modulates ("fine-tunes") the next process as far as <u>when, where, and how much</u> of the next will occur. We speak of determinants only in the sense that one stage determines initial conditions (when, where, how much) for an otherwise autonomous process of the immediate next stage. We don't think the egg has foreknowledge of the organization of the adult. For this reason we favor the word "cues" instead of "determinants" (since they don't <u>cause</u> the next process but just orient or delimit it) and we expect that a cue can affect any of a wide variety of cell activities (motility, cell shape, translation, secretion, etc.), depending on the stage, and not just gene expression.

Acknowledgements: This work was supported by USPHS grant GM19363.

References:

Black,S.D. & Gerhart,J.C.(1986) Dev.Biol.116: 228-240.
Coggins,L.(1972) J.Cell Sci.12: 71-93.
Danilchik,M.V. & Gerhart,J.C.(1986) Dev. Biol.in press
Gerhart,J.C. et al (1986) BioScience 36: 541-549.
Gimlich,R.L.(1986) Dev.Biol.115: 340-352.
Gimlich,R.L. & Gerhart,J.C.(1984) Dev.Biol. 104: 117-130.
Keller R.E. et al (1985) J.Emb.Exp.Morph. 89S: 185-209.
Mane,M. & Elinson,R.(1980) Roux's Arch.198: 73-76.
Nieuwkoop,P.D.(1977) Curr.Top.Dev.Biol.11: 115-132.
Opresko,L.et al (1982) Cell 22: 47-57.
Scharf,S.R. & Gerhart,J.C.(1980) Dev.Biol. 79: 181-198.
Scharf,S.R. & Gerhart,J.C.(1983) Dev.Biol. 99: 75-87.
Suzuki,A.S.et al (1984) Dev.Growth Diff.26: 81-94.
Vincent,J.-P.et al (1986) Dev.Biol.114: 484-500.
Wiley,C.C.et al (1985) J.Emb.Exp.Morph.89S: 1-15.

EARLY EVENTS IN TISSUE DETERMINATION IN THE AMPHIBIAN EMBRYO

H. R. Woodland, E. A. Jones & C. Wilson
Department of Biological Sciences, University
of Warwick, Coventry CV4 7AL, England.

From the earliest days of experimental embryology, amphibians have been a favourite material for analysing development. This is chiefly because amphibian embryos lend themselves to manipulative experiments, especially grafting and more recently, micro-injection. These two techniques are the theme of the experiments described below.

Our current understanding of the first steps in tissue diversification in Xenopus is summarised in Fig 1. In vertebrates the only available probes for differentiation events are terminally differentiated products. The early events of differentiation can only be revealed indirectly by functional assays of commitment to particular pathways of development. These are of two main types. Most simply tissue fragments are placed in the neutral environment of a saline solution to differentiate in isolation; this has been called a "specification" assay (1). Alternatively, the tissue is grafted into embryonic sites different from that which it normally occupies, to provide a more stringent test of its commitment to a particular developmental pathway.

MARKERS OF CELL DIFFERENTIATION

In analysing the events of early development it is very useful to have molecular markers of development. This supplements tissue identification in individual grafting experiments, but becomes indispensible where extreme treatments are employed, such as blocking cell division with drugs. We have used three monoclonal antibodies which identify the major tissue types of each of the three allocation events outlined in Fig 1.

Fig 1. (A) Flow diagram of the main steps in forming the major embryonic tissues. (B) and (C) show diagrams of the two inductive events (arrows) shown in (A). (B) Mesodermal induction. (C) Endodermal induction.

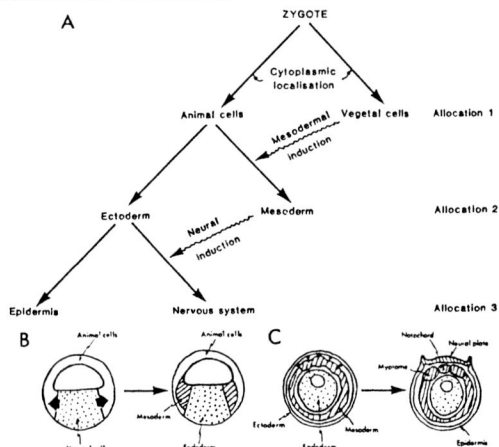

Monoclonal 2F7 identifies all of the variety of epidermal cells in the amphibian larva (2). 5A4 reacts with the myofibrils of striated muscle. Lastly 2G9 reacts with most or perhaps all cells of the nervous system. Such markers, as well as cloned probes like actin mRNA sequences, are invaluable in answering such questions as, when does mesoderm and nervous system formation occur? what limits the extent of these tissues? and so on.

The Mesoderm/Ectoderm Allocation

The early amphibian embryo consists of two kinds of cells (excluding the germ cell lineage), animal and vegetal (3). In isolation the animal cells become epidermis, expressing a variety of epidermal markers, including 2F7 binding (Jones and Woodland, 1986). This therefore represents the free-wheeling status of these cells, changes in the environment being necessary to divert them into other pathways. However, experiments involving cell disaggregation and blocking cell division suggest that to allow even this free-wheeling differentiation both simple cell interactions and cell divisions are necessary, up to but not after the early blastula stage (2).

It is now well established that the mesoderm is generated from animal cells by an inductive stimulus from the endoderm (eg 4, 5, 6). Using the muscle marker to identify mesoderm in grafted tissue, we have charted the period over which the animal cap cells are responsive to mesodermal induction and vegetal cells can induce mesoderm. In the mass of the embryo both disappear in the mid-gastrula (stage 10½) which is when cell rearrangements render the termination of this process essential (teleologically speaking!). However, the mesodermal induction stimulus persists in the posterior neural plate through to the late neurula stage (stage 24). This part of the neural plate is where the tail mesoderm arises (7) but we still do not know when these cells are actually induced to form mesoderm. It is possible that the mesodermal inducer has a quite different function in this region, since for other reasons, it has been implicated, in a double gradient model, in establishing the posterior nature of the developing CNS (8).

Our results suggest that mesodermal induction begins at a very early stage of development, probably between the 32- and 64-cell stages, long before transcriptional activation at the mid-blastula transition (9). Our current working hypothesis is that the extent of the mesoderm is limited by a barrier created by the induced cells. This requires that the mesodermal cells are not inductive, which is certainly true at later stages, and thus an equatorial band of mesoderm would be formed wherever responsive cells border on inductive vegetal cells.

The Nervous/Epidermal Allocation

The possession of early markers of both nervous and epidermal tissue classes has enabled us to devote particular attention to this third tissue allocation event. It was established long ago by Spemann and his associates (10) that ectoderm in contact with dorsal mesoderm is induced to become nervous system. In addition, the classical literature contains reports (not using Xenopus) that nervous system can itself induce more nervous system, making it complicated to envisage just how the neural plate is so precisely limited in extent. We find that in Xenopus post-cranial regions only notochord and myotomes are inductive and we have been completely unable to demonstrate any ability of neural plate to induce more nervous system (eg by transplanting it to ventral positions of blastulae and gastrulae). Thus in post-cranial regions at least, the extent of the neural plate seems to be simply limited to the regions of ectoderm that overlie the notochord and myotomes in the early neurula. Very soon the neural plate rolls up and new ectoderm overlies these regions. However, we find that the ectoderm has entirely lost responsiveness to neural induction by this stage, even though the stimulus persists at least until the early neurula, but rather is fully committed to the epidermal pathway of development. Thus, in these regions, the extent of the neural plate is apparently established in a simple way.

Neural induction begins as soon as the ectoderm becomes underlain by dorsal mesoderm in the early gastrula (stage 10). This is earlier than a recent report (11), but these authors examined a region of ectoderm that is not underlain at the earliest stages of gastrulation. In any case neural induction begins very early and proceeds in trunk and brain regions for only a short period up to the end of gastrulation.

REGULATION OF INJECTED ACTIN GENES IN TRANSGENIC FROGS

The ultimate aim of the research described above is to isolate the genes which direct early development and to unravel their function. To do this it is vital to be able to re-introduce genes into the embryo with an expectation that they will be correctly regulated. Very recent work indicates that this is feasible in Xenopus. Correct temporal regulation has been shown for a gastrula-specific gene (12), and we have investigated spatial regulation using a cardiac actin gene (13). The cardiac actin gene is expressed only in heart muscle in the adult animal, but in all striated muscle in the embryo and tadpole (13, 14). Its transcripts appear at the end of gastrulation and rapidly increase to become one of the most abundant mRNAs in the neurula.

We isolated a cardiac actin gene from X. borealis. This is sufficiently different from its X. laevis homologue that its transcripts can be detected in an X. laevis embryo (Fig 2). The gene was injected into the cytoplasm at the one cell stage and the embryos dissected one day later, at the neurula stage. As can be seen (Fig 2) nearly all expression of the injected gene is found in the somites, just like expression from the endogenous gene. Very low levels of transcripts are seen in other regions, but this is also true of the chromosomal gene, and so it does not necessarily represent inappropriate expression.

Fig 2. Expression of an X. borealis cardiac actin gene injected into an X. laevis zygote. The embryos were dissected into the regions shown, at the neurula stage. A, injected embryos (0.5 ng DNA/embryo); U, uninjected controls. The RNA was extracted and hybridized to a probe spanning the 5' boundary of the second exon, then subjected to S1 nuclease digestion and electrophoresis. The bands generated by tadpole RNA from X. borealis (b), X. laevis (1) and a 1:10 mixture of the two (b/1) are shown on the left. (From ref. 13).

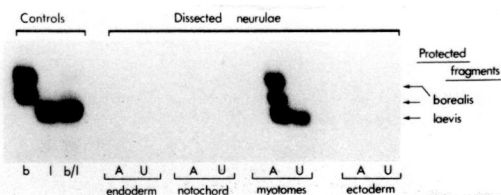

In order to conduct a properly controlled experiment we fused the two 5' exons of the actin gene to the last exon of a mouse globin gene. The expression of this "actbin" construct was compared to that of a "hisbin" construct which consisted of the 3' sequences of the mouse globin gene fused to the 5' upstream and transcribed regions of an X. laevis H3 histone gene. Whereas only 10% of hisbin expression occurred in the myotomes, 90% of the actbin was found in this region. Since both kinds of injected gene were fairly evenly distributed through the embryo, the localised expression represented correct regulation that was specific to the actin gene.

Apart from the fact that these experiments show proper regulation of injected genes in an animal well adapted to experimental embryology they have a number of other features of interest. The first is that the assays are very quick to perform. Correct spatial expression can be demonstrated within one day of injecting the zygote, and even single embryos provide abundant material for analysis. The results are consistent from embryo to embryo, probably because most expression occurs from the many thousands of unintegrated genes which are spread evenly through the embryo. Although the results need to be supplemented with data from other genes it seems likely that such transformed Xenopus embryos have special advantages for the analysis of tissue-specific gene expression.

REFERENCES

(1) Slack, J.M.W. (1983) From Egg to Embryo. Determinative events in early development. Cambridge University Press.
(2) Jones, E.A. and Woodland, H.R. (1986) Cell 44, 345-355.
(3) Woodland, H.R and Jones, E.A. (1985) Nature 318, 102-104.
(4) Nieuwkoop, P.D., (1973) Advances in Morphogen 10, 1-39.
(5) Dale, L., Smith, J.C. and Slack, J.M.W. (1985) J. Embryol. Exp. Morph. 89, 289-313.
(6) Gurdon, J.B., Fairman, S., Mohun, T.J. and Brennan, S. (1985) Cell 41, 913-922.
(7) Pasteels, J. (1939) Academie Royale de Belgique Classe des Sciences Bulletin 25, 660-666.
(8) Toivonen, S. and Saxen, L. (1968) Science N.Y. 159, 539-540.
(9) Newport, J. and Kirschner, M. (1982) Cell 30, 675-686.
(10) Spemann, H. (1938) Embryonic Development and Induction. Yale University Press, New Haven.
(11) Jacobsen, M. and Rutishauser, U. (1986) Developmental Biology, 116, 524-531.
(12) Krieg, P.A. and Melton, D.A. (1985) EMBO J. 4, 3463-3471.
(13) Wilson, C., Cross, G.S. and Woodland, H.R. (1986) Cell, in press.
(14) Mohun, T. J., Brennan, S., Dathan, N., Fairman, S. and Gurdon, J. B. (1984) Nature 311, 716-721.

DEVELOPMENTAL REGULATION OF METAL BINDING PROTEINS
IN ARTEMIA SALINA. R.A.Acey, B.N.Yoshida and E.
Maldonado, Department of Chemistry, California State
University, Long Beach, CA 90840

INTRODUCTION

Metallothioneins(MT) are a unique class of low
molecular weight metal binding proteins found in a
variety of pro- and eukaryotic organisms(1). Since
heavy metals are potent inducers of MT, the protein
has been implicated in metal detoxification. However,
there is an increasing awareness of a fundamental
role of MT in metal homeostasis and cell differentiation during development(2).A popular system for
studying embryonic development is the crustacean
Artemia salina. Previous studies have revealed the
presence of a MT-like cadmium binding protein (MBPI)
in Artemia embryos(3). We present evidence here that
MBPI is developmentally regulated. Moreover, MBPI is
associated with significant amounts of zinc, and
during the course of development, a second zinc binding activity appears.

METHODS

Dormant encysted embryos of Artemia were steri-
lized and cultured as described elsewhere(4). Embryos
were homogenized in 10 mM Tricine, 250 mM sucrose,
5 mM $MgCl_2$, 0.5 mM EDTA, 1 mM DTT, pH 8.0. The medium
was supplemented with 1 mM PMSF and 10 ug/ml soybean
trypsin inhibitor. The crude homogenate was filtered,
homogenized in a Dounce homogenizer, and centrifuged
for 90 minutes at 40,000 rpm. The supernatant was
dialyzed exhaustively under nitrogen against 50 mM
Tris HCl, 0.02% (w/v) sodium azide, pH 8.0. Equal
amounts of protein from various stages of development
were applied to a G-75 column and eluted with dialysis
buffer. Eluate was collected in tubes containing DTT.
Column fractions were assayed for zinc by flame atomic
absorption spectrophotometry. In vitro metal binding
activity was determined by preincubating supernatant
with limiting amounts of ^{109}Cd (or ^{65}Zn) prior to
fractionation and monitoring the column eluate for
radioactivity. Total cadmium binding activity was
measured by the method of Eaton and Toal(5).

RESULTS

In cytosols obtained from encysted embryos, a low
molecular weight zinc binding activity was clearly
evident (Fig.1). The observed activity is super-
imposable on the cadmium binding activity previously
referred to as MBPI. While endogenous cadmium level
in cytosol from developing embryos is below our level
of detection, significant amounts of zinc are associated with the proteins eluting in the void volume
and the MBPI peak (Fig.2A). Interestingly, during the
next 24 hours of embryonic development, there is a
significant decrease in the amount of zinc associated
with MBPI and concomitant emergence of a third zinc
binding activity (ZnBP) (Fig.2B). Little change is

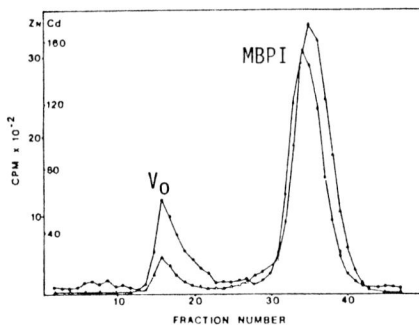

Fig. 1 In vitro metal binding activities ▲ ^{109}Cd
● ^{65}Zn. Fractionated on a 1.0 cm. x 40 cm. column

evident in the amount of zinc associated with protein eluting in the void volume. However, after 48
hours of development, the amount of zinc in the void
volume is diminished while ZnBP and MBPI are associated with increased amounts of zinc. Moreover, the
cadmium binding capacity of MBPI increases with age
of the embryo(Table I).

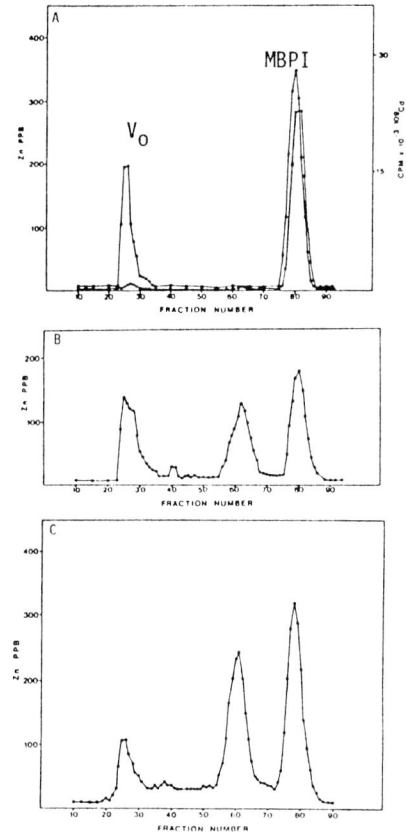

Fig. 2 Endogenous Zinc ● in cytosols from A)dormant cysts,
B) 24 hour embryos and C) 48 hour embryos.
Fractionated on a 2.5 cm. x 100.0 cm. column.
■ = ^{109}Cd levels

Table I. Cadmium binding capacity of MBPI

Hrs. of Development	CPM ^{109}Cd bound*
0	2727
24	8115
48	19080

*Standardized to 220 nm absorbance

DISCUSSION

It is apparent from our results that Artemia is
an interesting system for the study of metal homeostasis and gene expression. The fact that the cadmium binding activity associated with MBPI is increasing throughout early development while that of
zinc fluctuates,may be because it is composed of more
than one protein. An attractive postulate for the
appearance of ZnBP would be metal flux from a metal
storage ligand (MBPI?) to specific metalloproteins.
Other possible explanations would include genetic
expression, changes in turnover rates, or altered
rates of metal exchange as a result of differences
in metal uptake or compartmentalization.

REFERENCES
(1) Yutaka, K. and Kagi, J.H.R.,(1978) Trends Bio-
    chem Sci. 3, 90 - 93.
(2) Nemer, M. et. al.,(1984) Dev. Biol.102,471-482
(3) Thall, A. and Acey, R.(1985) Fed. Proc.44,1461.
(4) Bagshaw, J.C. and Acey, R.(1979)in Biochemistry
    of Artemia Development(Bagshaw, J.C. and Warner,
    A.H., eds.) pp. 136-149, University Microfilm.
(5) Eaton,D.L. and Toal, B.F.(1982)Toxicol. Appl.
    Pharm. 66, 134-142.

MONOCLONAL ANTIBODY TO THE GLUCOCORTICOID RECEPTOR BINDS TO A LYMPHOMA CELL MEMBRAIN PROTEIN: CORRELATION TO GLUCOCORTICOID INDUCED CELL LYSIS

B. Gametchu

Department of Human Biological Chemistry and Genetics, The University of Texas Medical Branch, Galveston, Texas 77550

Introduction. Glucocorticoids have long been known for their ability to cause lysis of lymphocytes (1) and this property is well exploited in medicine where the steroid hormones in combination with other drugs, have been used for the treatment of certain types of leukemias and lymphomas (2). The mechanism of lymphoid cell killing by glucocorticoids is not well understood; however it is known that the effect is mediated through a receptor (3). In general, patients with high numbers of glucocorticoid receptor (GR) show more favorable response to glucocorticoid treatment than patients with less receptor sites/cell, indicating a correlation between GR and clinical prognosis (4,5). However, this correlation is not perfect, since patients with high receptor sites sometimes, but not always, respond to the steroid therapy. Therefore one of the major questions in this field is, why do some receptor positive cells fail to respond?

Materials, Methods and Results. Glucocorticoid-lysis sensitive S-49 mouse T-lymphoma cells were studied for immunocytochemical localization of the GR by direct immunofluorescence in which the purified monoclonal anti-GR antibody was conjugated with fluorescein isotheocyanate (6). The results showed two very novel and interesting observations: First, only 53.0 ± 9.5% of the clonal S-49 cells contained specific immunofluorescing GR antigen (Fig. 1). Second, there was marked intensity of perimembrane specific fluorescence in positive cells. From this, it appeared that the GR not only resides in the cytoplasm and nucleus (as

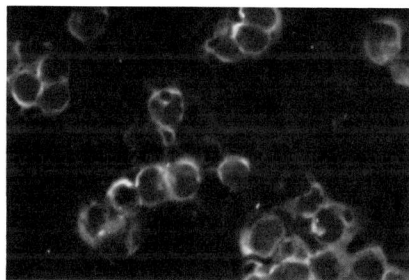

previously shown), but might also be in the plasma membrane. On the basis of this plasma membrane bound immunoreactive receptor, the cells were separated into membrane receptor-enriched (100%) and membrane receptor-deficient (38.9 ± 2.5%) cells by adsorption of live cells through their membrane receptor to monoclonal antibody coated-tissue culture flasks. Membrane isolation, followed by immunoblot analysis showed that the plasma membrane antigen-enriched cells contained five immunoreactive protein bands four of which were larger than the nuclear or cytoplasmic receptors and ranging in size from 100 to 145 kDa (Fig. 2). Furthermore, glucocorticoid sensitivity experiments were done by growing the cells in the presence and absence of 10^{-6} M concentration of dexamethasone (7). The results showed that the membrane adsorption selected, immunoreactive receptor-enriched cells, were 100% lysed by the effect of glucocorticoids, whereas the membrane receptor deficient cells showed only partial lysis (Fig. 3). These results indicate a striking correlation between glucocorticoid induced cell lysis and the presence of anti-GR monoclonal antibody reactive

plasma membrane receptor suggesting that the lysis response is at least in part mediated by plasma membrane GR or antigenically related protein.

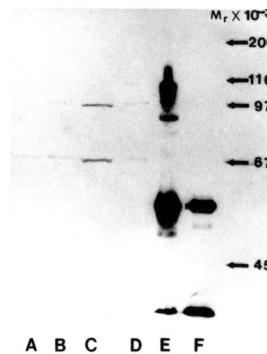

Fig. 2

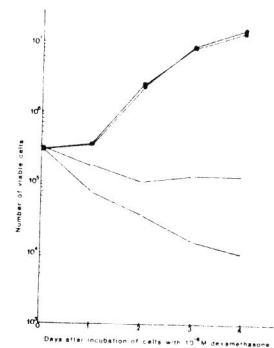

Fig. 3

Lanes AB, and CD are nuclear and cytosolic preparations containing 50 and 100 μg of total protein, respectively. Lane E, membrane receptor immunoprecipitate with monoclonal antibody to GR. Lane F membrane receptor immunoprecipitate with nonspecific mouse IgG.

○, ●, Dexamethasone-treated and untreated (control) cells, respectively, from specific membrane antigen-enriched group. □, ■, Dexamethasone treated and untreated (control) cells from specific membrane antigen-deficient group, respectively.

REFERENCES

(1) Dougherty, R.F., White, A. (1945) Am. J. Anat. 77, 81.
(2) Goldin, A., Sandberg, J., Henderson, E. et al. (1971) Cancer Chemother. Rep. 55, 309-507.
(3) Thompson, E.B. et al. (1983) in Steroid Hormone Receptors: Structure and Function (Eriksson, H. and Gustafsson, J.A., eds.) pp. 171-194.
(4) Konior-Yarbro, G.S., Lippman, M.E., Johnson, G.E. and Leventhal, B.G. (1977) Cancer Res. 37, 2688-2695.
(5) Lippman, M.E., Halterman, R.H., Leventhal, B.G., Perry, S. and Thompson, E.B. (1973) J. Clin. Invest. 52, 1715-1725.
(6) Gametchu, B. and Harrison, R.W. (1984) Endocrinology 114, 274-279.
(7) Huet-Minkowski, Gasson, J.C. and Bourgeois, S. (1981) Cancer Res. 41, 4540.

Acknowledgements: This research was partly supported by the PHS grant number CA17701 awarded by the National Cancer Institute.

IMMUNOLOCALIZATION OF UBIQUITINATED HISTONE H2A IN A SOLUBLE FRACTION OF MYOTUBE CHROMATIN AFTER MICROCOCCAL NUCLEASE DIGESTION

M. Hussin, A. L. Haas[*], B. A. Dawson and J. Lough

Departments of Anatomy & Cellular Biology and Biochemistry[*], Medical College of Wisconsin, Milwaukee, WI. 53226

INTRODUCTION: Histones H2A and, to a lesser extent, H2B, are transiently modified by the peptide ubiquitin (1). Although the function of histone ubiquitination is unknown, a vital role is implied by this protein's conservation during vertebrate evolution. Two reports have indicated that ubiquitinated histone H2A (uH2A) is localized in the active chromatin of chicken erythrocytes (2) and Drosophila cells (3). We have recently demonstrated a dramatic decline in uH2A content when myotubes form during chicken myogenesis (4), paralleling dramatic reductions in DNA transcription. Here we report preliminary observations indicating that uH2A resides in the active chromatin of differentiating myotube cells.

METHODS: Cultured myoblasts were treated with ara-C to obtain pure myotube cultures (4). One-half of the cultures was treated with 10 mM Na-butyrate for the hour preceding harvest to enhance enrichment of active chromatin (5). At day 4, nuclei were isolated and digested with 25 units/ml MNase at $15^{\circ}C$ until 4% of the DNA was acid-soluble. A low-speed supernatant (S_0) was saved and the nuclei were extracted in 100 mM NaCl and pelleted to obtain a second supernatant (S_{100}) and pellet (P) fractions. Protein and DNA from each fraction were analyzed using Western (4) and Southern blots, respectively.

RESULTS: Figure 1 shows a typical digestion pattern of myotube nuclei prior to fractionation, indicating that Na-butyrate treatment, which we previously demonstrated to cause hyperacetylation of myotube histones (6), predisposes chromatin to MNase digestion:

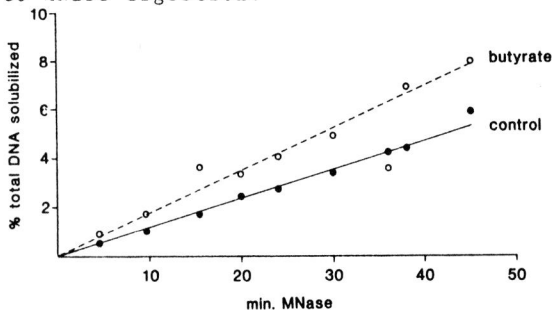

Figure 2 (below) shows the content of DNA in each chromatin fraction for a typical experiment, revealing that in most cases the release of chromatin to the S_0 and S_{100} supernatants is modestly increased by Na-butyrate pretreatment (shaded bars) as compared with controls (open bars). The protein:DNA ratios of the supernatant fractions gives no indication of selective protein extraction by salt fractionation (not shown).

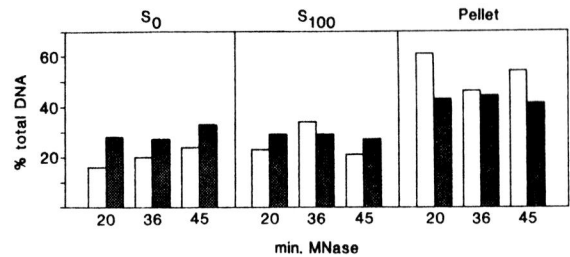

Figure 3 is a Western blot immuno-probed for ubiquitinated proteins in each fraction. Note that the S_{100} contains most of the uH2A in the nucleus along with a larger, unknown ubiquitinated protein designated "x". Also, free ubiquitin (Ub) is present only in the S_0 and S_{100}. Finally, butyrate treatment results in slightly less uH2A in the S_{100} with corresponding subtle increases in the S_0.

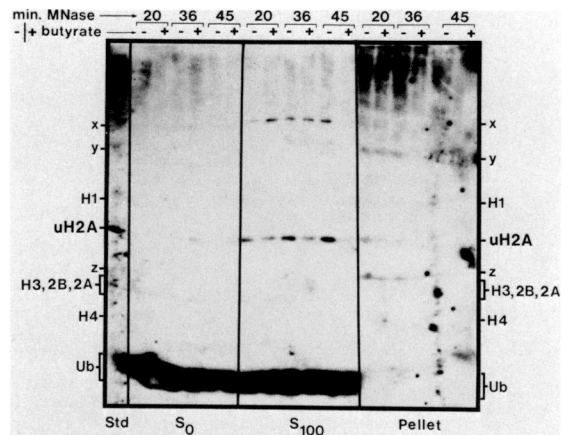

We have probed the DNA from each fraction for enrichment of the skeletal alpha actin gene using a plasmid (90C6) containing a genomic insert of the 3' untranslated region of the gene (7). Preliminary exposure of the blot (not shown) displays most of the hybridization as a "smear" in the supernatant fractions with an unexplained band in the pellet.

REFERENCES:
(1) Goldknopf, I.L. & Busch, H. (1977) P.N.A.S. 74, 864-868.
(2) Goldknopf, I.L., Wilson, G., Ballal, N.R. and Busch, H. (1980) J. Biol. Chem. 255, 10555-10558.
(3) Levinger, L. and Varshavsky, A. (1982) Cell 28, 375-385.
(4) Wunsch, A.M., Haas, A.L. and Lough, J. In Press, Developmental Biology
(5) Ferenz, C.R. and Nelson, D.A. (1985) Nuc. Acids Res. 13, 1977-1995.
(6) Lough, J. (1981) Cell Biol. Int. Rep. 5, 691-698.
(7) Fornwald, J.A., Kuncio, G., Peng, I. and Ordahl, C.P. (1982) Nuc. Acids Res. 10, 3861-3876.

ACKNOWLEDGEMENTS: Supported by grants from the NIH (A.L.H, J.L) and a Predoctoral Fellowship from the American Heart Association, Wisconsin Affiliate (B.A.D.)

TISSUE SPECIFIC EXPRESSION AND HYPOMETHYLATION OF RAT PROLACTIN GENE DURING GESTATION AND LACTATION.

Vipin Kumar and Debajit K. Biswas.

Laboratory of Pharmacology, Harvard School of Dental Medicine and Harvard Medical School, Boston, MASS 02115, USA.

Over the last few years, extensive studies have been carried out by several investigators on the sequence organization of eukaryotic genes and on their transcriptional activities. Modification by site specific hypermethylation and hypomethylation of gene sequences has been implicated with the transcriptional behavior of these genes (1). An inverse relationship i.e. the activation of the gene by reduced methylation and inactivation by increased methylation of bases has been postulated (2). Tissue specific expression of globin, ovalbumin, conalbumin, immunoglobulin and crystalline genes have been correlated to the degree of hypomethylation of the internal "-C-" residues of the "-CCGG-" sequences in these genes. However, it is not clear as yet whether hypermethylation or hypomethylation of bases is a passive circumstantial event or a causative factor in the process of inactivation or activation of the genes. All the in vivo studies and the limited experimental data originated from the studies involving transfer of hypermethylated and hypomethylated gene sequences suggest that such a modification may probably be one of the causative factors responsible for the regulation of expression of the genes.

Serum level of prolactin increases in pregnant adult mammals during the gestation period and more prominently so during lactation (3). The higher levels of prolactin in the serum of pregnant adults during gestation may be correlated mostly to the increased secretion of the hormone from the pituitary gland (4). Whereas the increased hormone level during lactation can be correlated to the increased synthesis of the protein hormone in the pituitary gland. The possible role of DNA methylation on the transcriptional behavior of rPRL gene of rat pituitary gland during gestation and lactation has been experimentally verified in this investigation.

Level of PRL-Specific mRNA Sequences During Gestation and in Lactation.

Northern blot analysis (5) of RNA isolated from pituitary gland of adult female rats at different stages of gestation reveals that the level of PRL message in pituitary gland of pregnant and nonpregnant rats is not significantly different from each other. However, the level of mRNA-PRL sequences is significantly higher in the pituitary gland of lactating animals. PRL-specific mRNA cannot be detected in the liver of these rats at any stage during gestation or lactation.

DNA Hypomethylation and Transcriptional Behavior of Prolactin Gene in Pituitary Gland and Liver of Pregnant and Lactating Rats.

The level of rPRL gene sequences in the pituitary gland of pregnant and lactating rats has been found to be same. Thus the observed increased level of rPRL gene expression is not mediated via the amplification of the gene. Results of this investigation reveal that the methylation pattern of rPRL gene in the pituitary gland is altered in a fashion which can be correlated to the stage specific expression of the gene during gestation and lactation. Such a modification of the gene structure is very much tissue specific because methylation pattern of rPRL gene in liver of the same pregnant and lactating rats is not affected. The results on the methylation pattern of internal "-C-" residues of the five "-CCGG-" sequences in the rPRL gene and its flanking sequences in pituitary gland, during gestation and lactation is reported in this abstract. These results reveal that, the three "-CCGG-" sequences in the coding region of the rPRL gene are hypomethylated concomitant to the increased expression of the gene. The methylation pattern of the internal "-C-" residues of the two distal "-CCGG-" sequences at the 5' and 3' flanking regions of the gene is not affected. The same "-CCGG-" sites in rPRL gene in liver in which the expression of the gene cannot be detected, are hypermethylated. These results demonstrate the role of hypomethylation of the specific regions of a gene in the tissue specific transcriptional activation of the gene during a physiological event in the life cycle of the adult animal.

REFERENCES:

1. Razin, A., Cedar, H. and Riggs, A.D. (1984). DNA Methylation, Biochemistry and Biological Significance. Springer-Verlag, New York, NY.

2. Ceder, H. (1984). DNA Methylation and Gene Expression. In: DNA-Methylation, Biochemistry and Biological Significance. Ed. by Razin, A., Ceder, H. and Riggs, A.D., Springer-Verlag, New York, NY, pp 147-164.

3. Wilson, D.W., Hanes, S.D., Pichler, M.H. and Biswas, D.K. (1983). 5-Bromodeoxyuridine-Induced Amplification of Prolactin Gene in GH Cells is an Extrachromosomal Event. Biochemistry, 22, 6077-6083.

4. Linzer, D.I.H. and Talamantes, F. (1985). Nucleotide Sequence of Mouse Prolactin and Growth Hormone mRNAs and Expression of these mRNAs During Pregnancy. J. Biol. Chem. 260, 9574-9579.

5. Maniatis, T., Fritsch, E.F. and Sambrook, J. (1982). Molecular Cloning. Cold Spring Harbor laboratory, Cold Spring Harbor, New York, NY.

PROOPIOMELANOCORTIN GENE EXPRESSION AND PEPTIDE PRODUCT SECRETION IN THE DEVELOPING RAT PITUITARY GLAND.

D.I. Lugo and J.E. Pintar

Dept. of Anatomy and Cell Biology of Columbia University, 630 W. 168th Street, NY, NY 10032.

Introduction

Pituitary gland development has previously been investigated immunocytochemically, biochemically, and ultrastructurally (1,2,5,6,8). While immunocytochemical studies have shown that the anterior lobe of the pituitary gland begins storage of proopiomelanocortin (POMC) peptide products prior to the intermediate lobe (2), biochemical characterization of peptide products at early stages of development (e13-e15) (6) have detected an intermediate-like pattern of POMC proteolytic cleavages at these ages. Although morphological studies have identified secretory granules in the anterior lobe as early as e14 (8), the time at which POMC cells acquire the capability to release peptides and respond to physiological regulators is not known.

Therefore we have investigated the initiation of POMC gene expression, peptide accumulation and peptide secretion in the developing pituitary gland. Here, we compare the results of in-situ hybridization experiments that determine the onset and location of POMC gene expression with immunocytochemical studies that determine sites of peptide accumulation. Further, we have used the reverse hemolytic plaque assay (4) to determine whether basal and stimulated release of POMC peptides occurs from the times at which POMC synthesis begins.

Materials and Methods

For the immunocytochemical and in-situ hybridization studies, animals (Sprague-Dawley rats e10-e21 and neonates) were fixed with 4% paraformaldehyde pH7-0.1% DEP by either immersion (e10-e15) or perfusion (e16-e21, neonates), frozen sectioned at 8 um using a cryostat and mounted onto poly-L-lysine coated slides. In-situ hybridization experiments were carried out using a cRNA probe specific for exon 3 of the POMC gene originally provided by J.L. Roberts and hybridized according to already published procedures (7). Immunocytochemical studies were performed utilizing an antiserum specific for B-endorphin kindly provided by A. Liotta (3). For the reverse hemolytic plaque assays, pituitary glands of different ages (e13-p9) were dissected, dispersed using 0.03% collagenase and assayed (4).

Results

(1) POMC gene expression and peptide accumulation

In-situ hybridization detected POMC mRNA containing cells as early as e13 on the ventralmost aspect of the pituitary; a similar location of cells containing POMC-peptide products was observed by immunocytochemical methods. The number of cells expressing the POMC gene as well as containing POMC peptides increase in this ventral aspect as it proliferates. By e15, labelled cells begin to appear in the middle posterior region of the intermediate lobe prior to immunocytochemical detection of POMC peptides in this region, while corticotrophs are detected throughout the developing anterior lobe. By e17 most of the intermediate lobe contains labelled cells and by e19 POMC expression and peptide localization resembles very much the adult gland with corticotrophs scattered throughout the anterior lobe, while nearly all the cells in the intermediate lobe are labelled.

(2) POMC peptide secretion

The reverse hemolytic plaque assay demonstrated basal secretion from POMC containing cells as early as e13.5. Moreover, secretion by these corticotrophs could not be stimulated by addition of secretagogues (CRF 10^{-8}M). By e15.5 not only could basal secretion be detected but corticotrophs appeared to be responsive to secretagogues (CRF 10^{-8}M). At later pre-natal ages, dissection of the pituitary gland into the anterior and neurointermediate lobes was possible. In these experiments, (e21-p10) we were able to observe not only secretagogue-stimulated (CRF 10^{-8}M) secretion of anterior lobe corticotrophs but also of intermediate lobe melanotrophs.

Conclusions

By utilizing in-situ hybridization we have unequivocally demonstrated that the cells of the presumptive anterior lobe begin POMC-mRNA synthesis prior to the intermediate lobe. Taken together with previous biochemical data (6) this suggests that the intermediate-lobe-like pattern of POMC proteolytic processing characteristic of the fetal pituitary occurs in anterior lobe POMC producing cells. The possible transient expression of an adult intermediate lobe kallikrein is presently being investigated.

Our experiments utilizing the reverse hemolytic plaque assay provide the first direct evidence that a functional release system in POMC cells is present during early stages of cell differentiation. The implications of such early secretory activity remains to be elucidated. In addition we suggest that fetal POMC cells may be responsive to physiological regulators such as CRH prior to development of a functional hypothalamic-hypophyseal portal system.

References

1. Allen, R.G., Pintar, J.E., Stack, J., and Kendall, J.W. Develop. Biol. 102:43-50 (1984).
2. Dupouy, J.P. Inter. Rev. of Cytology 68:197-249 (1980).
3. Liotta, A.S., Yamaguchi, H., and Krieger, D.T. J. Neurosci. 1:585-595 (1981).
4. Neill, J.D. and Frawley, L.S. Endocrinology 112:1135-1137 (1983).
5. Pintar, J.E. In: Gene Expression and Cell-Cell Interaction in the Developing Nervous System, Jean M. Lauder and Phillip G. Nelson, eds., Plenum Publ. Corp., New York, pp. 51-64, 1984.
6. Pintar, J.E., Krieger, D.T. and Liotta, A.S. Ontogeny of pituitary B-endorphin and MSH-sized peptides during fetal rat pituitary development (submitted).
7. Pintar, J.E. and Lugo, D.I.: Localization of Peptide Hormone Gene Expression in Adult and Embryonic Tissues. In: In-Situ Hybridzation: Applications to Neurobiology, (Valentino et al.), Oxford University Press, New York, 1986 (in press).
8. Stoeckel, M.E., Hindelang-Gertner, C. and Porte, A. Cell Tiss. Res. 198:465-476 (1979).

POSSIBLE ROLE FOR A NUCLEASE-SENSITIVE COMPONENT IN THE REGULATION OF TRANSLATION IN EARLY DEVELOPMENT OF THE SEA URCHIN Lytechinus pictus

Elizabeth N. Mandley and Alina C. Lopo·
Division of Biomedical Sciences, University of California, Riverside, CA 92521-0121

Sea urchin eggs, as do all animal eggs, contain large amounts of messenger RNA as well as all the other necessary components for protein synthesis; however, they are translationally repressed [1]. This repression of protein synthesis is released at fertilization, when there is a gradual but rapid increase in the rate of protein synthesis [2]. Because there is only limited transcription during the early development of the sea urchin embryo, the regulation of gene expression during this period is accomplished almost exclusively via translational regulation. The sea urchin is a particularly useful model to study this problem because of the dramatic activation of protein synthesis and because material is inexpensive and easily obtained.

We are using two approaches to understand this classic problem of gene expression in early development: purification of components of the translational apparatus, with an emphasis on the initiation factors [3], and the use of in vitro translation systems (IVTSs) from unfertilized eggs and early embryos to compare the details of the mechanims of translation. In vitro translation systems have been developed from unfertilized eggs of Lytechinus pictus [4] and from unfertilized eggs and embryos of Strongylocentrotus purpuratus [5,6]. These IVTSs have proven useful in providing support for the idea that there are at least dual [5,7,8] and likely multiple levels of translational control operating at fertilization [5-9]. Here we present preliminary evidence for translational control by a nuclease-sensitive component.

MATERIALS AND METHODS

IVTSs from Lytechinus pictus unfertilized eggs were prepared according to modifications of the methods of Winkler and Steinhardt [4] or Lopo et al. [5]. Nuclease treatment of the IVTSs was as follows. The lysate was hand-thawed and the desired volume was dispensed into sterile tubes. The lysate was adjusted to final concentrations of 8.76mM $CaCl_2$, 0.3 mg/ml creatine phosphokinase, and 0.3 units/ul micrococcal nuclease. Following incubation at $0°$C for 60 min, the reaction was stopped by addition of EGTA (to a final concentration in the IVTS of 30mM; the original IVTS was 10mM EGTA). The IVTS was warmed to the appropriate translation temperature [4,5; and Mandley and Lopo, unpublished], and the translation assay was carried out according to Lopo et al. [5]. Protein synthetic activity was determined by measuring incorporation of radiolabeled amino acid into protein.

RESULTS

When Lytechinus pictus (Lp) or Strongylocentrotus purpuratus (Sp) unfertilized egg IVTSs are supplemented with globin or TMV mRNA, the mRNAs are translated, yet there is no stimulation of translation [5,7,8]. We extended these results by supplementing Lp unfertilized egg IVTS with total RNA, polyA(+) RNA, and polyA(-) RNA from unfertilized eggs. In all cases there was either no stimulation, or a slight depression (polyA+) of translation (Table 1).

Table 1

Control	1.00
Control + TMV (0.1 mg/ml)	1.05
Control + globin mRNA (0.1 mg/ml)	0.89
Control + total RNA (0.1 mg/ml)	1.05
Control + polyA(+) RNA (0.005 mg/ml)	0.77
Control + polyA(-) RNA (0.1 mg/ml)	1.04

We then treated the Lp unfertilized egg IVTS with low concentrations of micrococcal nuclease. This type of treatment is the standard procedure for preparing messenger RNA-dependent lysates from rabbit reticulocyte and other mammalian systems. Rather than eliminate all endogenous translational activity (as we expected), these nuclease-treated IVTS (NT-IVTS) had approximately 30-40% increased translational activity over the untreated controls (Table 2). Autoradiograms of SDS-polyacrylamide gels revealed that similar translation products were synthesized by both control and NT-IVTSs, suggesting that there was no selective degradation of specific messages by the nuclease treatment.

Table 2

Control	1.00
Nuclease-treated	1.34
NT + total RNA (0.1 mg/ml)	1.16
NT + polyA(+) RNA (0.005 mg/ml)	2.45
NT + polyA(-) RNA (0.1 mg/ml)	2.32

When the NT-IVTS were supplemented with polyA(+) RNA, a 2.5-fold stimulation over the non-nuclease treated control was observed (Table 2). Addition of polyA(-) RNA resulted in approximately similar stimulation (2.32). However, addition of total RNA resulted in only minimal stimulation (1.16). Finally, in preliminary experiments, TMV RNA also appears to stimulate the NT-IVTS approximately 2-fold (data not shown).

The results presented here suggest that unfertilized eggs may contain a nuclease-sensitive component that is involved in repressing translation. The limited nuclease treatment of the IVTS may result in at least partial degradation of this component, thereby lifting part of the block on translation, and thereby permitting partial stimulation of protein synthesis by both endogenous (NT-controls) and added message. The stimulation following nuclease treatment is modest; however, this is consistent with evidence that we and others have presented [5,7-9] that there are multiple controls operating to limit translation in the unfertilized egg, and that these operate in concert at fertilization to achieve the translational activation seen following fertilization. Thus, a nuclease-sensitive component may be acting to inhibit only a portion of the process.

At this time we do not have any information on the nature of this nuclease-sensitive component. However, in recent years considerable attention has focused on small nuclear and cytoplasmic RNAs involved in regulating the activity of many cellular process [10], including translation [11]. With this in mind, we are continuing studies on the nature of this component.

REFERENCES

1. Davidson, E.H., Hough-Evans, B.R. and Britten, R.J. (1982) Science 217:17-26.
2. Epel, D. (1967) Proc. Natl. Acad. Sci. 57:899-906.
3. Lopo, A.C., Lashbrook, C.C., Infante, A.A., Infante, D. and Hershey, J.W.B. (1986) Arch. Biochem. Biophys, in press.
4. Winkler, M.M. and Steinhardt, R.A. (1981) Develop. Biol. 84:432-439.
5. Lopo, A.C. and Hershey, J.W.B. (1985) Fed. Proc. 44:1801.
6. Lopo, A.C., Lashbrook, C.C. and Hershey, J.W.B. (1987a) Developmental Biology, submitted.
7. Lopo, A.C., MacMillan, S. and Hershey, J.W.B. (1987b) Developmental Biology, submitted.
8. Winkler, M.M., Nelson, E.M., Lashbrook, C. and Hershey, J.W.B. (1985) Develop. Biol. 107:290-300.
9. Hille, M.B., Danilchik, M.V., Colin, A. and Moon, R.T. (1985) In Belle W. Baruch Library in Marine Science, Vol. 12: Symp. Cell. Molec. Mech. Develop. (Sayer, R. and Showman, R., eds.).
10. Turner, P. (1985) Nature 316:105-106
11. Winkler, M.M., Lashbrook, C., Hershey, J.W.B., Mukherjee, A.K. and Sarkar, S. (1983) J. Biol. Chem. 258:15141-15145.

SEA URCHIN MATERNAL AND EMBRYONIC U1 RNAS ARE DISTINCT CLASSES

M.A. Nash*, S.E. Kozak[+], L.M. Angerer[+],
R.C. Angerer[+], H. Schatten[++], G. Schatten[++],
and W.F. Marzluff*.

Departments of Chemistry* and Biological Science[+],
Florida State University, Tallahassee FL 32306 and
Department of Biology[++], University of Rochester,
Rochester, NY 14627

Small nuclear RNAs (snRNAs) play a key role in RNA biosynthesis in eucaryotic cells since they are essential for reactions in RNA processing. Maternal snRNAs are required during oogenesis for processing of maternal mRNAs, and the most abundant snRNAs, U1 and U2, are present in large amounts in mature sea urchin eggs (1). Additional U1 and U2 RNAs are apparently required for processing of embryonic transcripts synthesized after fertilization, because synthesis of these RNAs begins around the 32-cell stage and continues at high rates through cleavage, declining only after the rate of cell division decreases markedly (2).

In the experiments presented here we examined localization of U1 RNA and snRNP antigens in sea urchin oocytes, eggs and embryos using both cytological and biochemical techniques. We have compared the subcellular distribution of total U1 RNA to that of molecules newly synthesized after fertilization and have found that these two classes of molecules do not equilibrate.

Localization of U1 RNA by in situ hybridization. The distribution of U1 RNA was determined directly by hybridizing ^{3}H-labeled antisense (signal) and sense (control) transcripts to tissue sections of ovaries, eggs or embryos of the sea urchins, S. purpuratus and L. pictus (3). Sections of ovary revealed that labeling was primarily over nuclei of both small and large oocytes. No labeling of nucleoli or cytoplasm was detected. However, after oocyte maturation, a drastic shift in subcellular localization of U1 RNA into the cytoplasm was seen. Haploid pronuclei were essentially unlabeled. Grain densities over the cytoplasm of eggs hybridized with antisense and sense probes gave signal/noise ratios from 8-18 for different egg preparations and different experiments, while signals over nuclei were indistinguishable from background. We concluded that all the mature egg U1 RNA resides primarily, if not exclusively, in the cytoplasm. This cytoplasmic localization was shown to be maintained throughout early cleavage. Nuclear accumulation of U1 RNA was not detectable until 16-cell stage. At this stage a subset of micromere nuclei were labeled to a level 3-fold higher than that of surrounding cytoplasm. Conversely, micromeric cytoplasm stained twofold lower than did the nuclei. As development proceeded, labeling became progressively nuclear. At the latest stages studied (gastrula and pluteus larva) all nuclei labeled to approximately uniform intensity, and little signal was detected in the cytoplasm. Since the cytoplasm of micromeres was labeled to the same extent whether or not the nuclei were labeled, we conclude that the nuclear accumulation is most likely due to new synthesis of U1 RNA.

Localization of anti-RNP antigen. The snRNAs are found in ribonucleoprotein particles (snRNPs), each class of which contains a common set of proteins recognized by sera of the Sm specificity or anti-RNP specificity. The latter sera are specific for antigens found only on U1 RNP. These human sera react efficiently with sea urchin snRNPs. U1 snRNPs were localized by indirect immunofluorescence microscopy. In most respects the distributions observed were very similar to those found for U1 RNA (essentially identical data were obtained with anti-SM and anti-RNP sera). In large oocytes RNP antigen was present at highest concentration in germinal vesicles, though significant fluorescence was observed in the cytoplasm of some oocytes. Fluorescence in unfertilized mature eggs was primarily cytoplasmic, with little detectable signal in pronuclei. After fertilization high cytoplasmic concentrations of RNP antigen were seen throughout early development. Antigen was not seen to accumulate in micromere nuclei. A progressive shift in the location of the signal was seen starting in early blastula stage, progressing from cytoplasm to nuclei. By gastrula, the antigen was localized predominately within nuclei.

Biochemical analysis of U1 RNA content of nuclei and cytoplasm. An S1 nuclease protection assay was used to determine the relative amounts of U1 RNA in nuclear and cytoplasmic fractions of later stage L. variegatus embryos. In morulae most of the U1 RNA was cytoplasmic, at blastula stage more than 50% remained cytoplasmic and not until gastrula was the majority of U1 RNA found in the nucleus. It became imperative at this point to test whether the RNP (and SM) antigens and U1 RNA found in the cytoplasm existed together as RNP. This hypothesis was confirmed by precipitating egg U1 RNA (detected by our S1 protection assay) with antiserum. The ability to precipitate U1 RNA with anti-RNP serum indicated that the U1 RNA is in a particle similar to that of nuclear U1 RNP. Yet our previous data indicated that these particles remained cytoplasmic during early embryonic development. To further examine this point, distributions of total and newly synthesized U1 RNA were compared. Embryos of S. purpuratus were labeled with [32]PO$_4$ from 16-cell to blastula stage, and RNA was extracted from nuclear and cytoplasmic fractions and displayed on polyacrylamide gels. Total U1 RNA of a parallel unlabeled culture was measured by the S1 protection assay. The bulk of the total U1 RNA was shown to be cytoplasmic; more than 95% of the newly synthesized U1 RNA synthesized between 16-cell and blastula was found in the nuclear fraction. These data lead to the conclusion that newly synthesized and maternal U1 RNAs do not equilibrate during early development. Whether or not the maternal U1 RNA functions in the cytoplasm or is simply inactivated at germinal vesicle breakdown is not known.

There is a considerable difference in the embryonic strategies of U1 RNA expression in the three most widely studied species of animals, i.e., mouse, Xenopus and sea urchin. The large amount of U1 RNA present in mature oocytes and eggs of sea urchins may not function during development. Synthesis during early development is rather slow, resulting in only a 3-4 fold increase by pluteus (1). Xenopus, on the other hand, has a marked decrease in the concentration and absolute amount of U1 RNA in post vitellogenic oocytes, and has no further U1 RNA synthesis until the mid-blastula transition. In the mouse, a 10-fold increase in U1 RNA synthesis is seen during the first 4 cell divisions of the developing embryo (Lobo and Marzluff, unpublished results). As in sea urchins, U1 RNP is found in the cytoplasm of mouse eggs (Schatten and Schatten, unpublished results).

REFERENCES

1. Brown, D.T., G.F. Morris, N. Chodchoy, C. Sprecher and W.F. Marzluff. 1985. Nuc. Acids. Res. 13:537-556.
2. Nijhawan, P.N. and W.F. Marzluff. 1979. Biochemistry 18:1353-1360.
3. Cox, K.H., D.V. DeLeon, L.M. Angerer and R.C. Angerer. 1984. Dev. Biol. 101:485-502.

DIFFERENTIAL REPLICATION OF CELL TYPE SPECIFIC GENES OF PHYSARUM POLYCEPHALUM: IMPLICATIONS FOR CELLULAR COMMITMENT

G. Pierron[1], D. Pallotta[2] and H.W. Sauer[3]

[1]Institut de Recherches Scientifiques sur le Cancer, CRNS, Villejuif, France, [2]Dept. of Biology, Laval University, Quebec, Canada G1K 7P4, [3]Dept. of Biology, Texas A&M University, College Station, TX, 77843

INTRODUCTION - We have been exploring the idea that cell cycles enjoy a non-trivial relationship with cell differentiations during development. We wish to argue that selective transcription of genes is necessary but not sufficient to sustain the molecular dogma of "differential gene expression", and present some new experiments which indicate that differential gene replication may be required before a cell becomes differentiated. We had previously demonstrated the preferential activation of transcription units within early replicons (1). We now show that cell type specific genes replicate early when they are expressed but late when they are repressed.

MATERIALS AND METHODS - Physarum is the appropriate biological material to study differential genome replication. It is a multipotential eukaryotic developmental system with several completely alternative cell types (2). In its multinucleated plasmodium, all 10^8 nuclei not only divide at the same time (i.e. within 5 min., generation time 10 hours) but also traverse the S-phase of the mitotic cycle in natural synchrony, as revealed by high resolution flow cytometry (3). The timing of replication of individual genes can be precisely determined (4). Cell type specific cDNA clones have been detected by differential screening of 3 different cDNA libraries (5).

RESULTS - In the experiments probing the cell type specific pattern of gene replication, we have utilized several full length cDNA clones of genes exclusively expressed in the plasmodium, LAV 1.1, the amoeba, LAV 3.1 and the spherule, LAV 2.1. After nick-translation of isolated inserts these probes were hybridized to Southern blots of restricted total genomic DNA to visualize their respective restriction fragments. In the crucial experiment, these probes were hybridized to the fraction of the genome that had replicated during the first 40 min, and to the DNA which replicated during the remainder of the 3 hour S-phase. The results clearly show that the plasmodium specific gene has replicated early in the plasmodium, whereas the genes that are only active in the amoeba or the spherule replicate late in the plasmodium. Preliminary results with additional cell type specific cDNA clones confirmed this conclusion. As a control experiment the same DNA blot was hybridized consecutively with the different cDNA probes to rule out unequal DNA transfer. Also, a cloned highly repetitive transposon-like DNA element, known to be highly methylated in vivo (6), exclusively reacted with the late DNA. We have chosen the period of 40 min of early S-phase because we had previously shown that 3 of the actin loci replicated by that time, whereas ardA replicated thereafter (4). The chronology and synchrony of replication of the actin gene family has been confirmed with a different strain, containing allele pairs of different restriction fragment lengths. We have also cloned (7) and sequenced the late replicating ardA, confirmed

its actin amino acid sequence and detected the typical eukaryotic transcription initiation and polyadenylation signals (8). However, after comparing the ardA gene with the 3'-untranslated sequences of cDNAs for the early replicating ardB and C gene, we arrived at the intriguing conclusion that the ardA gene, though structurally intact, seems to be repressed in the plasmodium or the amoeba, perhaps because it replicates late.

DISCUSSION - Our results are consistent with the idea that expressed genes must be positioned in an early time compartment of S-phase. We postulate that cell type specific genes must change from a late to an early compartment before they can become activated. Although our results lead to a more complex picture regarding the interdependence of DNA replication and transcription in development, they may throw some light on how the genome is selectively programmed. We wish to point out that the potential activation of early genes, and overall repression of the late replicated genome, would require only a few transcription factors to explain complex gene activity pattern changes during development. It is worth mentioning that selective repression of the oocyte-specific multiple 5S rRNA genes in Xenopus during development (9) seems to correlate with the late replication of that gene cluster and a limiting concentration of the transcription factor TF IIIA specific for RNA polymerase III (10). The permanent expression of the somatic early replicating 5S rRNA genes in Xenopus is therefore consistent with the early replication of protein-coding, cell type specific genes in Physarum. Moreover, differential gene replication may be the molecular equivalent of cellular commitment, which was previously concluded from numerous observations by experimental embryologists (11).

REFERENCES

(1) Pierron, G., Sauer, H.W., Toublan, B. and Jalouzot. (1982) Eur. J. Cell Biol. 29, 104-113.
(2) Sauer, H.W. (1982) Developmental Biology of Physarum, Cambridge Univ. Press.
(3) Kubbies, M. and Pierron, G. (1983) Exp. Cell Res. 149, 57-67.
(4) Pierron, G., Durica, D.S. and Sauer, H.W. (1984) Proc. Natl. Acad. Sci. USA, 81, 6393-6397.
(5) Pallotta, D., Bernier, F., Hamelin, M., Martel, R. and Lemieux, G. (1985) in The Molecular Biology of Physarum polycephalum (Dove, W.F. et al., eds) pp. 315-327, Plenum Press, New York.
(6) Pearston, D.H., Gordon, M. and Hardman, N. (1985) The Embo. J. 4, 3557-3562.
(7) Nader, W., Edlind, T.D., Huettermann, A. and Sauer, H.W. (1985) Proc. Natl. Acad. Sci. USA, 82, 2698-2702.
(8) Nader, W., Isenberg, G. and Sauer, H.W. (1986) Gene, in press.
(9) Brown, D.D. (1984) Cell, 37, 359-365.
(10) Gilbert, D.M. (1986) Proc. Natl. Acad. Sci. USA, 83, 2924-2928.
(11) Sauer, H.W. (1980) Entwicklungs-biologie, Springer, Heidelberg.

ACKNOWLEDGEMENTS - We thank R. Flanagan for isolating cDNA inserts used in this study and G.L. Shipley for stimulating discussions. This work was partially supported by NIH grant DCB-8608017 to H.W.S.

EFFECTS OF MESODERM-INDUCING ACTIVITY ON ISOLATED
XENOPUS ANIMAL POLE REGIONS.

J.C. Smith and K. Symes.

Laboratory of Embryogenesis, National Institute for
Medical Research, The Ridgeway, Mill Hill, London
NW7 1AA, U.K.

The first inductive interaction in amphibian
development occurs at the blastula stage, when an
equatorial mesodermal rudiment is induced from the
animal hemisphere under the influence of a signal
from vegetal pole blastomeres (1). The molecular
basis of mesoderm induction is unknown, but we have
recently discovered that the Xenopus XTC cell line
(2) secretes a factor of relative molecular mass
16,000 which has the properties expected of a
mesoderm inducing factor (3). Thus, isolated
animal pole regions exposed to XTC conditioned
medium form muscle, notochord and other mesodermal
cell types instead of differentiating as epidermis.
Here we show that the earliest response to XTC
conditioned medium by animal pole explants is the
initiation of gastrulation-like movements. These
movements are accurate predictors of future mesodermal development and they begin when sibling
control embryos commence convergent extension (4),
at the mid-gastrula stage. The movements, occurring
at least 2½ hr prior to muscle-specific actin gene
activation (5), thus constitute the earliest known
marker of mesoderm induction. This should simplify
the task of coming to understand the intracellular
events between receipt of a mesoderm induction
signal and subsequent determination as a mesodermal
cell.

(I) CHANGES IN SHAPE OF ANIMAL POLE EXPLANTS
PREDICT FUTURE MESODERMAL DIFFERENTIATION.

Animal pole explants cultured in XTC conditioned
medium differentiate into mesodermal cell types
(Fig. 1a) while those cultured in medium lacking
inducing activity differentiate as epidermis (Fig.
1b). These results can be anticipated by the
appearance of the explants when sibling controls are
neurulae; induced explants transform into elongated
structures (Fig. 1c) while uninduced explants remain
as spheres (Fig. 1d).

Fig. 1. Mesodermal differentiation is predicted by
gastrulation-like movements. (a) An induced animal
pole explant differentiates into muscle (MUS),
mesenchyme (MES) and neural tube (NT). (b) an uninduced explant forms atypical epidermis. (c) an
induced explant whose sibling controls are early
neurulae. (d) an uninduced explant.

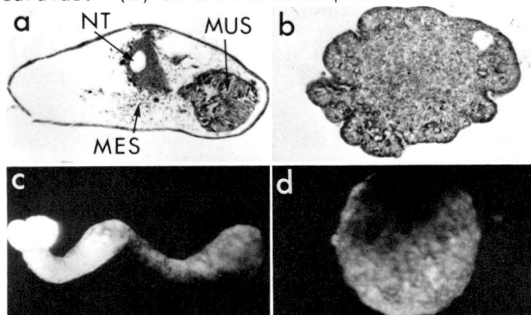

(II) GASTRULATION-LIKE MOVEMENTS START WHEN
SIBLING CONTROL EMBRYOS BEGIN CONVERGENT EXTENSION.

Animal pole explants exposed to XTC conditioned
medium at the mid-blastula stage were studied by
direct observation, by time-lapse video micrography
and by taking still photographs at 20 minute inter-

vals (Fig. 2). For the first three hours both
treated and control explants rounded up so that the
superficial pigmented cells almost completely
enclosed the unpigmented cells that had lined the
blastocoel cavity (2.3 and 3.3. hr). Between three
and four hours after exposure to XTC conditioned
medium, however, when control embryos were mid
gastrulae, the unpigmented cells still exposed to
the medium seemed to push away from the body of the
explant (4.3 and 5.3 hr). At the same time the
pigmented cell layer appeared to constrict and then
elongate (6.3 to 9.3 hr). The general appearance
of the explants and the rate of elongation (4um/
min) are very similar to isolated dorsal marginal
zones (4). Furthermore, the time the movements
begin depends upon the developmental stage of the
responding cells and not on the time at which the
cells were exposed to conditioned medium.

These results, together with those above,
suggest that the movements observed in isolated
animal pole explants are analogous to the
gastrulation movements of intact embryos. Since
the movements commence at the mid gastrulae stage,
they represent the earliest indication of mesoderm
induction yet reported.

Fig. 2. Photographs of induced animal pole explants
taken at hourly intervals from 2.3 to 9.3 hours.

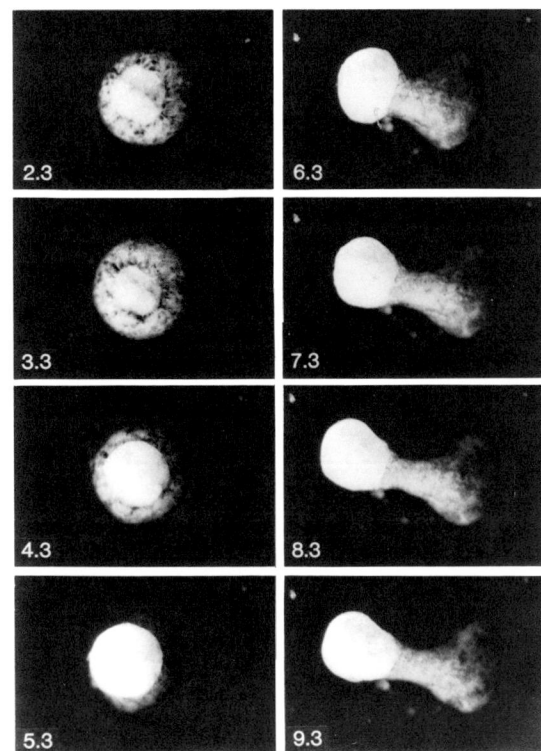

REFERENCES

(1) Dale, L., Smith, J.C. & Slack, J.M.W. (1985).
    J. Embryol. exp. Morph. 89, 289-313.
(2) Pudney, M., Varna, M.G.R. & Leake, C.J. (1973).
    Experientia 29, 466-467.
(3) Smith, J.C. (1987). Development, In Press.
(4) Keller, R.E., Danilchik, M., Gimlich, R. &
    Shih, J. (1985). J. Embryol. exp. Morph. 89,
    (Suppl.), 185-209.
(5) Gurdon, J.B., Fairman, S., Mohun, T.J. &
    Brennan,S. (1985). Cell 41, 913-922.

XENOPUS OOCYTES AND SOMATIC CELLS CONTAIN DIFFERENT SPECTRA OF PHOTOCROSSLINKED mRNP PROTEINS

Ruth E. Swiderski and Joel D. Richter

Worcester Foundation for Experimental Biology, 222 Maple Ave., Shrewsbury, MA 01545

One potential mechanism for differential translation during early development is that changing sets of proteins bind mRNAs and regulate their assembly onto polysomes. In Xenopus oocytes, greater than 90% of the available mRNA is stored as non-polysomal messenger ribonucleoprotein particles (mRNPs). As oocyte maturation, fertilization, and embryonic development proceed, most of the remaining maternal mRNAs are expressed. We have suggested that mRNA binding proteins might be involved in the modulation of this expression based on the results of an in vitro RNA-protein filter binding assay (1,2). We report here the use of UV irradiation to covalently cross-link poly(A)RNAs and their associated proteins in living Xenopus oocytes and somatic cells. We observed unique spectra of cross-linked proteins in these two cell types. Furthermore, oocytes and somatic (kidney) cells contained unique proteins bound to the poly(A) tail of the RNA. These observations suggest special roles for these proteins in the expression of oocyte and somatic mRNAs.

RESULTS

In order to determine the efficiency of cross-linking, oocytes were labeled with ^{3}H-uridine and then irradiated with increasing doses of 254 nm light. The partitioning of labeled RNA into the organic phase of a phenol/aqueous solvent system was then measured. At the maximum UV dose employed (10^6 ergs/mm^2), 85% of the newly synthesized RNA (predominantly 28S and 18S) was cross-linked to protein, thereby indicating the efficient nature of this procedure in transparent stage 1 oocytes.

Since we are particularly interested in those proteins which bind to poly(A)RNAs, we determined the UV cross-linking efficiency of several species of mRNA. Oocytes irradiated at increasing UV doses were homogenized and the RNA remaining in the aqueous phase was subjected to RNA gel blot analysis. The results show that greater than 90% of mRNAs for fibronectin, actin, ribosomal protein L15, and histone H4 are cross-linked at a UV dose of 10^6 ergs/mm^2. The blot also demonstrated that different mRNAs exhibited unique cross-linking efficiencies, indicating that perhaps unique sets of proteins are bound to each species of mRNA or that there is message-specific stoichiometry of similar sets of binding proteins. The maximum UV dose did not result in photolysis of poly(A)RNAs as determined by Proteinase K digestion of cross-linked RNPs followed by RNA gel blot analysis.

The spectrum of proteins which are cross-linked to poly(A)RNA in stage 1 oocytes as assessed by in vitro radioiodination and autoradiography is shown in Figure 1A. The major species are 93, 85, 72, 60, 56, 49, 43, 37, 35, 30, 23, 20, and 19 kD. The oocyte pattern was compared to the cross-linked poly(A)RNA proteins obtained from Xenopus kidney cells. In these cells, the major species are 78, 68, 56, 54, 43, 39, 34, 30, 25, 20, 19 kD. Only five proteins are common to both oocytes and somatic cells.

We addressed the possibility that ribosomal proteins present in polysomes were being cross-linked to poly(A)RNA. This seemed unlikely since the amount of poly(A) present on polysomes in stage 1 oocytes is less than 10%. While up to 90% of somatic cell poly(A)RNA is located on polysomes,

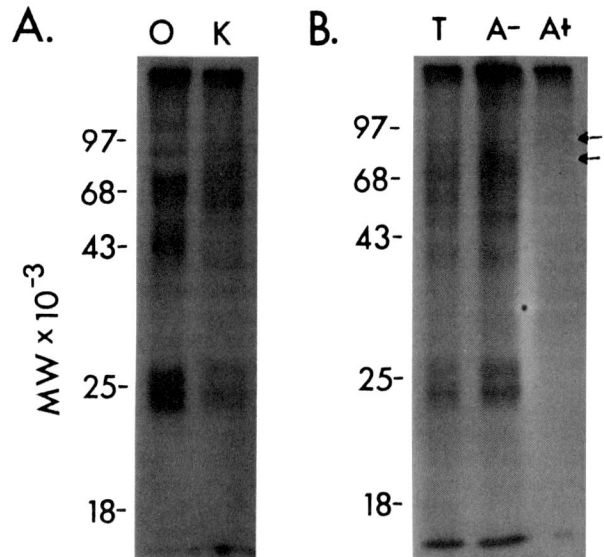

Figure 1. (A).Photocrosslinked mRNP proteins from stage 1 Xenopus oocytes (lane O) and kidney cells (lane K). (B). Iodinated poly(A)RNA binding proteins from kidney cells: total crosslinked proteins (lane T), proteins crosslinked to the non-adenylated fraction of mRNA (lane A-) and proteins associated with the poly(A) tail (lane A+).

we saw no enhancement of cross-linked proteins in the region of the gel where ribosomal proteins migrate. Second, when we monitored the mass of 28S and 18S rRNA present in the cross-linked fractions following exposure of the oocytes to increasing UV doses, we did not detect a coordinate increase in the amount of 28S and 18S rRNAs in RNP complexes.

The data of Figure 1A establish that oocytes and somatic cells contain different poly(A)RNA binding proteins. We have extended these studies by analyzing those proteins which are bound to the poly(A) tail of mRNAs. Total poly(A)RNPs were prepared by UV irradiation, oligo(dT)cellulose chromatography, and iodination. These RNPs were digested with RNAse A and T1, which preserved the poly(A) tail. The poly(A) tail was then re-chromatographed on oligo(dT) cellulose and subsequently digested with RNase T2 to hydrolyze the poly(A) tail. Figure 1B shows the RNPs of the kidney cells; total poly(A)RNPs, proteins bound to the non-adenylated portion of the RNA, and the poly(A)-associated proteins. The poly(A) tail binding proteins in kidney cells are 78 and 68 kD while those in oocytes are 72 and 60 kD (data not shown). Therefore, not only do oocytes and somatic cells contain different sets of total poly(A)RNA binding proteins, they also contain different sets of poly(A) tail associated proteins as well.

Our results suggest a developmentally regulated shift between oocyte and somatic poly(A)RNA binding proteins. The effect of such a switch may modulate translational regulation, RNA stability, or RNA localization in the oocyte or embryonic cells. We are continuing our investigation to determine when in the developmental timetable of oogenesis and early embryonic development this shift takes place.

REFERENCES
(1) Richter, J.D. and Smith, L.D. (1983) J. Biol. Chem. 258, 4864-4869.
(2) Richter, J.D. and Smith, L.D. (1984) Nature (London) 309, 378-380.
ACKNOWLEDGMENTS
This work was supported by grants from the NIH (JDR), a Cancer Research Scholar Award from the American Cancer Society, Mass. Division (JDR) and an NIH postdoctoral training grant (RES).

REGION-SPECIFIC EXPRESSION OF MOUSE HOMEOBOX GENES IN THE EMBRYONIC CENTRAL NERVOUS SYSTEM.

L.E. Toth, K.L. Slawin, J.E. Pintar* and
M.C. Nguyen-Huu.

Departments of Microbiology, Urology and
Anatomy and Cell Biology*, Columbia University
P&S, 630 West 168 th Street New York, New York
10032.

INTRODUCTION. Genes that specify the identity, polarity and number of body segments in Drosophila, termed homeotic and segmentation genes, share a characteristic 180 base pair sequence called the homeobox. Based on crosshybridization with this sequence, homeobox containing genes have been identified and isolated from the DNA of mammals including mice and humans. Although all mammalian homeobox containing genes thus far examined are expressed during embryogenesis, the sites of expression have not been well defined. Here, we report the results of in situ hybridization experiments that define the spatial pattern of expression during development of two mouse homeobox genes, M3 and M5. The M3 and M5 homeobox genes are located approximately 20 kb apart within the Hox-1 locus on mouse chromosome number 6 (1,2). Both genes are expressed in the mid gestational mouse embryo as shown by Northern analysis (3,4).

METHODS. Adjacent cryostat sections through e10.5, e11.5, e12.5 and e16.5 mouse embryos fixed for 24 hours in 4% formaldehyde were hybridized with ^{35}S labelled RNA probes synthesized, using SP6 RNA polymerase, from pGem vectors containing either a 224 bp EcoRI fragment of M3 or a 183 bp EcoRI- AccI fragment of M5 and washed to a final stringency of 0.2 X SSC at 50° C.

RESULTS. M3 and M5 transcripts were detected within the embryonic spinal cord of embryos on days 10.5 through 12.5 post coitum (see figure). In the 10.5 day embryo hybridization was detected throughout the embryonic spinal cord. At e 11.5 for M3 and e 12.5 for e 12.5 a region of maximal hybridization occurs in the upper cervical spinal cord. Anterior to this region, hybridization abruptly decreased and was not detected in the brain. Posterior to this region, expression decreased gradually, was lower in the thoracic CNS and was undetectable in the lumbar and sacral CNS. At e16.5, the expression of both genes was not detected by in situ hybridization. No hybridization was detected following hybridization using sense RNA probes (identical to M3 mRNA sequences). These results have been confirmed and extended by hybridization to both horizontal and transverse sections of entire embryos (data not shown).

DISCUSSION. This report represents the first demonstration of developmentally regulated and region-specific expression of mammalian homeobox genes within the mammalian embryonic central nervous system. Despite the different developmental stratagies seen with Drosophila and mouse embryos, this expression pattern is strikingly reminiscent of that exhibited by Drosophila homeobox genes within the ventral nerve cord of the fly embryo and larvae and is consistent with a possible role for mammalian homeobox genes in the determination of specific regions in the embryonic central nervous system and the body plan.

REFERENCES.

1. Duboule,D. Baron, A., Mahl,P. and Galliot,B. EMBO J. 5 1973-1980 (1986).

2. Patel et al, Manuscript in preparation.

3. Rubin, M.R., Toth, L.E., Patel,M.D., D'Eustachio,P., and Nguyen-Huu, M.C. Science 233 663-667 (1986).

4. Colberg-Poley, A.M., Voss, S.D., Chowdhuty,K., Stewart,C.L., Wagner,E.G. and Gruss,P. Cell 43 39-45 (1985).

This research was supported by NIH HD-19821 (CNH) and HS-18592 (JP).

10.5   11.5   12.5   16.5

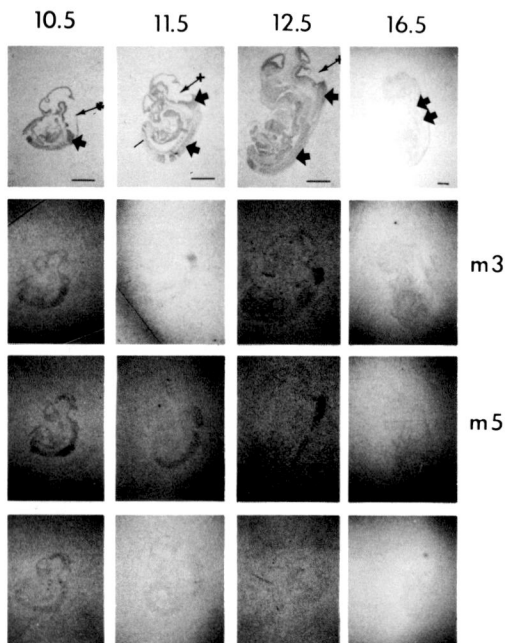

m3

m5

The localization of M3 and M5 transcripts in 10.5, 11.5, 12.5 and 16,5 day mouse embryos. (row 1) Light micrographs of hematoxylin -eosin stained sections. Horizontal bar = 1mm. bold arrow = spinal cord, starred arrow = 4th ventricle (row 2,3 and 4) X-ray autoradiographs of mid sagittal sections probed with M3 (row 2), M5 (row 3) and sense M3 RNA probes (row 4).

POLYTENE CHROMOSOMES IN TROPHOBLAST GIANT CELLS

S. Varmuza and J. Rossant
Mount Sinai Hospital Research Institute, Toronto,
Ontario, CANADA

Trophoblast giant cells, the first differentiated cells to arise in the mouse embryo, are characterized by their large size and excess DNA content. This has been shown to result from a cessation of mitosis, but not DNA replication, shortly after differentiation begins (1, 5, 6). Endoreduplication of giant cell chromosomes may lead to amplification of the genome by up to 500 fold. However, because giant cell chromosomes do not condense, their cytological state, ie whether polytene or polyploid, has not been convincingly determined, although several authors favor the former, based on cytological examination of the interphase chromatin. We have addressed this issue by examining the hybridization patterns of two DNA markers:
1) the major satellite sequence, which is present near the centromeres of all chromosomes (4), and
2) a transgenic insert which is present on the telomere of chromosome 3 (2).
Both of these markers are detectable by in situ hybridization with biotinylated probes.

Distribution of Satellite Sequences

Primary giant cells derived from blastocysts grown in tissue culture, and secondary giant cells (7.5 day ectoplacental cone outgrowths) or isolated secondary giant cell nuclei (10.5 day embryos) were fixed and hybridized with biotinylated M. musculus satellite DNA as described (4). The number of hybridization sites per nucleus were counted, and the mean frequencies determined. Giant cells from blastocyst outgrowths, ectoplacental cone outgrowths, or isolated giant cell nuclei contained 18.8+4.6, 17.5+5.0, or 19.9+3.9 hybridization sites. This number roughly approximates the haploid number of 20 chromosomes in the mouse. However, it is not unusual for centromeres in interphase nuclei to be "randomly or non randomly associated" (3).

Distribution of a Telomere Marker in Diploid and Polyploid Cells

Before using the transgenic marker as a test of ploidy on the giant cells, we first determined its usefulness by hybridizing diploid (blood) and polyploid (liver) cells from hemizygous and homozygous carriers with the biotinylated probe. Blood cells from hemizygous mice typically displayed one hybridization site per nucleus, while cells from homozygous mice contained two. This test proved so reliable that we routinely screened our mice in this way instead of analysing tail DNA during breeding of the line. Polyploid cells from hemizygous mice contained one, two, or four hybridization sites, while liver cells from homozygous mice contained two, four, or eight sites. Occassional deviations from the expected numbers were attributed to overlapping hybridization sites. Thus, the marker proved to be a reliable indicator of the number of marked chromosomes in the target cells.

Distribution of a Telomere Marker in Trophoblast Giant Cells

When giant cells from hemizygous embryos were hybridized with the biotinylated probe, a single hybridization site was observed. Very occassionally, two or more hybridization signals per nucleus were observed. Giant cells from homozygous embryos, on the other hand, contained two hybridization sites. The homozygous embryos were obtained by mating hemizygous individuals. The resulting F2 generation of embryos were all outgrown in tissue culture medium, and homozygous carriers were identified by the presence of two hybridization sites in diploid cells. We consistently observed that a significant proportion of giant cell nuclei contained diffuse hybridization signals. These were seldom observed in diploid cells. However, we have no explanation for the "fuzzy" staining at present.

Giant Cells Contain Polytene Chromosomes

The haploid number of centromere associated hybridization sites (satellite sequences) might be interpreted to mean that the chromosomes in trophoblast giant cells are polytene. However, the observation that interphase nuclei from other diploid tissues also contain haploid complements of centromere associated sequences renders this interpretation suspect. Thus, our finding that a telomeric marker for a single chromosome, which reliably indicates ploidy in other tissues, produces a single hybridization signal in giant cells is much more convincing evidence of polyteny in these cells. Because the homozygous giant cells yielded two hybridization signals, it is apparent that the two homologues are not closely apposed as they are in dipteran polytene chromosomes.

While our studies indicate that trophoblast giant cell chromosomes are polytene, although not in the classical sense, we cannot determine whether the centromeres are associated. One way to establish this would be to condense the chromosomes in vitro, perhaps with a Xenopus oocyte extract. Should this procedure prove successful, it may also be feasible to use such preparations for cytogenetic mapping and cloning of mouse genes.

References
1. Barlow, P. and Sherman, M. (1974) Chromosoma 47:119-131.
2. Lo, C. (1986) J. Cell. Sci. 81:143-162.
3. Manuelidis, L. (1985) Ann. N.Y. Acad. Sci. 450:205-221.
4. Rossant, J., Vijh, K.M., Grossi, C., and Cooper, M. (1986) Nature 319:507-511.
5. Snow, M. and Ansell, J. (1974) Proc. R. Soc. Lond. 187:93-98.
6. Zybina, E., Kudryavtseva, M., and Kudryavtsev, B. (1973) Tsitologiya 12:1081-1096.

DIFFERENTIATION

INVOLUCRIN - THE GENE AND THE PROTEIN

H. Green and M. Simon
Department of Physiology and Biophysics,
Harvard Medical School, 25 Shattuck Street,
Boston, MA 02115

The integument of terrestrial animals consists of a stratified epithelium whose principal cell type, the keratinocyte, is differentiated in two principal respects:

1) It possesses very abundant intermediate filaments composed of a family of keratins.

2) In the later stages of its terminal differentiation, the keratinocyte makes a highly insoluble envelope on the cytoplasmic side of its plasma membrane. At about the same time, the cell organelles undergo programmed destruction and the cell becomes a skeletal structure consisting almost entirely of filaments and envelope (1).

The evolution of the vertebrate epidermis took place in stages, summarized in Table 1.

Table I. EVOLUTION OF THE EPIDERMIS IN VERTEBRATES

Aquatic	Stratified epithelium consisting mainly of keratinocytes
Terrestrial (amphibia)	Add programmed cell death resulting in cornified layer of cells with cross-linked envelopes
Primates	Add involucrin to the envelopes (2,3)

The Cross-Linked Envelope The insolubility of the keratinocyte envelope is the result of protein cross-linking catalyzed by the enzyme transglutaminase (4-6). The enzyme introduces isopeptide bonds between the carboxamide groups of peptide bound glutamine residues and the E-amino groups of lysyl residues. The transglutaminase that catalyzes the formation of the envelope in the keratinocyte is predominantly bound to the plasma membrane (7,8). This enzyme has been extensively purified and is probably specific to epithelial cells, since monoclonal anti-sera to it do not react with the (soluble) transglutaminase of connective tissue cells (8). Like other transglutaminases, the keratinocyte enzyme is activated by Ca^{++}, and envelope polymerization can therefore be initiated prematurely by introducing Ca^{++} into the cell (2).

The cross-linked envelope is composed of a number of proteins. When the transglutaminase of human keratinocytes is activated, at least 6 proteins become unextractable by ionic detergents and are presumed to be cross-linked. Involucrin is the only cytosolic one: the others are membrane-associated before the cross-linking begins; three of these proteins (including involucrin) are not found in fibroblasts (9), and are therefore specialized products of the keratinocyte.

The Protein Involucrin This is a rather unusual protein. In hydrodynamic studies, it behaves like an elongated rod (2). It is stable at high temperature and may be purified extensively by simply heating a crude extract of keratinocyte cytosol to 100°; most cytosolic proteins aggregate and are precipitated and the involucrin may then be recovered from solution (10). When electrophoresis of cytosolic extracts is carried out in the absence of sodium dodecyl sulfate, most of the proteins

interact and do not migrate as discrete bands, but involucrin does migrate as a discrete band with nearly the same mobility as in the presence of sodium dodecyl sulfate (10). Evidently the negative charge carried by involucrin (pk 4.5) is nearly as great as in the presence of SDS.

Involucrin as preferential substrate of epidermal transglutaminase Involucrin is an amine-accepting substrate of transglutaminase. To examine the degree to which keratinocyte transglutaminase prefers involucrin as substrate to other cytosolic proteins, involucrin was separated from other cytosolic proteins by immuno-affinity chromatography. The cytosolic proteins, now depleted of involucrin, were then compared with the purified involucrin as substrate for transglutaminase and acceptor of labelled putrescine, using a membrane-containing particulate fraction of keratinocytes as a source of enzyme. Per microgram of protein, involucrin was labeled at least 80-fold more intensely than the average of other cytosolic proteins (8). Involucrin is also preferred to other cytosolic proteins in copolymerization reactions with proteins of the keratinocyte membrane, resulting in polymers insoluble in SDS. It is therefore of interest how the structure of the involucrin molecule makes involucrin a preferred substrate.

The Involucrin Gene The gene for human involucrin has been cloned and sequenced (11). It contains 2 exons. The first contains only a short 5' untranslated sequence. The second contains the entire coding region of 1.8 kb. This consists of 59 repeats of different

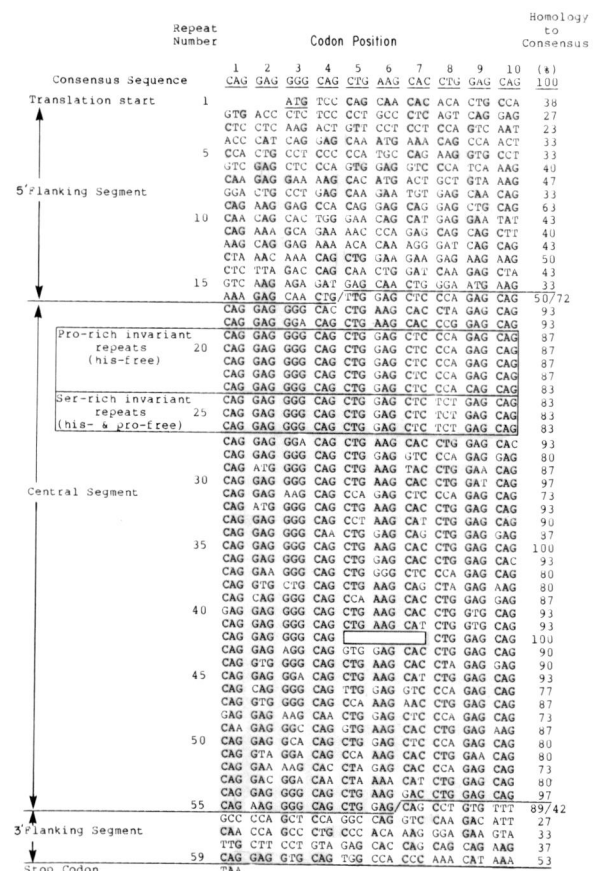

Figure 1. Coding region gene for human involucrin (from Eckert and Green, (11)).

variants of a 30 nucleotide sequence. These variant repeats may be grouped, as shown in Fig. 1. Analysis of these blocks of repeats strongly suggests that the gene was created de novo out of a primordial repeat of the triplet CAG (Fig. 2).

The first stage in the evolution of the gene appears to have been a series of repeats of $(CAG)_{10}$, creating that part of the gene corresponding to the present flanking segments. Considerable divergence of these repeats then followed, so that the 20 repeats in the flanking segments of human involucrin are on average only 47% homologous to $(CAG)_{10}$.

The next stage consisted of over 20 duplications of one of the divergent repeats of the flanking segment. This led to the central segment of the gene, a region about 80% homologous to a consensus departing significantly from $(CAG)_{10}$ (Fig. 2). This consensus contains only 3 CAG codons per repeat.

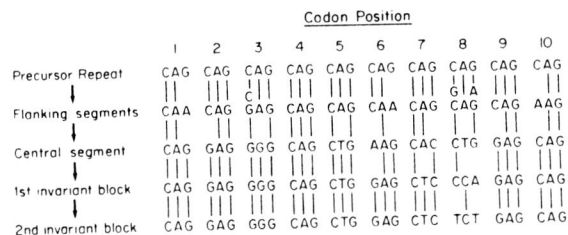

Codon Position

	1	2	3	4	5	6	7	8	9	10
Precursor Repeat	CAG	CAG	CAG	CAG	CAG	CAG	CAG	CAG	CAG	CAG
Flanking segments	CAA	CAG	GAG	CAG	CAG	CAA	CAG	CAG	CAG	AAG
Central segment	CAG	GAG	GGG	CAG	CTG	AAG	CAC	CTG	GAG	CAG
1st invariant block	CAG	GAG	GGG	CAG	CTG	GAG	CTC	CCA	GAG	CAG
2nd invariant block	CAG	GAG	GGG	CAG	CTG	GAG	CTC	TCT	GAG	CAG

Figure 2. Proposed evolution of the coding sequence of involucrin

The final stages were the creation of 2 invariant blocks. A divergent repeat of the central segment gave rise to the first invariant block. A divergent repeat of this block was then duplicated to give rise to the second invariant block. There has been essentially no divergence in the repeats of these recently created blocks; the 3 repeats of the second block are identical and the 5 repeats of the first block differ in only 1 nucleotide out of 150 (12).

The most essential amino acid of involucrin is presumably glutamine, since it alone is able to accept amines and form isopeptide bonds; participation of involucrin lysyl residues seems much less important in the polymerization reaction in vitro (8). There are 150 glutamine residues in involucrin and these are among the most conservative residues of the central segment. Two of the three glutamines of each repeat of the central segment are adjacent and the pair is bordered by glutamic acid residues. Such a simple repetitive sequence might favor multiple interactions with other proteins located in the cell membrane; an example of an interactive protein possessing multiple repeats has been studied earlier (12). The pattern of repetition in involucrin also suggests an evolutionary advantage, for if the enzyme were to recognize this sequence, involucrin could remain a preferred substrate while continuing to enlarge by the addition of successive repeats anywhere in the molecule. The principal result would be to add more mass to the involucrin molecule and hence to the envelope without disturbing the ability of involucrin to act as a transglutaminase substrate.

In recent years a number of other proteins have been found to contain repeats of a short amino acid sequence (13-19). In most cases the significance of the repeats is not known. Even in the case of a quite old protein, serum albumin, it is clear that the protein evolved from numerous repeats of 9 codons, although the repeats are now highly divergent and it is possible to trace their evolution only in broad outline (20). Evolution of proteins through duplication of short sequences is not a rare event; but as involucrin must be one of the most recently evolved proteins, the details of its evolutionary history are probably more clearly revealed. It seems likely that evolution of proteins from simple nucleotide sequences is still a current process.

References

1. Green, H. (1980). The Harvey Lecture, Series 74, 101-139.
2. Rice, R.H., and Green, H. (1979). Cell 18, 681-694.
3. Rice, R.H. and Thacher, S.M. (1986). in Biology of the Integument 2. Vertebrates. pp 752-761. J. Bereiter-Hahn, A. G. Matoltsy and K. S. Richard, eds. (Springer-Verlag).
4. Rice, R.H. and Green, H. (1977). Cell 11, 417-422.
5. Sugawara, K. (1979). Agric. Biol. Chem. 43, 2543-2548.
6. Rothnagel, J.A. and Rogers, G. E. (1984). Mol. Cell. Biochem. 58, 113-119.
7. Simon, M. and Green, H. (1985). Cell 40, 677-683.
8. Thacher, S.M. and Rice, R. H. (1985). Cell 40, 685-695.
9. Simon, M. and Green, H. (1984). Cell 36, 827-834.
10. Etoh, Y., Simon, M. and Green, H. (1986). Biochem. Biophys. Res. Comm. 136, 51-56.
11. Eckert, R.L. and Green, H. (1986). Cell 46, 583-589.
12. Waite, J.H., Housley, T.J., and Tanzer, M. L. (1985). Biochemistry 24, 5010-5014.
13. Pederson, K., Devereaux, J., Wilson, D. R., Sheldon, E. and Larkins, B. A. (1982). Cell 29, 1015-1026.
14. Ozaki, L.S., Svec, P., Nussenzweig, R.S., Nussenzweig, V. and Godson, G. N. (1983). Cell 34, 815-822.
15. Enea, V., Ellis, J., Zavala, F., Arnot, D. E., Asavanich, A., Masuda, A., Quakyi, I. and Nussenzweig, R.S. (1984). Science 225, 628-630.
16. Dame, J.B., Williams, J.L., McCutchan, T. F., Weber, J.L., Wirtz, R.A., Hockmeyer, W. T., Maloy, W.L., Haynes, J.D., Schneider, I., Roberts, D., Sanders, G.S., Reddy, E.P., Diggs, C.L. and Miller, L.H. (1984). Science 225, 593-599.
17. Kataoka, T., Broek, D., and Wigler, M. (1985). Cell 43, 493-505.
18. Allison, L.A., Moyle, M., Shales, M. and Ingles, C.J. (1985). Cell 42, 599-610.
19. Corden, J.L., Cadena, D.L., Ahearn, Jr., J. M. and Dahmus, M.E. (1985). Proc. Natl. Acad. Sci. USA 82, 7934-7938.
20. Alexander, F., Young, P.R., Tilghman, S.M. (1984). J. Mol. Biol. 173, 159-176.

GENERATION OF COMPLEX CONTRACTILE PROTEIN PHENOTYPES THROUGH PROMOTER SELECTION AND ALTERNATIVE PRE-mRNA SPLICING. B. Nadal-Ginard; R.E. Breitbart; A. Andreadis, M. Gallego; Y-T. Yu; G. Koren; G. White; P. Bouvagnet and V. Mahdavi. Howard Hughes Medical Institute. Laboratory of Molecular and Cellular Cardiology, Department of Cardiology, Children's Hospital and Department of Pediatrics, Harvard Medical School. Boston, MA 02115.

The regulated expression of structurally distinct, developmentally regulated and cell type-specific protein isoforms is a fundamental characteristic of eukaryotic cells. The molecular mechanisms responsible for generating this protein diversity might be broadly categorized into two main systems: those that select a particular gene among the members of a multigene family for expression in a particular cell and those that generate several different proteins from a single gene. This latter mechanism includes DNA rearrangement and alternative pre-mRNA splicing, each producing the differential use of intragenic sequences that lead to the production of multiple protein isoforms from a single gene. DNA rearrangement appears to be restricted to a very limited set of genes coding for immunoglobulins and T-cell receptors. In contrast, increasing numbers of genes in organisms ranging from _Drosophila_ to human, including their RNA and DNA viruses, are known to be alternatively spliced. This mode of regulation is particularly prevalent in muscle and includes myosin light chain (MLC), tropomyosin (TM) and troponin T (TnT) genes. The major constituents of the thick--myosin heavy chain (MHC) and MLC_S--and thin filaments--actin, TM, and troponins (C,T and I)--of vertebrate sarcomeres are each encoded by a multigene family of moderate size, ranging from six to eight members (1). With the exception of the tropomyosins (see below), the expression of each member of these families is regulated at the transcriptional level in a tissue and developmentally regulated manner. The restricted combinatorial use of the different members of these multigene families allows for the generation of a moderate number of qualitatively different sarcomere types that, at least in some cases, exhibit significantly different physiological characteristics (2). In addition, the potential for the production of different sarcomere is highly increased by the generation of multiple protein isoforms by individual MLC, TM and TnT genes. Therefore, in order to understand the mechanisms involved in generating sarcomere diversity, it is necessary to elucidate both, the elements responsible for the selective transcription of a given gene in a particular cell and developmental or physiological state as well as the factors involved in alternative splicing.

I. TRANSCRIPTION OF MYOSIN HEAVY CHAIN GENES IS REGULATED BY A MINIMUM OF TWO DISTINCT CIS-ACTING ELEMENTS THAT INTERACT WITH DEVELOPMENTAL-AND TISSUE-SPECIFIC TRANS-ACTING FACTORS. cDNA and genomic clones corresponding to seven MHC genes (embryonic, neonatal, adult fast IIA, adult fast IIB, slow/cardiac β, cardiac α and extraocular) have been isolated, mapped on the chromosome (3) and their responsiveness to thyroxine (T_3) analyzed (4). In addition, the complete nucleotide sequence of the embryonic gene has been obtained (5). In order to determine the sequences involved in the expression of these genes, we have produced mini-genes constituted by 5' upstream sequences, exons 1 to 3 and 37 to 41 as well as CAT gene constructs fused to the embryonic as well as cardiac α and β MHC gene upstream sequences. These constructs were used for transient gene expression assays in cultured HeLa cells, C_2 myoblasts and myotubes. The expression of these constructs is cell-type and differentiation stage-specific. None of the constructs was expressed in C_2 myoblast or in HeLa cells. Deletion analyses revealed that sequences located between -1141 and -671 of the embryonic MHC gene and between -650 and -320 of the cardiac β gene behave as position-, orientation-, and promoter-dependent elements. In addition, the promoter region of these three genes plays a different role in their regulation. In the case of the embryonic gene it is also

development and tissue-specific even in the presence of the SV40 enhancer. For the cardiac α gene its tissue-specificity is less dramatic, since it can be overcome by the SV40 enhancer. Interestingly, the expression of transfected and endogenous α -and β -MHC genes was not affected by T_3 in the C_2 myotubes. However, T_3 dramatically increased the level of expression of both transfected and endogenous α -MHC genes in primary cardiocytes. These results suggests that the tissue- and temporal-specific expression of these three MHC genes as well as their hormonal regulation by T_3 involves the interaction of two different 5' regulatory sequences with tissue-specific trans-acting factors. The expression of these factors, however is not limited to muscle cells. The presence of trans-acting factors in HeLa cell nuclear extracts that specifically interact with the embryonic MHC promoter was demonstrated by competition studies in an in vitro transcription system that mapped a binding site between -21 and -110. The interaction of this sequence with trans-acting factors from both HeLa cells and C_2 myotubes was examined in a protein-DNA binding mobility shift assay. Both nuclear extracts demonstrated specific binding to this sequence. Although the precise binding site has not yet been mapped it does not correspond to the CAAT and TATAA boxes of the promoter.

Taken together, the results outlined in this section demonstrate that the developmental and hormonal regulation of the MHC genes is complex and involves unusual but very specific enhancer-like elements that modulate the activity of their corresponding promoter which in turn are also cell-type and developmental-stage specific as well as hormone responsive. These cis-acting elements interact with trans-acting factors some of which are specific for particular muscle cell types while other factors appear to be expressed in muscle and non-muscle cells.

II. THE POWER OF ALTERNATIVE SPLICING TO GENERATE CONTRACTILE PROTEIN DIVERSITY COVERS A BROAD RANGE.
The generation of protein diversity by alternative splicing appears to be a common mechanism used by muscle cells. The power of this mechanism to produce multiple proteins from a single gene covers a broad range. It is relatively low in the MLC1/3 gene where two proteins are produced, intermediate in the α -TM gene which has the capacity to produce eight different proteins, and extremely high in the TnT gene which has the capacity to encode 64 different proteins.

The **MLC1/3 gene** is one of the best documented examples of **mutually exclusive** splicing produced through the alternative use of two different promoters (6,7). Vertebrate fast skeletal muscle contains two alkali MLC isoforms (MLC1 and MLC3) that differ from each other at their amino terminus (exons 1 and 2, respectively) and at additional aminoproximal sequences encoded by exons 4 and 3, respectively. The structural organization of the gene has been evolutionarily conserved (7) with the MLC1 and MLC3 promoters located 10kb apart. The isoform-specific exons are arranged in a sequence that necessitates alternative splicing. In the longer transcript the exclusion of the MLC3-specific exon 2 can easily be explained by its lack of consensus splice acceptor sequence. On the other hand, no known splicing mechanism can account for the mutually exclusive use of the MLC3-and MLC1-specific exons 3 and 4. The wide separation of the transcription initiation sites for MLC1 and MLC3 give rise to two mRNA precursors of significantly different size (20kb and 10kb, respectively) and sequence. These sequences could play a role in determining the pattern of alternative splicing of isoform-specific exons (see below).

The α -TM gene produces smooth and striated α -TM isoforms by the use of a single promoter region, two intragenic regions containing alternatively spliced "isotype switch exons" and two poly(A) addition sites located ~6kb apart (8). Contrary to expectation, this gene is not exclusively expressed in muscle cells, but it can be detected at significant levels in all cells and tissues so far analyzed, with the exception of liver. Seven different mRNA isoforms have been detected by S1-nuclease mapping. Thus, the α-TM gene produces striated (cardiac and skeletal), smooth and several non-muscle specific isoforms from a single promoter region by alternative splicing. As far as we know, this is the first example of a housekeeping gene generating differentiation-specific products by alternative splicing.

The organization of the fast **Troponin T** gene and its functional capacity to encode multiple TnT isoforms by alternative splicing from a single primary transcript has been reported (9,10,11). Among the 18 exons of this gene, three types of splicing are exhibited. Each of exons 4 to 8, near the 5' end of the primary transcript may be individually included or excluded from the mature mRNA in a **combinatorial** fashion to generate as many as 32 different sequences within the amino terminal region of the protein. Of these possible combinations, every one that has been probed in S1-nuclease analysis has been detected. Exons 16 and 17, in contrast, are alternatively spliced in a **mutually exclusive** manner to encode different α internal and β peptides, respectively, near the carboxyl terminal end. These two exons increase to 64 the number of distinct but related TnT isoforms that this gene can generate. The remaining exons are **constitutively** spliced. Alternative splicing of the two "isotype switch" regions is tissue-specific and developmentally regulated, arguing for the existence of controlling trans-acting factors.

The three genes described in this section, demonstrate the power and diversity of specifically regulated alternative pre-mRNA splicing. These genes constitute a favorable system in which to elucidate those elements involved in generating the different forms of alternative splicing and those that must distinguish between alternative and constitutively spliced exons.

III. USE OF MINI GENE CONSTRUCTS TO DETERMINE THE RELATIVE ROLE OF CIS-AND TRANS-ACTING ELEMENTS IN ALTERNATIVE SPLICING

A multitude of in vitro prepared mini genes constructs containing MLC1/3, TnT and α-TM gene sequences have been prepared. In order to express these constructs in non-muscle as well as undifferentiated and differentiated muscle cells, all these constructs were driven by strong and promiscuous promoters, such as SV40 and the LTR of retroviruses. In addition, stable transformants expressing each of the constructs at high level were isolated and the pattern of splicing of their transcripts determined by S1-nuclease mapping.

The picture that is emerging from these experiments is an intriguing one. Muscle contractile protein gene transcripts are alternatively spliced both in non-muscle and undifferentiated muscle cells, but the patterns of splicing differ according to the construct analyzed and, in most cases, is different from the pattern exhibited in differentiated muscle cells.

Three patterns of splicing, that charaterize different mechanisms, can be detected. **a) Cis-regulated splicing** with little or no influence of trans-acting cell-specific factors. This pattern is exhibited by the MLC 1/3 gene. In this case, the donor of exon 1 (MLC1-specific) and exon 2 (MLC3-specific) appear to determine whether exon 3 (MLC3-specific) or exon 4 (MLC1-specific) are utilized. In fact, the selectivity of the donor sites is so stringent that when the appropriate exon is removed from the construct, the alternative one is not utilized. This behavior appears to be the same in all cells tested, independently of whether they are myogenic or not, differentiated or undifferentiated. **b) Trans-regulated splicing.** This pattern of splicing is exhibited by constructs that include exons 4 to 8 of TnT either "in toto" or in different combinations. These constructs are alternatively spliced in each cell type tested, but the pattern of splicing is specific for each cell type. In general, most combinatorially spliced exons are excluded in non-muscle cells and myoblasts, although different combiniantions are included at low levels of efficiency. During the induction of differentiation there is a dramatic and progressive

increase in the number of combinations of TnT mRNAs generated by the constructs until they attain the basic pattern of the endogenous gene in fully differentiated cells. These results clearly indicate that there are trans-acting factors that are induced during differentiation and trigger the inclusion of a number of exons. **c) Cis-and trans-regulated exons.** Although it is clear that, as is the case of the combinationally spliced exons of TnT described in the previous section, there are cis-acting elements that also play a role, the interplay between these two elements is more clearly demonstrated by the α-and β-exons at the 3' end of the TnT gene. These two exons are spliced in a mutually exclusive manner in all muscle cells. This is also the case for all the mini-genes tested so far. Yet, in both muscle and non-muscle cells the mini-gene constructs always include the β exon while excluding the α. This result is less surprising when it is noted that β is the embryonic/fetal exon and that α is only expressed in adult tissues. More interestingly, the α exon is not included in the mRNA from any of the constructs even when the β exon has been deleted. In none of the cell types used so far, have we been able to detect the inclusion of the α-exon. These results strongly suggest that in the absence of specific trans-acting factors the α exon is not recognized by the splicing machinery of the cell and this exon is constitutively excluded. Therefore, exclusion of the exon is not a simple matter of competition with the β exon for the splice sequences of the common flanking exons. Trans-acting factors present only in adult muscle cells, however, make this exon recognizable to the splicing system and able to successfully compete with the β-exon. **d) Alternative splicing of constitutive exons.** Of particular interest is the finding that TnT exons 3 and 9 that are constitutively spliced in muscle cells (11), are alternatively spliced when the TnT mini genes are expressed in non-muscle or undifferentiated muscle cells, while they are uniformly incorporated in myotubes. The switch from alternative to constitutive splicing of these exons occurs progressively during myogenesis. Therefore, the distinction between alternative and constitutively spliced exons is not dependent only on the intrinsic properties of the exon (cis) but can be influenced by the cell environment (trans).

CONCLUSIONS: Taken together the results summarized here demonstrate that different splice junctions in a gene transcript are not equivalent but express different affinities for each other even when they have a canonical splice junction sequence. These relative affinities are dependent on the nucleotide sequences flanking the exons and, in some cases, by the differentiated state as well as cell type where the transcript is expressed. Therefore, the relative affinity among splice sites can be modulated by cis- and trans-acting elements. At least some of the trans-acting elements are expressed in a tissue-specific and developmentally regulated manner. Moreover, the distinction between alternative and constitutively spliced exons appears not to be an intrinsic property of the exons but is greatly influenced by the characteristics of the neighboring splice sites as well as the trans-acting factors available in the cell where the transcript is expressed.

REFERENCES 1. Nadal-Ginard et al., (1982) In muscle development: Molecular and Cellular Control, Ed. by M.L. Pearson and H. Epstein. Cold Spring Harbor Laboratory Press pp. 143-168. 2. Lompre, A.M., Nadal-Ginard, B. and Mahdavi, M. (1984) J. Biol. Chem. 259:6437-6446. 3. Leinwand, L.A., et al., (1983). Science 221:766-768. 4. Izumo, S., Nadal-Ginard, B., and Mahdavi, V. (1986) Science 231:597-600. 5. Strehler, E.E, et al., (1986). J. Mol. Biol. 190:291-317. 6. Periasamy, M. et al., (1984). J. Biol. Chem. 259:13595-13604. 7. Strehler, E.E. et al., (1985). Mol. and Cell. Biol. 5:3168-3182. 8. Ruiz-Opazo, N., and Nadal-Ginard, B. (1987). Submitted. 9. Medford, R.M. et al., (1984) Cell 39:409-421. 10. Breitbart, R., et al., (1985) Cell 41:67-82. 11. Breitbart, R. and Nadal-Ginard, B. (1986) J. Mol. Biol. 188:313-324.

C-FOS INDUCTION: A POSSIBLE NUCLEAR SWITCH IN GROWTH REGULATION

T. Curran[1], L. Sambucetti[1], K. Rubino[1], B.R. Franza[2], and James I. Morgan[3]

Departments of Molecular Oncology[1] and Neurosciences[3], Roche Institute of Molecular Biology, Nutley NJ 07110 and Cold Spring Harbor Laboratories[2], Cold Spring Harbor, NY 11724

Retroviral oncogenes (v-onc) were identified and isolated on the basis of their ability to induce aberrant cell growth. It was not surprising therefore to discover that their cellular homologues, the proto-oncogenes (c-onc) play key roles in the control of proliferation and differentiation. Several proto-oncogenes have been found to be related or identical to growth factors and their receptors (for review see ref.1). Recent studies have implicated the c-onc genes that encode nuclear proteins (such as c-fos and c-myc) in the control of cell growth and differentiation, as these genes are induced rapidly following stimulation of cells by growth factors and other agents (2-12). We have investigated the control of c-fos expression, and the properties of the fos gene product, in an attempt to understand the function of the fos proto-oncogene.

(I) **INDUCTION AND PROPERTIES OF THE FOS GENE PRODUCT.** Many laboratories have shown that c-fos and c-myc are induced following treatment of serum-deprived fibroblasts with polypeptide growth factors and serum (2-8). Generally, c-fos is induced with a rapid time-course; peak expression of mRNA occurs at 30 to 45 min post-stimulation and returns to basal levels after 1 to 2 h. Synthesis of the c-fos protein follows the mRNA profile; maximal levels are obtained at 1 to 2 h post-induction and decay occurs with a half-life of approximately 2 h. During the induction process, a number of changes occur; (a) c-fos transcription is activated and then repressed, (b) c-fos mRNA accumulates and is degraded, and (c) the c-fos protein is synthesized, modified, transported to the nucleus, complexed with cellular proteins and subsequently degraded. Thus, a complex series of events is triggered during induction, the consequence of which is a dramatic, but transient, elevation of c-fos protein levels prior to the onset of DNA synthesis. This process is very similar in the many other biological situations in which it occurs, although it sometimes precedes differentiation rather than mitogenesis.

The c-fos product is generally detected as a protein with an apparent molecular weight of 62 kilodaltons (kDa) on SDS-polyacrylamide gel electrophoresis (Fig. 1). This size is a consequence of post-translational modification since the unmodified protein migrates on SDS-polyacrylamide gels with an apparent molecular weight of 54 to 55 kDa. In PC12 cells the fos protein is complexed with a cellular protein of 40 kDa that is present in immunoprecipitates, whereas, in fibroblasts it is bound to a 39 kDa cellular protein. In addition, in both cell types, a series of antigenically related proteins, of 46 and 35 kDa, are also detected using fos-specific affinity-purified peptide antibodies (Fig. 1). These fos-related antigens are induced with a similar time-course to c-fos, they are nuclear and share many biochemical properties with the c-fos product. Both the c-fos and v-fos products can be extracted readily from nuclei using buffer containing 0.5% NP40 and 400 mM KCl. Fos protein complexes extracted from nuclei by this procedure were applied to a series of columns. The fos product was not retained on cellulose columns but was retained by phosphocellulose, although it eluted with low concentrations of salt (200 mM KCl). A tighter

association with double-stranded DNA cellulose was observed and elution was achieved only with 500 mM to 1 M KCl. Up to 85% of the fos protein complex was retained on DNA cellulose. In all assays, the v-fos protein complex behaved in a similar manner to the c-fos protein complex. In addition, the fos related antigens were also retained on DNA cellulose and eluted with 500 mM to 1 M KCl. To determine if fos proteins are also associated with DNA in vivo, isolated nuclei were treated with DNase,micrococcal nuclease and RNase. Both DNase and micrococcal nuclease released the majority of the fos protein from the nucleus, whereas, treatment with RNase did not release any fos protein. These data support the hypothesis that fos is involved in the regulation of gene expression.

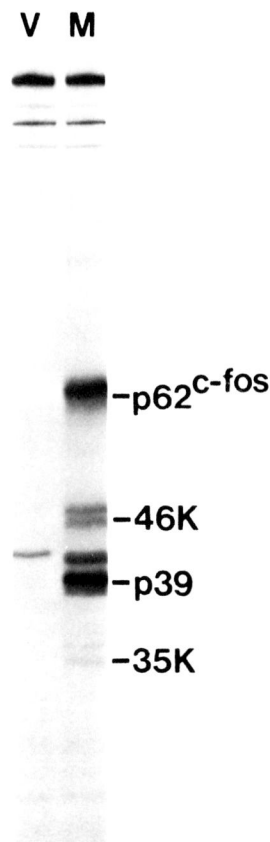

Fig. 1. **Detection of c-fos protein, FOS-related antigens and FOS-binding proteins in serum stimulated rat fibroblasts.** Rat fibroblasts were maintained for two days in medium containing 0.5% fetal calf serum. This was replaced with methionine-free medium supplemented with 20% dialyzed calf serum and the culture was incubated at 37°C for 30 min. Approximately 300 μCi of ^{35}S-methionine was then added and incubation was continued for a further 15 min. Cell lysates were prepared, treated with control antibodies (V) or FOS-specific peptide antibodies (M), and the immunoprecipitation products were analyzed on an SDS-polyacrylamide gel as previously described (15).

(II) CONTROL OF C-FOS EXPRESSION IN PC12 CELLS. We have focused our attention on the PC12 pheochromocytoma cell line in an attempt to elucidate some of the biochemical events that lead to c-fos expression. Two major pathways have been identified that are involved in c-fos activation. Treatment of PC12 cells with nerve growth factor (NGF) or fibroblast growth factor (FGF) leads to a series of alterations that include neurite outgrowth and changes in neurotransmitter synthesis (13,14). However, the earliest change at the level of gene expression yet identified is the induction of c-fos mRNA and protein (9,10). Thus, in this system, c-fos protein synthesis occurs as a prelude to a differentiation process. In addition we have found that depolarization of PC12 cells can evoke a strong c-fos induction (11). Treatment of cells with high levels of potassium ions leads to a calcium- and calmodulin-dependent c-fos activation. This process is distinct from that activated by NGF, as the NGF response does not require calcium or calmodulin. The depolarization pathway can also be triggered by veratridine, which causes depolarization by provoking an influx of sodium ions. Direct activation of the voltage-dependent calcium channel with the drug BAY K 8644 also induces c-fos, further emphasizing the role of this ion channel in the control of c-fos expression. Surprisingly, although this calcium- and calmodulin-dependent pathway is a potent activator of c-fos transcription, it leads to the synthesis of a partially modified fos protein (12). This pathway is not present in fibroblasts and, so far, it is specific for PC12 cells. One might speculate that activation of c-fos by agents that classically depolarize neurones implies that c-fos and the co-regulated genes, are excellent candidates for coupling excitation of the neurone to long term adaptive modifications of transcription.

Although there are two major pathways that mediate c-fos induction in PC12 cells, other pathways can lead to a modest increase in expression. Activation of protein kinase C by phorbol esters and activation of cAMP-dependent kinase by dbcAMP are two examples. However, although such signals only have a minor effect on c-fos expression in PC12 cells, they can have major effects in other cell types. Thus, c-fos induction may be regarded as an early nuclear marker for various signal transduction systems. However, the degree and specificity of the response may be dictated by the differentiated or physiological state of the stimulated cell.

(III) ROLE OF 3' SEQUENCES IN C-FOS REGULATION. Relatively little is known about the processes involved in the specific degradation of c-fos mRNA following induction. We have investigated the expression and stability of fos transcripts derived from SP6 vectors, using the rabbit reticulocyte lysate translation system. For this purpose, we have constructed vectors that encode portions of the v-fos and c-fos genes. Translation of v-fos mRNA yields a 55 kDa protein as expected. Interestingly, translation of c-fos mRNA (derived from a full-length cDNA clone) results in synthesis of the highly modified 62 kDa form of the c-fos protein that is present in serum-stimulated fibroblasts. The v-fos transcript exhibits a simple dose-dependent translational activity. In contrast, although the c-fos transcript is translated extremely efficiently at low concentrations, very little expression is detected at high concentrations. Analysis of RNAs recovered from lysates suggests that, at high concentrations, c-fos mRNA is rapidly degraded. Experiments using v-fos/c-fos chimeric transcripts indicate that the inhibitory activity is provided by the 3' untranslated region of the c-fos gene. In addition, pre-incubation of lysates with c-fos 3' sequences causes

an inhibition of expression of exogenous globin mRNA. The data are consistent with the hypothesis that the 3' end of the c-fos gene activates a nuclease in reticulocyte lysates.

REFERENCES

(1) Heldin, C-H and Westermark, B. (1984) Cell 37, 9-20.
(2) Kelly, K., Cochran, B.H., Stiles, C.D. and Leder, P. (1983) Cell 35, 603-610.
(3) Campisi, J., Gray, H.E., Pardee, A.B., Dean, M. and Sonenshein, G.E. Cell 36, 242-247.
(4) Cochran, B.H., Zullo, J., Verma, I.M. and Stiles, C.D. (1984) Science 226, 1080-1082.
(5) Greenberg, M.E. and Ziff, E.B. (1984) Nature 311, 433-438.
(6) Kruijer, W., Cooper, J.S., Hunter, T. and Verma, I.M. (1984) Nature 312, 711-716.
(7) Muller, R., Bravo, R., Burckhardt, J. and Curran, T. (1984) Nature 312, 716-720.
(8) Bravo, R., Burkhardt, J., Curran, T. and Muller, R. (1985) EMBO J. 4, 1193-1197.
(9) Curran, T. and Morgan, J.I. (1986) Science 229, 1265-1268.
(10) Greenberg, M.E., Greene, L.A. and Ziff, E.B. (1985) J. Biol. Chem. 260, 14101-14110.
(11) Morgan, J.I. and Curran, T. (1986) Nature 322, 552-555.
(12) Curran, T. and Morgan, J.I. (1986) Proc. Natl. Acad. Sci. USA (in press).
(13) Greene, L.A. and Tischler, A.S. (1976) Proc. Natl. Acad. Sci. U.S.A. 73, 2424-2428.
(14) Greene, L.A. and Rein, G. (1977) Nature 268, 349-351.
(15) Curran, T., Van Beveren, C., Ling, N. and Verma, I.M. (1985) Mol. Cell. Biol. 5, 167-172.

MUSCLE CELL DETERMINATION IN ASCIDIAN EMBRYOS

William R. Jeffery and Craig R. Tomlinson

Center for Developmental Biology, Department of Zoology, University of Texas, Austin, TX 78712.

Tadpole larvae of the ascidian Styela contain about 40 muscle cells located in three bands on each side of their tails. The muscle cells are derived from three sources. The primary source is decendants of the B4.1 blastomeres of 8-cell embryos (Fig. 1B). The B4.1 cells are characterized by yellow pigmentation (1). The yellow pigment is obtained from a localized egg cytoplasmic region, the yellow crescent, which segregates to the posterior vegetal region of the fertilized egg where it is inherited by the B4.1 cells (Fig. 1A). Secondary sources of muscle cells originate from descendants of the A4.1 and b4.2 blastomeres of 8-cell embryos (Fig. 1B).

The relative contributions of the B4.1, A4.1 and b4.2 cell lineages to muscle cells has been defined by horseradish peroxidase injection. In ascidian larva with 42 tail muscle cells, 28 anterior cells originate from B4.1 and 14 posterior cells originate from A4.1 and b4.2 (2,3). Ascidian muscle cells are thought to be specified by the action of determinants localized in the yellow crescent (1), but this region does not not enter all the muscle cell progenitors (Fig. 1). Here, we present an alternate hypothesis for muscle cell determination based on the accumulation of muscle-specific actin mRNA and protein.

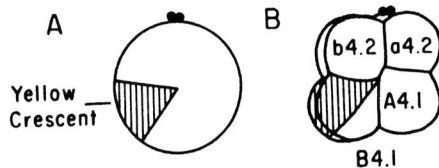

Fig. 1. A fertilized egg (A) and 8 cell embryo (B) of Styela.

ACTIN GENES, TRANSCRIPTS, AND PROTEINS. 2D gel electrophoresis indicated that there are three major actins in Styela (Fig. 2). Two basic actins are present in all adult and embryonic tissues while an acidic actin is present only in adult and larval muscle (Fig. 2). We assume that the basic actins are cytoskeletal actins, and the acidic actin is a muscle actin. Eggs and early embryos contain only the cytoskeletal actins. 2D gel electrophoresis of ^{35}S-methionine-labeled proteins indicated that the muscle actin first accumulates to detectible levels in the 128-cell embryo. The results suggest that Styela embryos synthesize a specific muscle actin.

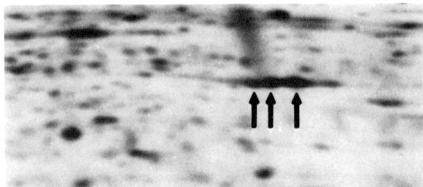

Fig. 2. A portion of a 2D gel showing the three actin isoforms (arrows) of a tadpole larvae. The muscle actin is on the right.

To obtain probes for actin transcripts, we screened a Styela muscle cDNA library with the coding region of a Drosophila actin gene. Positive clones were sequenced and compared to consensus amino acid sequences of mammalian cytoskeletal and muscle actins (4). A cDNA clone was selected for further analysis which shared 11 of 15 diagnostic amino acid positions with the muscle actins. Subclones of this cDNA were prepared which consisted of the coding (SpGA) and 3' non-coding (SpMA) regions of the corresponding mRNA (Fig. 3). In genomic Southern blots using several different restriction enzymes, SpGA recognized three bands and SpMA recognized only one band suggesting that Styela contains at least three actin genes, one of which is a muscle actin gene. In Northern blots of total muscle RNA, SpGA hybridized to three bands and SpMA hybridized to only one band suggesting that Styela muscle contains at least three actin mRNAs, one of which is a muscle actin mRNA. In translation selection analysis, SpGA selected mRNA which directed the in vitro translation of the cytoskeletal and muscle actins and SpMA selected mRNA which directed the translation of only the muscle actin. The results suggest that the cDNA clone corresponds to a muscle-specific actin mRNA and that in Styela there may be a single muscle actin gene, transcript, and polypeptide.

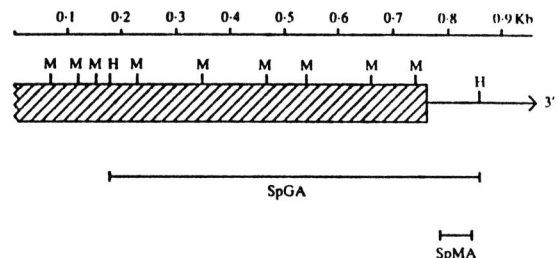

Fig. 3. The muscle actin cDNA clone showing SpGA and SpMA probes and Mnl I (M) and Hae III (H) sites.

TEMPORAL AND SPATIAL EXPRESSION OF THE MUSCLE ACTIN GENE DURING EMBRYOGENESIS. To determine the temporal expression of the muscle actin gene during embryogenesis, dot blots containing RNA extracted from eggs and embryos at various stages of development were probed with SpGA and SpMA. The muscle actin transcript was present in unfertilized eggs but gradually decreased in titer during the early cleavages until the late gastrula stage, when there was a significant increase in the level of muscle actin mRNA. Muscle actin transcripts continued to accumulate during the tailbud stage, when muscle cells begin to differentiate, and began to disappear at the tadpole stage, when muscle cell differentiation is complete. Thus, although there are some maternal muscle actin transcripts, these disappear prior to gastrulation and are subsequently replaced by much larger amounts of zygotic muscle actin mRNAs. The zygotic muscle actin transcripts are probably responsible for directing most of the muscle actin synthesis that occurs during muscle cell differentiation.

To determine the spatial distribution of zygotic muscle actin mRNAs, we subjected sections of embryos to in situ hybridization with radioactive RNA probes synthesized in vitro after SpGA and SpMA were cloned into transcription vectors. Beginning at the late gastrula or early neurula stage, zygotic muscle actin transcripts began to appear in some of the presumptive muscle cells and continued to accumulate during the tail bud stages (Figs. 4-5). There were no detectible levels of muscle actin mRNA in the epidermal, gut, brain, spinal cord, notochord, or mesenchyme cells. The muscle actin gene appears to be expressed specifically in presumptive tail muscle cells.

There were differences in the levels of muscle actin transcripts which accumulated in the anterior and posterior presumptive muscle cells in early tailbud embryos, however. Anterior tail muscle cells showed higher levels of muscle actin mRNA than posterior tail muscle cells. By the late tailbud stage these differences disappeared and all of the presumptive muscle cells showed similiar levels of muscle actin mRNA (Fig. 5). The results suggest that presumptive muscle cells derived from the A4.1 and b4.2 cell lineages may lag behind those derived from the B4.1 cell lineage in the timing of muscle actin gene expression.

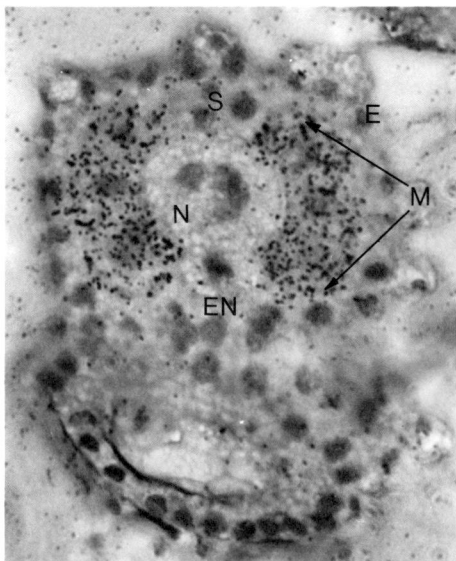

Fig. 4. A cross section through the head (below) and tail (above) of a late tailbud embryo hybridized in situ to determine the location of muscle actin mRNA. M:(band of three muscle cells). N:(notochord). S:(spinal chord). E:(epidermis). EN:(endoderm).

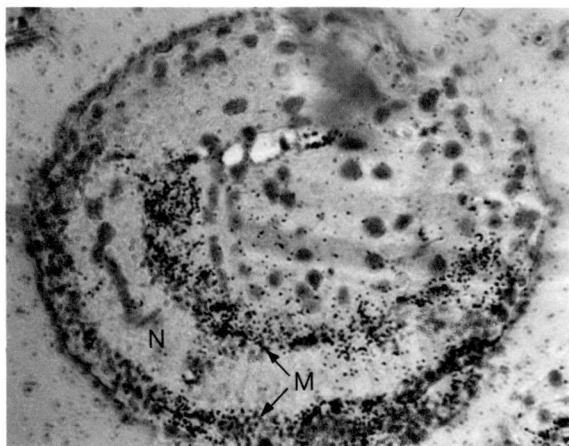

Fig. 5. A longitudinal section through a late tailbud embryo hybridized in situ to determine the accumulation of muscle actin transcripts. M:(two bands of tail muscle cells). N:(notochord).

MUSCLE ACTIN PROTEIN AND mRNA ACCUMULATION IN CLEAVAGE-ARRESTED EMBRYOS. Tissue-specific markers appear in cleavage-arrested ascidian embryos exclusively in those cells expected according to their normal developmental fates, a result inter- preted as evidence for specification of cell fate by segregated determinants (5,6). To examine the segregation pattern of muscle actin and its mRNA, ^{35}S-methionine-labeled proteins extracted from cleavage-arrested embryos were subjected to 2D gel electrophoresis or cleavage-arrested embryos were subjected to in situ hybridization with the SpGA and SpMA probes. Muscle actin accumulated to high levels in cleavage-arrested embryos, and electrophoresis of proteins from seperated B4.1 and A4.1/b4.2/a4.2 portions of 8-cell embryos indicated that muscle actin was concentrated in the B4.1 cells. Likewise, muscle actin transcripts accumulated exclusively in the B4.1 cell lineages of cleavage-arrested embryos (Fig. 6). Despite their contribution to tail muscle, the A4.1 or b4.2 cell lineages did not accumulate detectible muscle actin mRNA. These results suggest that segregated determinants may control muscle actin gene expression in the B4.1 cell lineage, but not in the A4.1 or b4.2 cell lineages.

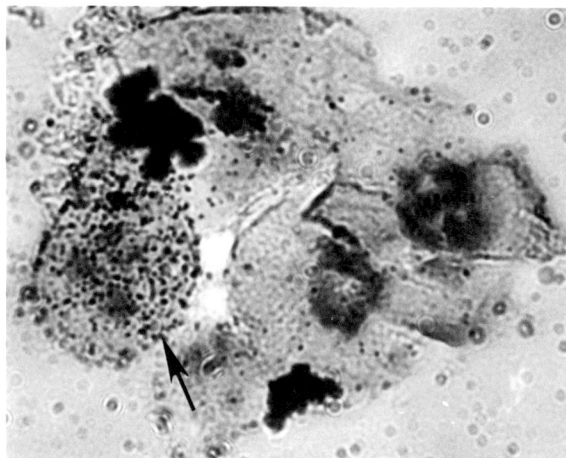

Fig. 6. A section through a cleavage-arrested 8- cell embryo showing muscle actin mRNA in a B4.1 cell (arrow).

DISCUSSION. Styela embryos synthesize a muscle- specific actin mRNA and protein during early development. In cleavage-arrested embryos, the actin and its mRNA accumulate exclusively in blastomeres that contain the yellow crescent. Some cells that lack the yellow crescent and do not accumulate muscle actin or its mRNA in cleavage- arrested embryos eventually express muscle actin mRNA and differentiate into muscle cells in normal embryos. These cells presumably do not depend directly on segregated determinants for muscle actin gene expression. It is proposed that yellow crescent determinants may have two functions in ascidian embryos. First, they may function to promote muscle cell determination directly in the B4.1 cells. Second, they may also cause B4.1 cells to develop the capacity induce some of the A 4.1 and b4.2 lineage cells to become muscle cells.

REFERENCES

(1) Conklin, E.G. (1905) J. Acad. Nat. Sci. Phila. 13, 1-119.
(2) Nishida, H., and Satoh, N. (1983) Dev. Biol. 99, 382-394.
(3) Nishida, H., and Satoh, N. (1985) Dev. Biol. 110, 440-454.
(4) Vanderkerckhove, J., and Weber, K. (1979) Differentiation 14, 123-133.
(5) Whittaker, J.R. (1973) Proc. Nat. Acad. Sci. U.S.A. 70, 2096-2100.
(6) Jeffery, W.R. (1985) Cell 41, 11-12.

STAGE-SPECIFIC ACTIVATION OF APOLIPOPROTEIN GENES IN F9 EMBRYONAL CARCINOMA CELLS

K. Basheeruddin, S.J. Madore, M. Sorci Thomas, and D.L. Williams

Department of Pharmacological Sciences, State University of New York at Stony Brook, N.Y. 11794.

We have previously examined the expression of the apolipoprotein (apo) E gene in mouse embryos and F9 embryonal carcinoma (EC) cells (1). F9 cells can be irreversibly differentiated in response to retinoic acid into parietal endoderm (PE)-like or visceral endoderm (VE)-like cells which resemble cells of the parietal or visceral yolk sac of the midgestation mouse embryo (2,3). The apo E synthesized by F9 cells was reactive with antibody raised against plasma apo E and comigrated on SDS polyacrylamide gel electrophoresis with newly synthesized hepatic apo E at an apparent molecular weight of 33,000. Upon differentiation to PE-like cells, apo E synthesis and secretion were down-regulated. In contrast, differentiation of F9 cells to VE-like cells up-regulated apo E synthesis. DNA-excess solution hybridization assays were used to measure apo E mRNA in F9 cells and mouse embryonic tissues. The changes in apo E expression in F9 PE-like and VE-like cells are due to changes in apo E mRNA abundance. Furthermore, the difference in apo E mRNA content of F9 VE-like and PE-like cells is very similar to the difference shown by visceral and parietal yolk sacs of the 12 day embryo. These data indicate that apo E expression in differentiated F9 cells reflects the developmental pattern of apo E expression in vivo. In the 12 day embryo, visceral yolk sac contained 5 times more apo E mRNA than liver. In the next 6 days the apo E mRNA content of visceral yolk sac declined while liver apo E mRNA increased 30 fold. These data are consistent with the hypothesis that visceral yolk sac function declines during late gestation while fetal liver assumes its functions (5). Additional evidence is now presented that F9 cells express other apolipoprotein genes and secrete lipoprotein particles. F9 cells appear to be a suitable cell culture model to study the developmental activation of apolipoprotein genes and the cellular components required for lipoprotein assembly.

Apo AI mRNA was measured in stem cells and differentiated F9 cells by dot blot analysis using a mouse apo AI cDNA probe labeled by the random priming method. F9 VE-like cells expressed 2-4 fold higher levels of apo AI mRNA than F9 stem cells. The VE-like cells contain 50% as much apo AI mRNA as adult liver. DNA-excess solution hybridization measurements with a single stranded apo AIV probe showed that apo AIV mRNA was not present in F9 stem cells and PE-like cells but was present in VE-like cells (0.07 pg/ug RNA) and visceral yolk sac (2.8 pg/ug RNA) of the 12 day embryo. These data indicate that at least 3 apolipoprotein genes are active in visceral yolk sac in vivo and are up-regulated during differentiation of F9 cells to the VE-like phenotype in culture. In other experiments, F9 cells were metabolically labeled with ^{35}S-methionine for 6 hours in lipoprotein deficient serum. Culture medium contained newly synthesized proteins that floated in the density range d 1.006-1.21 g/ml upon ultracentrifugaton. SDS polyacrylamide gel electrophoresis showed that newly secreted apo Al was most abundant in fractions of density 1.063-1.15 g/ml and 1.15-1.20 g/ml. Thus, the F9 cell may also mimic the visceral yolk sac endoderm in producing lipoprotein particles.

In order to study the activation of apolipoprotein genes in VE-like F9 cells, a series of expression vectors were constructed with the monkey apo E 5'-flanking sequences inserted next to the bacterial chloramphenicol acetyltransferase (CAT) gene. The apo E DNA from bp -797 to +31 promoted CAT expression. A DNA-excess solution hybridization assay was used to quantitate CAT mRNA levels directly in F9 cells. In transient transfection assays the apo E promoter was found to be 40% as active as a DNA fragment containing the Herpes simplex virus thymidine kinase promoter and enhancer elements from a polyomavirus host range mutant that is very active in F9 cells. These results suggest that the F9 cell will be useful to identify DNA sequences and protein factors necessary for the activation of apolipoprotein genes during embryological development.

References.

1. Basheeruddin, K., Stein, P., Strickland, S. & Williams, D.L. (1986) submitted for publication.
2. Strickland, S., Smith, K.K., & Marotti, K.R. (1980) Cell 21, 347-355.
3. Hogan, B.L.M., Taylor, A. & Adamson, E.D. (1981) Nature, 291, 235-237.
4. Williams, D.L., Newman, T.C., Shelness, G.S. & Gordon, D.A. (1986) Meth. Enzy. 129, 670-689.
5. Yeoh, G. & Morgon, E.H. (1984) Biochem. J. 144, 215-224.

THE HOMEOTIC COMPLEX OF TRIBOLIUM CASTANEUM

Richard W. Beeman[1] and Robin E. Denell[2]

U.S. Grain Marketing Research Laboratory, Agricultural Research Service, U.S. Department of Agriculture, Manhattan, KS 66502[1] and Division of Biology, Kansas State University, Manhattan, KS 66506[2]

In the past few years studies of the fruitfly, Drosophila melanogaster, have resulted in dramatic gains in our understanding of the organization and roles of genes which control developmental commitments in this organism. Previously, geneticists had isolated and described mutations affecting adult cuticular morphology such that regions develop segment-specific features inappropriate for their location. Some of the genes identified by such "homeotic mutations" were found to be members of two major clusters: the bithorax complex or BX-C (1), and the Antennapedia complex or ANT-C (2). BX-C mutations affect posterior thoracic and abdominal development, and map along the chromosome in the same order as the segments predominantly affected. Although some ANT-C mutations cause more anterior homeotic effects, there is no correlation with chromosomal order and not all genes are obviously homeotic. Two major conceptual and methodological advances then resulted in cascades of new information: a focus on the embryological effects of mutations causing more extreme abnormalities incompatible with survival to the adult stage (1), and the application of molecular biological approaches (3). This work has led to a consensus that genes within the two complexes play primary roles in controlling determinative commitments throughout the organism, and the recognition that in addition to their developmental significance each complex shows many interesting characteristics with respect to their organization and regulatory interactions.

The Diptera are very advanced insects, and show many special features of head and thoracic morphology in both larval and adult stages. Thus the extent to which the organization and function of these complexes represent general features of insect development as opposed to special adaptations has been unclear. In order to assess the generality of the developmental mechanisms described for Drosophila, we have initiated a study of the homeotic genes of the red flour beetle, Tribolium castaneum. This insect is more primitive with respect to details of its embryological development and anterior morphology. Ten mutations associated with pupal and adult homeosis in this organism had been described. In the first phase (4), the 8 extant mutations were subjected to recombination and complementation mapping. They define 6 genes, of which 5 map into a single cluster on linkage group 2. These mutations affect segments along the entire anterior-posterior axis from the maxilla (maxillapedia) to the caudal tip (extra urogomphi), and map in the same order on the chromosome as that of the segments homeotically transformed. These results suggest the exciting hypothesis that this Tribolium cluster (denoted the Homeotic complex or HOM-C) represents the juxtaposed equivalent of the ANT-C and the BX-C. We speculate that a single complex represents the more primitive condition, and that the spatial separation of the ANT-C, as well as its greater organizational complexity and genetic diversity, reflect the evolutionary history of the Diptera.

Past work on the homeotic mutants of Drosophila provide a paradigm for our analysis of the structure and function of the HOM-C. That is, we intend to saturate this region with new mutations associated with adult effects and/or recessive lethality. The effects of mutations on embryonic and larval phenotypes will also be examined. Finally, the organization of the HOM-C and its constituent genes will be explored by molecular biological studies.

As a first step, we constructed a viable and fertile stock homozygous for four HOM-C mutations. A cross of gamma ray-treated wild type beetles to this stock allowed the recognition in the F_1 of new mutations which fail to complement these existing variants to give visible adult phenotypes. A total of 25 new mutations were isolated from approximately 39,000 progeny, and include both dominant and recessive alleles. Future mutagenesis efforts will be aimed at using a deficiency to recover recessive lethal mutations.

The adult phenotypes of the new mutations have been studied, and include several novel homeotic transformations of head, thoracic, and abdominal features. Complementation analysis indicates that one variant causing abdominal homeosis may be a deficiency. These and additional mutations will be examined for their effects on earlier stages of the life cycle, with particular emphasis on those which result in late embryonic or larval lethality. First instar larvae have thoracic legs, antennae, eye spots, and mandibles, all entirely lacking in Drosophila larvae. These structures as well as a wealth of setae and sense organs provide segment-specific markers which will be extremely valuable in scoring homeotic transformations, and an analysis of mutant embryos and larvae is underway.

Many of the genes of the BX-C and ANT-C include a 180 bp region (called the Antennapedia-type homeo box) which has been highly conserved during animal evolution (5). This sequence is translated to provide a domain of the protein encoded by each gene; this peptide bears partial homology to known DNA binding proteins (6), suggesting a transregulatory function. We have shown that the Tribolium genome includes several restriction fragments with sequence homology to this homeo box. We are currently testing to see if Tribolium has homology to the non-homeo box regions of Antennapedia and Ultrabithorax cDNAs. In the future we intend to molecularly clone the genes of the HOM-C to allow a comparison with their putative Drosophila homologues. The integrated developmental genetic and molecular study we have outlined should provide a much broader view of the genetic control of developmental decisions during insect embryogenesis.

REFERENCES

(1) Lewis, E.B. (1978) Nature 276, 565-570.
(2) Kaufman, T.C., Lewis, R., and Wakimoto, B. (1980) Genetics 94, 115-133.
(3) Bender, W., Akam, M., Karch, F., Beachy, P.A., Peifer, M., Spierer, P., Lewis, E.B., and Hogness, D.S. (1983) Science 221, 23-29.
(4) Beeman, R.W. (1986) Manuscript submitted.
(5) McGinnis, W., Garber, R.L., Wirz, J., Kuroiwa, A., and Gehring, W.J. (1984) Cell 37, 403-408.
(6) Laughan, A. and Scott, M.P. (1984) Nature 310, 25-31.

DIFFERENTIAL REGULATION OF ETS LOCI DURING MURINE HEPATIC REGENERATION

Narayan K. Bhat, Robert J. Fisher, Shigeyoshi Fujiwara, Richard Ascione and Takis S. Papas

Laboratory of Molecular Oncology, National Cancer Institute, Frederick, MD 21701-1013

INTRODUCTION

Avian erythroblastosis virus, E26, is a replication-defective virus containing tripartite oncogene, Δgag-myb^E-ets-Δenv (1). We have previously shown that in higher mammals the cellular homologue of the v-ets oncogene consists of two distinct proto-oncogenes, ets-1 and ets-2, localized on two different chromosomes (2) and both gene loci are transcriptionally active (3). In certain acute human leukemias both ets genes are translocated to different chromosomes (1). In murine tissues, the ets-2 gene is expressed as a single mRNA species (4.2 kb), unlike the multiple ets-1 transcripts. The ets-2 mRNA is more abundant in young proliferating tissues, when compared to terminally differentiated adult tissues. To examine the possibility that the ets-2 gene products are associated with cell proliferation, we studied its expression in regenerating liver, an in vivo model for cell proliferation.

RESULTS AND DISCUSSION

Partial hepatectomy induced ets-2 mRNA, peaking about four hours post-surgery, whereas sham operated liver did not induce the ets-2 message at all (Fig. 1). The DNA synthesis peaked well after the ets-2 transcripts, around 48 hours after partial hepatectomy. We were unable to detect ets-1-specific transcripts at several time points spanning one round of DNA synthesis, even while ets-2 mRNA was at its peak, suggesting that both ets-1 and ets-2 gene loci are subject to differential regulation. To place ets-2 mRNA induction in the sequence of expression of other genes involved in cell proliferation, we studied the expression of several select genes (Fig. 1) following partial hepatectomy: myc, ras^H, fos, actin, and tubulin genes which are involved in cell proliferation, all were induced by partial hepatectomy. These induction results are in agreement with the concept of an orderly expression of several genes, many of which are nuclear proto-oncogenes, during the transition from G$_0$ to G$_1$ phase of the cell cycle. The sequence of nuclear oncogene induction thus appears to be fos→myc→ets-2.

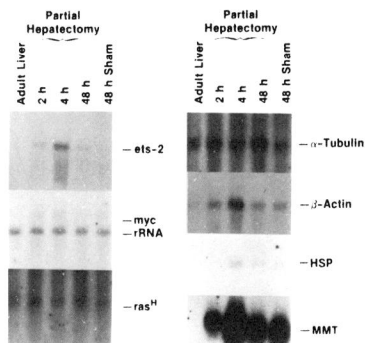

Fig. 1. Gene expression during liver regeneration. Poly A$^+$ RNA (15 µg) were size fractionated and filters were probed sequentially with the ^{32}P-labeled nick-translated probes. Transcripts detected by respective probes are indicated on the side.

These nuclear proto-oncogenes, expressed during the early phase of the cell cycle, are superinduced in the presence of protein synthesis inhibitors. Cycloheximide superinduces ets-2 mRNA by more than 15-fold (Fig. 2, lanes 2 and 5); whereas only low levels of ets-2 mRNA was detectable in sham operated animals injected with saline (Fig. 2, lanes 1 and 4). Partial hepatectomy amplifies (20 to 40-fold) the effect of cycloheximide (Fig. 2, lanes 3 and 6). This superinduction effect was not observed for the ets-1 gene, which reaffirms the differential regulation of ets-1 and ets-2 gene expression. Expression of the housekeeping genes (actin, tubulin), and stress-related protein genes (MMT and heat shock protein) are not observed to be superinduced. By contrast, myc, fos, ornithine decarboxylase (markers for cell proliferation) are superinduced by cycloheximide. This superinduction could be due either to stabilization of mRNA or the additive effects of hormones and growth factors released following surgery, which activate receptors and thereby activate the signal transduction processes.

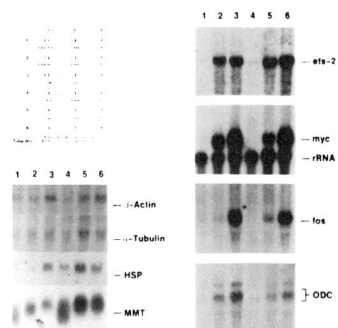

Fig. 2. Superinduction of ets-2 mRNA by cycloheximide. Left panel (top): Protocol used to superinduce ets-2 mRNA. Either 0.15 ml saline (SL) or 0.15 ml cycloheximide (CHX) in saline (100 mg/kg body weight) was injected i.p. at the indicated time. SH - sham operation; H - hepatectomy; S - sacrifice. Left panel (bottom), right panel: 15 µg of poly A$^+$ RNA from each sample (as described on left panel (top) was analyzed by Northern blot analysis. Transcripts detected by respective probes are indicated on the side.

CONCLUSIONS

In liver tissue, both ets-1 and ets-2 gene expression are differentially regulated. The ets-2 gene transcription is induced by partial hepatectomy and does not require de novo protein synthesis. The ets-2 mRNA induction pattern is similar to that of the other nuclear oncogenes, fos and myc. The expression of the ets-2 gene is intrinsically linked with the transition of cells from the G$_0$ to the G$_1$ phase of the cell cycle.

REFERENCES

(1) Papas, T. S. et al. in The Oncogene Handbook (Reddy, E. P. et al., eds.), Elsevier, New York, in press.
(2) Watson, D. K. et al. (1986) Proc. Natl. Acad. Sci. USA 83, 1792-1796.
(3) Watson, D. K. et al. (1985) Proc. Natl. Acad. Sci. USA 82, 7294-7298.

DEVELOPMENTAL CHANGES OF RAT OVARIAN AROMATASE

Mark E. Brandt, Kim M. Wilson, Stephen Zimniski, and David Puett

REPSCEND Labs, Department of Biochemistry, University of Miami, Miami, Florida, 33101

INTRODUCTION

Aromatase is a cytochrome P-450 enzyme that is required for sexual differentiation and for the development and maintenance of the normal reproductive cycle (1). This microsomal enzyme catalyzes the conversion of the androgens androstenedione and testosterone to estrone and estradiol, respectively, the latter representing the most potent estrogen. The purpose of this study was to examine the developmental changes in rat ovarian aromatase as animals undergo sexual maturation and to compare these changes with those measured in the normal estrous cycle and in gonadotropin-stimulated immature animals.

METHODS

In order to study the normal developmental changes occurring in ovarian aromatase activity during sexual maturation, immature female Sprague-Dawley rats were maintained and sacrificed at various ages until day 45. Changes in aromatase levels were measured in mature rats (55-70 days old) exhibiting at least two regular 4-day estrous cycles. Groups of animals were sacrificed on each day of the cycle. For the aromatase induction studies, immature rats (22 days old) were injected with a supraphysiologic dose (20 IU) of pregnant mare's serum gonadotropin (PMSG), a hormone that mimics endogenous gonadotropins, and sacrificed every 12 hours thereafter. Ovaries and blood were removed from the animals immediately after sacrifice for the preparation of a microsomal fraction by differential centrifugation and serum, respectively. Aromatase activity was determined by following the production of ^{3}H-water from 1β-^{3}H-androstenedione (2,3), and the results are expressed as fmol estrogen produced per min per mg of microsomal protein. The serum estradiol (E_2) levels were measured on ether extracts by radioimmunoassay under conditions which gave a sensitivity of about 10 pg per ml of serum.

RESULTS

Serum E_2 levels did not vary between days 31-45 of age (Fig. 1A), a period when ovarian aromatase activity was dramatically changing (Fig. 1B). In the mature cycling rat, ovarian aromatase activity and serum E_2 were low during estrus and metestrus, and then rose to maximum levels on proestrus. Thus, the changes observed in ovarian aromatase in the developing animal between about 34-45 day of age (cf. Fig. 1B) may reflect the initiation of the estrous cycle; in fact, on day 42 the aromatase activity was comparable to that observed in the proestrus adult. Administration of PMSG to immature rats led to a significant increase in the ovarian enzyme activity, which achieved a maximal value in about 72 hours. At this time, the induced aromatase activity was at least 5-fold higher than that determined in proestrus, while the serum E_2 level was about 10-fold higher than that measured in proestrus.

DISCUSSION

Ovarian aromatase activity was found to be extremely low in the immature animal and increased dramatically during a relatively short period of development to levels comparable to those in the adult. The time scale for this increase is also similar to that found in the normal estrous cycle and in immature rats stimulated with PMSG. In contrast, serum E_2 levels are not, in all cases, tightly coupled with ovarian aromatase activity, thus suggesting the possible importance of other regulatory factors. In conclusion, these results demonstrate that the adult pattern of ovarian aromatase activity is established in a relatively short period during sexual maturation in the rat. Supported by NIH (AM33973 and CA43226) and by an American Cancer Society Florida Division Summer Fellowship to K.M.W.

REFERENCES

1) Ojeda, S.R., Urbanski, H.F., and Ahmed, C.E. (1986) Recent Progress in Hormone Research 42: 385-440.

2) Rabe, T., Rabe, D., and Runnebaum, B. (1982) J. Steroid Biochem. 17: 305-309.

3) Zimniski, S.J., Brandt, M.E., Melner, M.H., Covey, D.F., and Puett, D. (1985) Cancer Res. 45: 4883-4889.

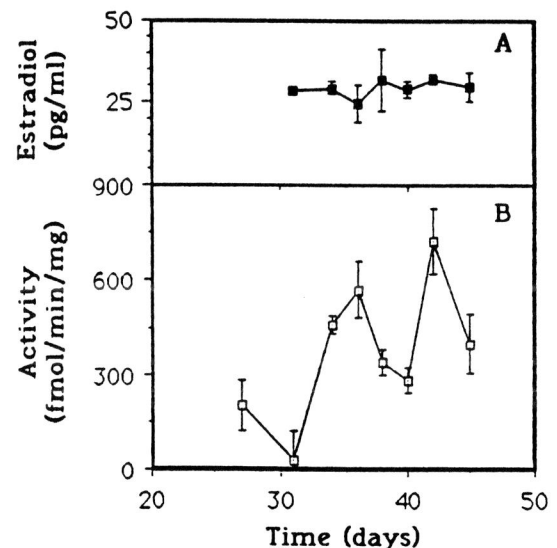

Figure 1. Serum E_2 levels (A) and ovarian aromatase activity (B) in the maturing rat. There are no significant differences between the E_2 values; in contrast, aromatase activity changes significantly. These enzyme assays were performed using 92 nM 1β-^{3}H-androstenedione and 80 μg microsomal protein/ml. Other studies were performed using ^{3}H-testosterone and ovarian microsomal fractions from animals between 22-25 days of age; the aromatase activity was low in these cases (data not shown).

EXPRESSION OF A HUMAN FAST SKELETAL MUSCLE MYOSIN
LIGHT CHAIN TWO GENE IN THE RAT MYOBLAST LINE, L8

C. Cisar and J. Calvo

Section of Biochemistry, Molecular and Cell
Biology, Cornell University, Ithaca, NY 14853

INTRODUCTION

Myosin is a major protein component of the
contractile apparatus of muscle cells. It is a
hexameric protein consisting of two myosin heavy
chains and four myosin light chains. There are
several types of myosin light chain. One of these,
myosin light chain two (MLC2), occurs in three
different forms: fast skeletal muscle, slow
skeletal muscle and cardiac muscle MLC2. The
expression of these genes is controlled in a
tissue- and developmental stage-specific manner.

The rat myoblast cell line, L8, has been used
as a model system for the study of control of gene
expression during myogenesis. The cells can be
induced to differentiate in vitro (1).

We have characterized a human fast skeletal
muscle MLC2 gene. Four of the seven exons have
been sequenced as well as 570 bp of upstream DNA.
The expression of the human fast skeletal muscle
MLC2 gene was analyzed after transfection into L8
cells. The human gene was regulated in a normal
manner in the rat cells.

MATERIALS AND METHODS

The DNA sequence was determined by the method
of Maxam and Gilbert (2). The cDNA primer
extension and S1 mapping techniques used were those
of de Banzie et al. (3). L8 cells were transfected
by calcium phosphate precipitation (4) and the RNA
isolated as described in Nudel et al. (5).

RESULTS

A lambda phage (1-1-1) containing the human
fast skeletal muscle MLC2 gene on a 15 kb genomic
DNA fragment was kindly provided by U. Nudel.

Exons 1 (AA 1), 2 (AA 2-29), 3 (AA 30-55) and 6
(AA 118-133) of the human fast skeletal muscle MLC2
gene have been sequenced as well as 570 bp of
upstream DNA. The intron/exon boundaries were
determined by comparison with the DNA sequence of
the rat fast skeletal muscle MLC2 gene (5) and by
location of splice recognition consensus sequences
(6). Exon 1 of the human fast skeletal muscle MLC2
gene is transcribed, but only the last three
nucleotides (ATG) are translated. The length of
exon 1 (62 bp) was determined by cDNA primer
extension. The transcription initiation site was
determined by S1 mapping using human myotube RNA
and agrees with the site determined by sequence
analysis and the cDNA primer extension experiment.

Three different DNAs were used to transfect L8
cells by calcium phosphate precipitation (Fig. 1).
(a) The lambda phage, 1-1-1, contains 15 kb of
human genomic DNA. It includes the complete human
fast skeletal muscle MLC2 gene plus approximately
5 kb of upstream and downstream DNA. (b) Plasmid
p6.5H3CAT contains a 6.5 kb Hind3 fragment from
phage 1-1-1 inserted into the vector pSVOCAT (7).
This plasmid includes the complete human fast
skeletal muscle MLC2 gene with 2.5 kb of upstream
DNA and approximately 500 bp of downstream DNA.

(c) Plasmid p3.4H3CAT contains a 3.4 kb Hind3
fragment from phage 1-1-1 inserted into the vector
pSVOCAT. This plasmid includes exon 1 of the human
fast skeletal muscle MLC2 gene with 2.5 kb of
upstream DNA and approximately 800 bp of the first
intron. The human DNA inserts were in the same
orientation as the CAT gene. RNA isolated from
pre- and post-differentiated transfected cells was
analyzed by S1 mapping using a human specific
probe. RNA from post-differentiated transfected
cells gave rise to a band of the size expected for
transcripts from the human fast skeletal muscle
MLC2 promoter. No such band appears in the lanes
containing RNA from pre-differentiated transfected
cells.

DISCUSSION

We have characterized a human fast skeletal
muscle MLC2 gene (Figure 1). A comparison with the
sequence of the rat fast skeletal muscle MLC2 gene
(5) reveals that the coding exons of the two genes
are highly homologous (>80% homology). In
addition, the positions of the intron/exon
boundaries are conserved. There is also striking
homology between the two genes in the 350 bp
immediately upstream of the initiation of
transcription sites (73% homology).

A 3.4 kb Hind3 human DNA fragment containing
exon 1 and approximately 2.5 kb of upstream DNA and
800 bp of the first intron, is sufficient for
regulated expression in the rat myoblast line, L8.
The L8 cell line can be used to study the
regulatory region/s of the human fast skeletal
muscle MLC2 gene.

Fig. 1. Human DNA used to transfect the L8 cell
line. (a) lambda phage 1-1-1 human DNA insert (b)
p6.5H3CAT human DNA insert (c) p3.4H3CAT human DNA
insert. E=EcoR1, H=Hind3; e1, e2, e3, e6 and e7
are exons 1, 2, 3, 6 and 7 of the human fast
skeletal muscle MLC2 gene.

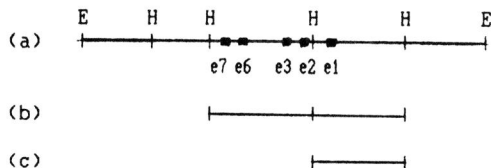

REFERENCES

(1) Yaffe, D. and Saxel, O. (1977) Differentiation
7, 159-166.
(2) Maxam, A. and Gilbert, W. (1980) in Methods in
Enzymology 65 (Grossman, L. and Moldave, K.,
eds.) pp. 499-560, Academic Press, New York.
(3) de Banzie, J. S., Sinclair, L. and Lis, J. T.
(1986) Nucleic Acids Res. 14, 3587-3601.
(4) Melloul, D., Aloni, B., Calvo, J., Yaffe, D.
and Nudel, U. (1984) EMBO J. 3, 983-990.
(5) Nudel, U., Calvo, J. M., Shani, M. and Levy,
Z. (1984) Nucleic Acids Res. 12, 7175-7186.
(6) Mount, S. M. (1982) Nucleic Acids Res. 10,
459-472.
(7) Gorman, C. M., Moffat, L. F. and Howard, B. H.
(1982) Mol. Cell. Biol. 2, 1044-1051.

THE TRANSFORMING POTENTIAL OF DIFFERENT INTRACELLULAR LEVELS OF ONCOGENIC p21

Sally J. Compere, Whitehead Institute, 9 Cambridge Center, Cambridge, MA 02142

INTRODUCTION

The c-Ha-ras-1 oncogene encodes a 21kd protein (p21)containing a single amino acid alteration from the normal cellular p21(1). In addition to an alteration in the structural gene, many human tumors express high levels of p21(2), suggesting a growth advantage for those cancer cell containing increased levels of the mutant protein. Furthermore,very high levels of ras potentiate both transformation and establishment of primary rat embryo fibroblasts in tissue culture (establishment is not observed with lower levels of the altered ras proteins)(3). These results suggest that careful quantitation of an oncogenic protein must accompany a definition of its properties.In this report the effect of various levels of the transforming oncogene from the EJ bladder cell line on mouse fibroblasts is analyzed.

RESULTS

To determine which cellular phenotypes are affected as a primary response to p21 production, cell lines were established in which the human transforming gene encoding p21 is under the control of the metallothionein-1 (MT-1) promoter(150 bp) (Fig.1a). The transcription of the MT-I promoter is enhanced by the presence of heavy metals (4). NIH3T3 cells were cotransfected with the MT-1/ras fusion gene and pSV2neo. Colonies of cells (labeled SBEJ #) resistant to G418 were clonally isolated and analyzed in the presence and absence of zinc for the expression of ras RNA (Fig.1b) and protein (Fig.1c).Transformed cells were scored for anchorage-independence(Fig.2) and morphology, tumor cells being refractile and not limited to monolayer growth. RNA analysis demonstrated increases in MT-1/ras RNA in response to incremental increases in zinc concentration (Fig.1b). Analysis of p21 proteins in many of these isolates (Fig.1c) demonstrates that

Fig. 1. Construction and Characterization of An Inducible Ras Oncogene.
a) Schematic of the MT-I promoter fused at an endogenous BglII site to a site just 5' of the AUG of p21.
b) RNA was measured according to Ref.5.
c) Radiolabeled· p21 proteins were immunoprecipitated according to ref.1.

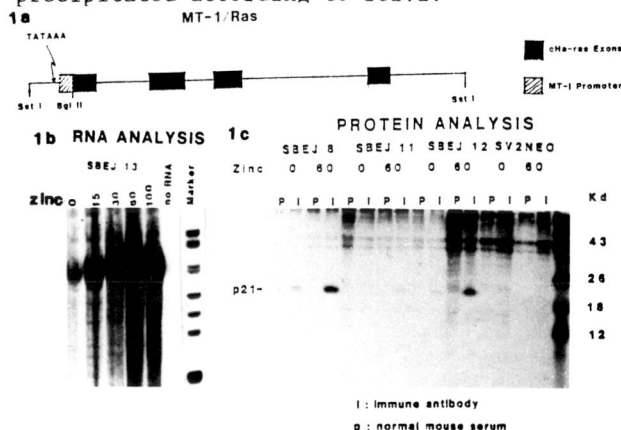

I : immune antibody
p : normal mouse serum

zinc induces an increase in the intracellular concentration of p21. A correlation can be inferred between the amount of p21 present and the ability of the cell to grow in soft agar (Fig.2). Cells containing the MT-1/ras gene exemplify three phenotypes: 1) SBEJ 11 cells exhibit morphological changes such as multilayer growth upon the addition of zinc,but are never highly refractile in appearance nor capable of anchorage independent growth even though detectable amounts of p21 are induced (Fig.1c), 2) many cells (e.g. SBEJ 13,8) are transformed in the absence of zinc and become more transformed in the presence of zinc and 3) rarely cells are obtained (e.g. SBEJ 14) which are highly transformed in the absence of zinc, with the addition of zinc appearing toxic.

DISCUSSION

This report demonstrates that a threshold level of expression of the aberrant protein is required for transformation as very low levels of p21 do not confer anchorage independent growth or a greatly transformed morphology. Furthermore,increasing the intracellular level of ras leads to an increasingly transformed phenotype. The prevalence of high concentration and mutation of ras in a diversity of human tumors suggests that both increased expression and mutation are required for potent transforming capabilities. In many systems such as mouse skin, experimental induction of a tumor requires multiple stimuli, such as initiators and promoters.As tumors progressively acquire more malignant phenotypes, they may reflect a progressive mutation and amplification of the oncogene .

This assay system has tested the quantity of p21 necessary for transformation. Further experiments will determine the _in vivo_ requirements of cells for tumor formation. Retroviral vectors carrying ras are being introduced into early mouse embryos to directly test when expression of the oncogene contributes to the genesis of cancer.

Fig. 2. Anchorage Independent Growth
Cells were seeded into agar +/-zinc.

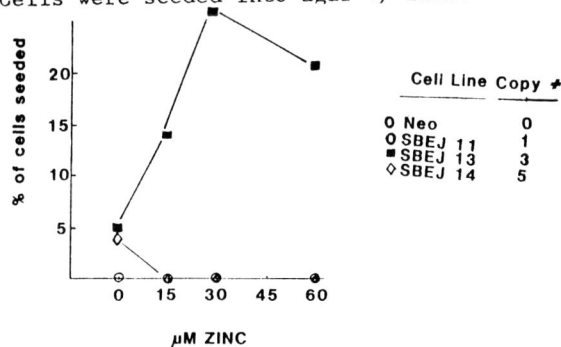

μM ZINC

REFERENCES

1) Tabin et al., (1982) Nature 300, 143-149.
2) Tanaka et al., (1986) Can. Res.46,1465-70.
3) Spandidos, D.A. and Wilkie, N.M. (1984) Nature 310, 469-475.
4) Brinster et al. (1982) Nature 296, 39-42.
5) Melton et al.,(1984) NAR 12,7035-7056.

ACKNOWLEDGEMENTS ·

SJC was a fellow of the Jane Coffin Childs Fund for Biomedical Research. I would like to thank R. Weinberg for his support and advice.

EXPRESSION OF THE ALPHA-FETOPROTEIN GENE IN TRANSGENIC MICE

Reid S. Compton, Sally A. Camper, Robb Krumlauf, Robert E. Hammer, Ralph Brinster and Shirley M. Tilghman

Department of Molecular Biology, Princeton University, Princeton, N.J. 08544

INTRODUCTION

Alpha-fetoprotein (AFP) is the major serum protein in the developing mammalian fetus. It is produced at high levels in the visceral endoderm of the yolk sac and in the fetal liver, and to a lesser extent in the gastrointestinal tract (1). Its levels are reduced several orders of magnitude in the neonatal liver as the result of a decrease in its transcription rate (2). The AFP gene in the mouse is under the control of at least two unlinked, trans-acting loci, termed raf and Rif (3). Raf is a cell autonomous function (4) which acts in liver to regulate the adult basal level of AFP mRNA. The Rif gene product determines the degree of inducibility of AFP mRNA during liver regeneration in response to injury.

To begin to map the regulatory domains in the AFP gene which are responsible for its tissue-specific pattern of expression, its post-natal repression and its reinduction during liver regeneration, we have generated a series of transgenic mice carrying varying amounts of the AFP gene 5' flanking region linked to an internally deleted AFP structural gene (5). These initial studies demonstrated that tissue specificity of the exogenous gene was achieved with 7 kb of 5' flanking DNA. We have extended these studies to include an analysis of the ability of the endogenous gene to respond to both raf and Rif.

MULTIPLE ELEMENTS DIRECT TISSUE SPECIFICITY IN TRANSGENIC MICE

Transient expression studies had indicated that there were at least 4 independent regulatory elements within the proximal 7 kb of the 5' flanking domain of the AFP gene; three enhancers located at approximately −7, −5 and −2.5 kb and a tissue-specific promoter at −0.1 kb (6). When constructs carrying these elements, either separately or together in various combinations, were introduced along with the AFP minigene into fertilized mouse eggs, we found that all were expressed only in the appropriate tissues, the visceral endoderm, liver and gut. The levels of expression, however, depended upon the number of enhancer elements present. That is, with all three elements, expression was observed at high levels in all three tissues, including the gut where the level of the endogenous mRNA is very low. When the enhancers were separately used in conjunction with the minigene, different levels were observed in the three tissues, suggesting that the enhancers have different biological activities in the three tissues.

THE AFP MINIGENE IS DEVELOPMENTALLY REGULATED IN TRANSGENIC MICE

Several transgenic mouse lines were generated from founder animals, and the activities of the introduced genes were examined in liver and gut

during development. All constructs, independent of the number of enhancers present, exhibited a decrease in transcriptional activity after birth, arguing that the regulatory signals for the decline were present in each enhancer, or alternatively in the common promoter region. Likewise, all constructs tested were able to reinitiate transcription during liver regeneration.

THE AFP MINIGENE RESPONDS TO RAF IN TRANSGENIC MICE

To identify the region of the AFP regulatory domain which responds to the two trans-acting loci, raf and Rif, we performed backcrosses between transgenic founder animals, which were generated from (C57BL/6 x SJL)F1 x (C57BL/6 x SJL)F1 fertilized eggs, to BALB/cJ mice. The latter carry the rare, recessive raf^b allele which results in hereditary persistence of AFP mRNA (7). By examining transgenic progeny we determined that the minigene carrying 7 kb of 5' flanking DNA was under the influence of the raf gene. Similar progeny as well as transgenic mice carrying a variety of deletions in the 5' flanking region are being tested for their ability to respond to both the raf and Rif genes during liver regeneration.

In the course of these studies, we examined the decrease in the levels of AFP and minigene mRNAs in the gut after birth. We observed that unlike the liver, the very low levels of AFP mRNA in adult gut are not under the influence of the raf gene. This observation, coupled with the demonstration that the raf gene acts in liver in chimeric mice (4), argues that the raf gene product is liver-specific in its action. This is supported by our observation that the raf gene regulates a second independent locus, termed H19, in the liver but not in muscle, where H19 is also produced (8).

REFERENCES

(1) Tilghman, S.M. (1985) in Oxford Surveys on Eucaryotic Genes, (N. MacLean, Ed.) pp. 160-206, Oxford Univ. Press, Oxford
(2) Tilghman, S.M. and Belayew, A. (1982) Proc. Nat. Acad. Sci. USA 79, 5254-5257
(3) Belayew, A. and Tilghman, S.M. (1982) Mol. Cell. Biol. 2, 1427-1435
(4) Vogt, T.F., Solter, D. and Tilghman S.M. (1986) submitted for publication
(5) Krumlauf, R. Hammer, R.E., Tilghman, S.M. and Brinster, R.L. (1985) Mol. Cell. Biol. 5, 1639-1648
(6) Godbout, R., Ingram, R.S. and Tilghman, S.M. (1986) Mol. Cell. Biol. 6, 477-487
(7) Olsson, M., Lindahl, G. and Ruoslahti, E. (1977) J. Exp. Med. 145, 819-827
(8) Pachnis, V., Belayew, A. and Tilghman, S.M. (1984) Proc. Nat. Acad. Sci. USA 81, 5523-5527

REGULATION OF ALPHA-FETOPROTEIN GENE EXPRESSION DURING EMBRYONAL CARCINOMA CELL DIFFERENTIATION

Patrick Coughlin and Joel Schindler

Department of Anatomy and Cell Biology, University of Cincinnati College of Medicine, 231 Bethesda Avenue, Cincinnati, Ohio 45267-0521

INTRODUCTION

Exposure of Nulli-SCC1 murine embryonal carcinoma (EC) cells grown in monolayer culture to hexamethylene-bisacetamide (HMBA) induces differentiation into a visceral endoderm-like phenotype with concurrent expression of alpha-fetoprotein (AFP). These HMBA treated EC cells demonstrate two unique characteristics; 1) AFP specific mRNA exhibits a bimodal temporal response to HMBA, initiated by an extremely rapid burst of AFP mRNA synthesis and 2) an extremely long lag period exists between the appearance of intracellular AFP and the secretion of AFP into the medium.

RESULTS

1) AFP mRNA Kinetics

Total poly A^+ RNA (1,2) from untreated Nulli-SCC1 EC cells and both retinoic acid (RA) and HMBA induced cultures was examined for AFP transcripts by Northern blot analysis. A 900bp mouse AFP cDNA representing the 3' end of the mRNA transcript (pBR322-AFP2) was used as a molecular probe (3). AFP specific mRNA was not detected in either uninduced EC cells or in RA induced cells. In contrast, poly A^+ RNA isolated from HMBA treated cells at various times after induction displayed interesting results. As early as one hour post induction, an AFP signal was detected. At three hours, this signal decreased in intensity, but was still present. Between three and 48 hours of HMBA exposure, AFP cDNA failed to hybridize with an appropriately sized 2.2 kb mRNA species. However, after six hours of exposure to HMBA, isolated mRNA contained five new species which hybridize with the cDNA, all of which are significantly larger than the 2.2 kb mRNA ordinarily seen. At 12 hours, three faint bands were present. At 24 hours, the same banding pattern as the six hour sample was seen, except that each band had an increased intensity. By 48 hours, the typical band at 2.2 kb was reestablished. To confirm that equal amounts of RNA were loaded onto each lane of the gels, filters were stripped and reprobed with a constitutively expressed murine APRT cDNA. The resulting band of approximately 1Kb was seen at equal intensity in all the samples tested.

To establish that the mRNA which hybridized to our AFP cDNA did code for AFP, we performed a series of in vitro translation experiments. Poly A^+ RNA samples from time points noted were used as template for the synthesis of AFP. Synthesized products were immunoprecipitated with anti-mouse AFP antiserum, and separated by SDS/PAGE. Only poly A^+ RNA from HMBA treated samples that had detectable levels of the appropriate 2.2Kb mRNA could synthesize material that was precipitable with anti-AFP antiserum.

2) Phenotype Specific AFP Protein Synthesis

Total cellular protein labeled with ^{35}S-methionine was immunoprecipitated with an anti-mouse AFP antiserum. After 48 hours of treatment, only HMBA induced cells synthesized an immunoreactive protein which comigrated with purified AFP on polyacrylamide gels. Unlabeled AFP competed with this labeled protein for antibody binding sites, suggesting that the labeled protein was, in fact, AFP. No such protein was identified in either uninduced cells or cells induced with RA. The use of a polyclonal antiserum resulted in extraneous bands on autoradiographs despite extensive washing of the precipitates. These extraneous bands were not competed with cold AFP. AFP was seen in cell extracts 3 hours after induction and remained present in the cells until the termination of the experiment at 48 hrs. In culture media, no AFP was observed until 48 hours of treatment. As in the previous experiments, neither uninduced cells nor cells induced with RA were positive for AFP.

DISCUSSION

It is of particular interest that AFP specific mRNA can be detected in induced EC cells as early as one hours after exposure to HMBA. To our knowledge, this is the most rapid induction of a phenotype-specific gene during EC cell differentiation yet reported. The bimodal nature of this gene expression is of particular interest. By both in vitro translation procedures and continuous labeling experiments, it seems the AFP specific mRNA detectable early in differentiation does code for a protein that is immunoprecipitable with anti-AFP antibody.

The appearance, from 6 to 24 hours of induction, of high molecular weight mRNA that hybridize with our AFP cDNA was unexpected. These large transcripts could be the consequence of several different events. It is possible that improper transcription initiation occurs at a cap site upstream from the authentic AFP cap site. It is equally possible that these large mRNA transcripts are the result of improper termination of transcription or inaccurate splicing of AFP primary transcripts. Interestingly, the high molecular weight transcripts appear at times when no authentic 2.2Kb AFP mRNA is detectable. It is unclear why such transcripts consistently appear with such temporal regularity, but in vitro translation results indicate that the high molecular weight mRNA can not serve as template for the synthesis of AFP.

The considerable lag between the appearance of AFP in the cells and its detection in media poses an interesting cell biological question. Within the limits of our detection, it seems AFP remains cell associated from 3 hours until its detection in the medium by 48 hrs. Interestingly, these times correlate with our observed detection of authentic AFP mRNA. It is possible that the intracellular accumulation of AFP is a signal for the termination of specific AFP transcription. As the cell begins to secrete AFP into the medium, presumably influencing intracellular levels of the protein, accurate transcription of the AFP gene is reinitiated.

We have previously shown that Nulli-SCC1 EC cells, when treated in monolayer cultures with HMBA, can differentiate into a visceral endoderm-like phenotype, with concurrent expression of AFP (4). Previous studies have shown that AFP expression is regulated at the transcriptional level (5). In this communication, we report some interesting observations about the temporal expression of the AFP gene in this model experimental system.

REFERENCES

1. Cathala, G., J-F. Savouret, B. Mendez, B.L. West, M. Karin, J. Martial, and J. Baxter. 1983. DNA 2(4):329-335.
2. Aviv, H., and P. Leder. 1972. Proc. Natl. Acad. Sci. U.S.A. 69:1408-1412.
3. Tilghman, S.M., D. Kioussis, M.B. Gorin, P.G. Ruiz, and R.S. Ingram. 1979. J. Biol. Chem. 254:7393-7399.
4. Schindler, J., R. Hollingsworth, and P. Coughlin. 1984. Differentiation 27:236-242.
5. Tilghman, S.M., and A. Belayew. 1982. Proc. Natl. Acad. Sci. U.S.A. 79:5254-5257.

76

EXPRESSION OF THE MULTIDRUG-RESISTANCE-ASSOCIATED P-GLYCOPROTEIN MULTIGENE FAMILY

Kathryn L. Deuchars, Farida Sarangi, Jane A. Endicott, Peter F. Juranka and Victor Ling

The Ontario Cancer Institute and the Department of Medical Biophysics, University of Toronto, Toronto, Ontario, Canada, M4X-1K9.

The overexpression of a 170 kDa plasma membrane glycoprotein, P-glycoprotein, is thought to be responsible for mediating multidrug resistance[1]. This phenotype is characterized by a pleiotropic resistance to a wide spectrum of structurally and functionally unrelated drugs[1]. Drug uptake studies have shown that multidrug resistant cells have a reduced net intracellular accumulation of the drugs involved[2], and we have recently proposed that this may be the result of P-glycoprotein acting in the plasma membrane as an energy-dependent export pump[3].

Southern blot analysis with a 630 bp cDNA fragment of P-glycoprotein, pCHP1, detected multiple bands amplified in multidrug resistant mutants[4]. This indicates that the overexpression of P-glycoprotein in multidrug-resistant cells is a result of gene amplification. The multiplicity of bands further suggests that P-glycoprotein is encoded by a multigene family.

P-glycoprotein cDNA clones were isolated from an Okayama and Berg pcD cDNA library constructed from a wild-type Chinese hamster ovary cell line. Four classes of transcripts have been identified by DNA sequence analysis. They appear to be transcribed from two different P-glycoprotein genes. Both genes produce two different transcripts by using alternative polyadenylation sites. Restriction maps of overlapping clones that represent 2.8 kb of Gene I and 2.4 kb of Gene II are shown in Figure 1. A full-length clone is expected to be 4.7 kb. The coding regions have 90.6% homology, indicating that they are related isoproteins. The non-coding regions, however, show only 37.4% homology, indicating that they are distinct family members. This supports the multigene family hypothesis.

Figure 1. Restriction enzyme map of the two P-glycoprotein genes. Letters represent restriction enzymes as follows: A, Aha3; B, BanI; D, DdeI; H, HinfI; Ha, HaeI; N, NcoI; R, RsaI; S, Sau3A; St, StuI; T, TaqI. The arrows mark the four alternate polyadenylation sites, A_n represents the poly-A tail and the hatch marks indicate the 3'-untranslated regions.

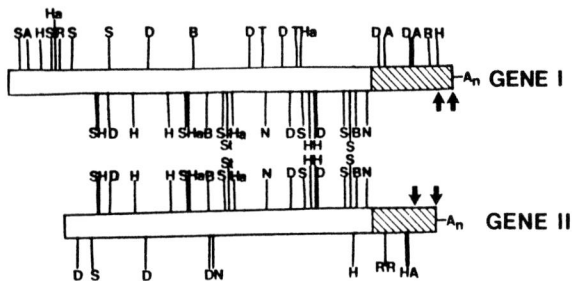

A total of 24 clones have been isolated. Table 1 shows the distribution of these clones between the four classes of transcripts. The short form of gene II appears to predominate (14 of 24).

Table 1. Distribution of alternately polyadenylated transcripts between two gene family members.

Gene Family	Number of Clones Isolated	Polyadenylation Site Used	
		First	Second
I	8	3	5
II	16	14	2

We can speculate on the role of multiple P-glycoprotein genes in the multidrug resistant phenotype. The complexity of this phenotype may reflect a synergism of the individual functions of each member. Differential expression of the members may result in the variability in cross-resistance profiles between multidrug-resistant cell lines. Gene-specific probes have been constructed from the 3'-untranslated regions that can be used to address the above questions, and to analyze the tissue-specific and temporal expression of the P-glycoprotein multigene family in the developing and adult organism.

The P-glycoprotein DNA and protein sequences are strongly conserved between species. The hamster pCHP1 probe hybridizes at high stringency to both mouse and human P-glycoprotein sequences, and monoclonal antibodies specific for P-glycoprotein recognize common epitopes shared between hamster, mouse and human P-glycoprotein[4,5]. Furthermore, functional conservation is indicated by the fact that, in transformant studies, hamster P-glycoprotein can be expressed and can function in the mouse cell[6]. This suggests an important role for P-glycoprotein in the normal cell. As in the multidrug-resistant cell, it may act as a pump, for some as yet unidentified compound(s).

References
1. Gerlach, J.H., Kartner, N., Bell, D.R. and Ling, V. (1986) Cancer Surveys 5: 25-46.
2. Riordan, J.R. and Ling, V. (1985) Pharmac. Ther. 28: 51-75.
3. Gerlach, J.H., Endicott, J.A., Juranka, P.F., Henderson, G., Sarangi, F., Deuchars, K.L. and Ling, V. (1986), submitted.
4. Riordan, J.R., Deuchars, K., Kartner, N., Alon, N., Trent, J. and Ling, V. (1985) Nature 316: 817-819.
5. Kartner, N., Evernden-Porelle, D., Bradley, G. and Ling, V. (1985) Nature 316: 820-823.
6. Deuchars, K.L., Du, R.-P., Naik, M., Evernden-Porelle, D., Kartner, N., Van der Bliek, A.M. and Ling, V. (1986), submitted.

This work was supported by the National Cancer Institute of Canada (NCI) and the U.S.A. National Institutes of Health Grant CA37130. K.L.D. is the recipient of an NCI Studentship, J.A.E. is the recipient of a Medical Research Council (MRC) of Canada Studentship, and P.F.J. is an MRC Fellow.

HUMAN ets-1 AND ets-2 PROTEINS: IDENTIFICATION AND INTRACELLULAR LOCALIZATION

S. Fujiwara, R. J. Fisher, A. Seth*, N. K. Bhat, T. S. Papas

Laboratory of Molecular Oncology, *Litton Bionetics Basic Research Program, NCI-Frederick Cancer Research Facility, Frederick, MD21701-1013

Avian erythroblastosis virus E26 has 2 distinct transforming genes, v-myb^E and v-ets (1). The human homologue of the v-ets is composed of 2 separate genes, ets-1 and ets-2 (2). Antibodies specific to the ets-1 and the ets-2 protein domains were prepared by using synthetic oligo-peptides and bacterially expressed proteins as antigens. The ets-1 protein was identified as a 51Kda cytoplasmic protein and the ets-2 protein as a 56Kda nuclear protein. The ets-1 protein is found primarily in lymphocytic lineage and the ets-2 protein is found in a wide variety of cell types.

Ets-1 protein (Figure 1)

An ets-1 specific peptide (^{229}TESYQTLHPISSEEL^{243}) was synthesized with the amino acid sequence of a region in the viral ets-1 domain (1). Antibodies to the peptide recognized the E26 p135$^{gag-myb-ets}$ and also the ets-1 domain of the p135 protein expressed in E. coli. This antibody immunoprecipitated a 51Kda protein from cell lines derived from the lymphocyte lineage. Subcellular fractionation showed the cytoplasmic localization of the protein.

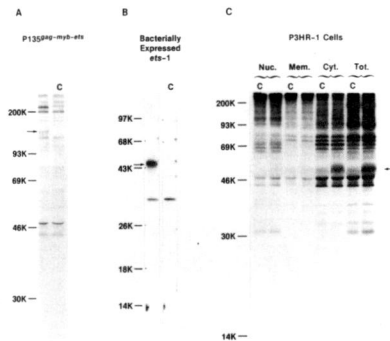

Figure 1 Identification and subcellular locali-zation of the human ets-1 protein by anti-peptide antibody. A: Immunoprecipitation of the p135 (C: competition by the peptide). B: Immunoblot detection of the bacterially expressed viral ets-1 protein domain. C: Immunoprecipitation of subcellular fractions of the human Burkitt lymphoma derived cells P3HR-1.

Ets-2 protein (Figure 2)

An ets-2 specific peptide (^{410}FKLSDPDEVARRW^{422}) was synthesized with the amino acid sequence of a region in the viral ets-2 domain (1). In addition, a Pst-Pst fragment of the viral ets-2 domain (nucleotide No. 1505-1923) (1) and a cDNA clone of the human ets-2 gene (cDNA-14) (2) were expressed in E. coli and antisera were prepared against them. Antibodies to the ets-2 peptide recognized the p135 and also the human ets-2 protein expressed in E. coli. This antibody identified a 56Kda protein from the Colo 320 DM and various other human cell lines. Two other antisera directed against the bacterially expressed proteins also recognized a similar 56Kda protein. Peptide mapping using the Staphylococcal V8 protease proved that the 56Kda proteins immunoprecipitated by the 3 different antibodies were identical.

Subcellular fractionation demonstrated the 56Kda protein in the cell nucleus.

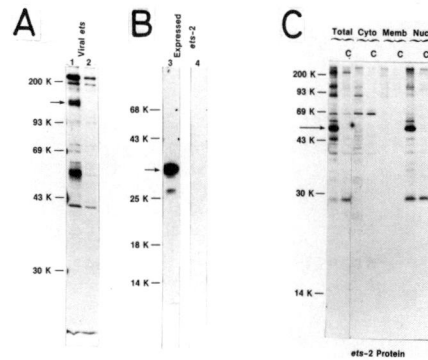

Figure 2 Identification and subcellular locali-zation of the human ets-2 protein by anti-peptide antibody. A: Immunoprecipitation of the p135 (lane 1) (lane 2: competition by the peptide). B: Immunoblot detection of the bacterially expressed human ets-2 protein (lane 3) (lane 4: competition). C: Immunoprecipitation of subcel-lular fractions of the human Colo 320 DM cells.

Chicken ets protein

By contrast to the mammalian ets genes, the chicken has a single ets gene where the ets-1 and the ets-2 domains are contiguous. Immuno-precipitation of subcellular fractions from the chicken lymphnode cells with the antiserum against the bacterially expressed human ets-2 protein identified a 55Kda protein predominantly in the cytoplasm.

Discussion

Antibodies prepared against the ets-1 and the ets-2 oligopeptides detected the E26 viral product, p135. Furthermore, these 2 antibodies recognized the ets-1 and ets-2 specific protein domains expressed in E. coli. The expression of the 51Kda protein reactive to the ets-1 specific antibody showed a good correlation with the amounts of the ets-1 mRNA in various human cell lines. Similarly, the 56Kda protein iden-tified by the ets-2 specific antibody was shown to be highly expressed in the Colo 320 DM cells and other human cell lines which express high levels of the ets-2 mRNA's. Additionally, 2 other antisera directed against bacterially expressed ets-2 domains also recognized the same protein. Thus, we identify the 51Kda and the 56Kda proteins as the products of the human ets-1 and ets-2 genes, respectively. Interest-ingly these 2 proteins showed distinct intra-cellular localizations; the ets-1 product was found in the cytoplasm while the ets-2 protein was detected in the nucleus. The chicken c-ets protein was found in the cytoplasm, as previously reported (3, 4).

References

(1) M. F. Nunn, P. H. Seeburg, C. Moscovici, P. H. Duesberg, Nature (London) 306, 391 (1983).
(2) D. K. Watson, M. J. McWilliams-Smith, M. F. Nunn, P. H. Duesberg, S. J. O'Brien, T. S. Papas, Proc. Natl. Acad. Sci. U.S.A. 82, 7294 (1985).
(3) J. Ghysdael, A. Gegonne, P. Pognonec, D. Dernis, D. Leprince, D. Stehelin, Proc. Natl. Acad. Sci. U.S.A. 83, 1714 (1986).
(4) J. H. Chen, Mol. Cell. Biol. 5, 2993 (1985).

STRUCTURE OF TWO DEVELOPMENTALLY REGULATED *DICTYO-STELIUM DISCOIDEUM* UBIQUITIN GENES

R. Giorda and H.L. Ennis

Roche Institute of Molecular Biology, Department of Biochemistry, Roche Research Center, Nutley, NJ 07110 USA

During synchronous germination of *Dictyostelium discoideum* spores, the differential expression of certain genes and their products results in the accumulation of proteins which may be involved in this developmental program. In previously reported work we have identified proteins that are developmentally regulated and specific for spore germination, and have cloned a number of mRNAs which are specifically expressed during germination (1,2). In the present study we report on the sequences of two genomic clones homologous to one of these cDNA clones named pLK229. The sequences of the deduced proteins are almost identical to human ubiquitin, a 76 amino acid protein whose sequence is conserved in many organisms (3).

DEVELOPMENTAL REGULATION OF pLK229 mRNA AC-CUMULATION. RNA blot analysis of RNA from vegetative cells and germinating spores indicates that pLK229-specific mRNA is developmentally regulated. Poly(A)$^+$ RNA was isolated from vegetatively growing cells, dormant spores and 1.5 h and 3 h germinating spores and 1 μg from each sample was size fractionated on an agarose-formaldehyde gel. The RNAs were transferred to nitrocellulose sheets and hybridized to nick-translated pLK229 plasmid DNA. The results show that spore RNA contains 6 species of mRNA which hybridize to pLK229 plasmid DNA: 1900, 1400, 1100, 840, 580 and 500 nt. The 1900 to 840 nt pLK229-specific mRNA are at highest concentration in spores and their levels decrease during germination. None of the 1900 to 840 nt mRNA is present in growing cells. The 580 and 500 nt species are present in highest amount in vegetative cells.

SEQUENCE OF GENOMIC AND cDNA CLONES. Southern blots indicate that the pLK229-specific genes constitute a multigene family comprising at least 7 genes. Two genomic sequences homologous to pLK229 were isolated by screening a plasmid and a λgt 10 phage library. Clone p229, isolated from the plasmid library, contains an insert of about 5.2 kb whereas clone λ229, isolated from the λgt 10 library, contains 2.8 kb of *Dictyostelium* DNA. The sequence of p229 is identical to that of the cDNA. The coding region of p229/pLK229 is 1143 nt long and codes for a polypeptide of 381 amino acids. This polypeptide is composed of five tandem repeats of the same 76 amino acid sequence (228 nt). In addition to an open reading frame of 687 nt (3 repeats of the 228 nt unit), clone λ229 contains 1741 nt of 5' and 444 nt of 3' non-coding sequences. It has a typical *Dictyostelium* "TATA" box (TATAAATA), the usual T-stretch, and CAAA and CATT boxes in the putative 5'-untranslated region. A curious feature is a pair of perfect ten nucleotide long repeats (TAATTGTATA) also in the 5'-non-coding region. There is no significant open reading frame 5' to the first coding ATG. Although the repeats themselves differ in up to 34 out of the 228 nucleotides, it is noteworthy that the amino acid sequence of each repeat is identical except for a substitution of a threonine for alanine in position 28 of the second, third, and fifth repeat of p229.

A computer search for homology to known proteins revealed that each 76 amino acid repeat is identical to human and bovine ubiquitin, except for a substitution of glycine for proline at position 19 and asparagine for threonine at position 22 (3).

The last repeat unit in each gene contains one additional amino acid at the carboxyl terminus, asparagine for p229 and leucine in the case of λ229. This terminal amino acid seems to vary among the ubiquitins found in other organisms and is apparently cleaved off as it is not found in the isolated protein (3).

DISCUSSION

As is indicated by its name, ubiquitin, a 76 amino acid protein, is found in all eukaryotic cells, either free or covalently bound to cellular proteins (3). The protein is required for ATP-dependent intracellular protein degradation and apparently targets protein degradation by covalent linkage (4). Ubiquitin also binds to histone H2A (5). The fact that ubiquitin mRNA is developmentally regulated (and by inference the synthesis of the protein) in *Dictyostelium* development argues for a function for ubiquitin in this process. Since ubiquitin is known to be necessary for the decay of a large number of proteins (4), perhaps this indicates that more protein degradation occurs during spore germination than during other developmental stages; or that ubiquitin is necessary for the degradation of specific proteins that are unique to spore germination. Ubiquitin is known to form a covalent bond with histone H2A, modifying chromosomal structure by inducing a disorganization of nucleosomes (5). Since actively transcribed genes have been correlated with loss in nucleosomal organization, ubiquitin might, in this way, be important in the regulation of gene expression in *Dictyostelium* development.

REFERENCES

(1) Dowbenko, D.J. and Ennis, H.L. (1980) *Proc. Natl. Acad. Sci. USA* 77, 1791-1795.
(2) Kelly, L.J., Kelly, R. and Ennis, H.L. (1983) *Mol. Cell. Biol.* 3, 1943-1948.
(3) Schlesinger, D.H., Goldstein, G. and Niall, H.D. (1975) *Biochemistry* 14, 2214-2218.
(4) Ciechanover, A., Finley, D. and Varshavsky, A. (1984) *J. Cell. Biochem.* 24, 27-53.
(5) Levinger, L. and Varshavsky, A. (1982) *Cell* 28, 375-385.

TRANSCRIPTIONAL REGULATION OF RIBOSOMAL PROTEIN GENES DURING MOUSE MYOBLAST DIFFERENTIATION

S.A. Harris, M.G. Agrawal, and L.H. Bowman

Department of Biology, University of South Carolina, Columbia, SC 29208

The terminal differentiation of myoblasts into muscle fibers is associated with a dramatic decrease in the rate of cell division and provides an excellent system for studying the developmental regulation of ribosome formation. We have previously shown that the rate of mature rRNA formation is reduced 2-3 fold following the differentiation of mouse MMDZ14 myoblasts into fibers. This reduction is regulated solely by a decrease in the rate of rDNA transcription in fibers (1). Analysis of the metabolism of ribosomal proteins (r-proteins) and r-protein mRNAs shows that a major mechanism regulating the formation of r-proteins following myoblast differentiation is the coordinate decrease in the rates of transcription of r-protein genes.

r-PROTEIN SYNTHESIS. The synthesis and stability of 16 large subunit and 11 small subunit r-proteins were measured in mouse MMDZ14 myoblasts and fibers to determine the level at which their accumulation is regulated. Pulse labelling experiments indicate that the rates of synthesis of these r-proteins are coordinately reduced 2.0 fold in fibers. Furthermore, no detectable turnover of r-proteins was observed during a 4 hour chase in either myoblasts or fibers. Therefore, r-protein accumulation is solely regulated by coordinate changes in their rates of synthesis during mouse myoblast differentiation.

STEADY STATE r-PROTEIN mRNA LEVELS. To determine if the reduced rate of r-protein synthesis is regulated by a decrease in r-protein mRNA concentrations, the steady state levels of three r-proteins mRNAs (S16, L18, and L32) were measured. Northern blots containing RNA from myoblasts and fibers were hybridized to cDNA plasmids for these r-proteins (2). These experiments demonstrate that myoblasts contain 1.5-2.0 fold more of each r-protein mRNA on a per nucleus basis.

TRANSCRIPTIONAL REGULATION. Nuclear run-on transcription experiments were performed to determine if the decreased concentrations of r-protein mRNAs in fibers are regulated at the level of transcription. The radioactivity incorporated into r-protein transcripts in the nuclear transcription reactions was measured by hybridization to filters containing S16, L18, and L32 cDNA plasmids as well as S16 and L32 intron sequences. These experiments indicate that the transcription rates of these r-protein genes decrease 3-5 fold following myoblast differentiation. Thus, the reduced rate of ribosome formation in fibers is due, in part, to parallel changes in the rate of transcription of both rRNA and r-protein genes.

TRANSFECTION ANALYSIS. In order to identify the sequences necessary for the transcriptional regulation of the L32 gene, we have subcloned two different fragments of the 5' region of L32 gene in front of the galK protein coding sequences in a eukaryotic expression vector (DSP). One construct contains 1.1 kb of the 5' nontranscribed region plus 11 nucleotides of transcribed sequences. The other construct contains only 0.45 kb of the 5' nontranscribed region and the same 11 nucleotides of transcribed sequences. These constructs were transfected into mouse myoblasts, and stable transformants were selected on the basis of the expression of the XGPRT gene which is also present in the DSP vector. Colonies were pooled from 2 independent transfections for each construct. The structure and steady state levels of the L32-galK transcripts from each transformant were analyzed in myoblasts and fibers by northern blot and S1 nuclease mapping techniques. The 5' end of the L32 galK transcript maps to the previously identified L32 transcription start site in each transformant (3). Furthermore, the size of the L32-galK mRNA expressed in the transformants suggests that these transcripts initiate at the L32 transcription start site and end at the polyadenylation site past the galK coding sequence. However, in contrast to the endogenous L32 mRNA whose concentration decreases 1.5-2.0 fold following myoblast differentiation, the steady state levels of the L32-galK mRNA increase 2-4 fold in fibers in all transformants from both constructs. We are currently measuring the transcription of the L32-galK gene in nuclear run-on experiments. Analysis of these experiments will determine whether the difference in accumulation of endogenous L32 mRNA and L32-galK mRNA during myoblast differentiation is due to differences in the transcription of these genes or in the processing or stability of their transcripts.

REFERENCES
(1) Bowman, L.H. (1986) Devel. Biol. (in press).
(2) Meyuhas, O. and Perry, R.P. (1980) Gene 20, 113-129.
(3) Dudov, K.P. and Perry, R.P. (1984) Cell 37, 457-469.

ACKNOWLEDGEMENTS
We thank R.P. Perry, K.P. Dudov and O. Meyuhas for the r-protein clones, and Mitch Reff and John Trill at Smith Kline and French Laboratories for the galK eukaryotic expression vector, DSP1BZDGH.

DEVELOPMENTAL PATTERNS OF GENE EXPRESSION IN RODENT BRAIN DETECTED BY IN SITU HYBRIDIZATION

G.A. Higgins, F.E. Bloom, R.J. Milner and M.C. Wilson

Research Institute of Scripps Clinic, La Jolla, CA 92037.

We are interested in identifying novel gene products which can be used as specific markers for CNS cell populations during development. Previously, we have isolated and begun to characterize individual cDNA clones from whole mouse and rat brain poly A+ mRNA, and from hippocampal and amygdala specific cDNA populations generated by subtractive hybridization (1,2,5). In the present study, we have used cDNA probes for in situ hybridization to examine the cellular sites of expression of these "brain-specific" mRNAs during the development of the rodent CNS.

RNA BLOTTING

The developmental time courses of expression of the mRNAs examined in this study is summarized in TABLE 1. Variation in the developmental timing of these mRNAs is apparent: rapid prenatal onset (Hcb D4), late prenatal onset with gradual postnatal expression (MuBr 8), postnatal onset (Acb F6) and postnatal onset with highest abundance at P20 (1B236; see (3)).

TABLE 1: Developmental Expression of Novel Brain Transcripts Determined by RNA Blotting

mRNA	E17	P2	P10	P20	Adult
Hcb D4	**	***	***	***	***
MuBr 8	-/*	*	**	***	***
Acb F6	-	-	*	***	***
1B236	-	-	**	***	*

E:Embryonic; P:Postnatal

IN SITU HYBRIDIZATION

In situ hybridization has been used to examine the cellular patterns of gene expression during development for these novel mRNAs.

(1) MuBr 8 mRNA: Cell-type Specific Marker for Neuronal Development

This novel 2.2 kb mRNA, originally generated from a whole mouse brain cDNA library (1), is expressed in many different neuronal subpopulations in the adult CNS. In the hippocampal formation, this mRNA is unique in that it is expressed at high abundance specifically by CA3 pyramidal neurons, and provides a marker for this subpopulation of neurons throughout development.

(2) 1B236 mRNA: Developmental Shift in Cellular Expression Revealed by In Situ Hybridization

The cDNA clone p1B236 hybridizes to mRNAs expressed in rat brain, but not liver or kidney (2), and has been well characterized as to its anatomical distribution (4) and developmental expression (3). In situ hybridization shows that this mRNA is expressed at highest abundance within oligodendrocytes undergoing myelinogenesis at early postnatal times (P15-P25), but that its expression switches to circumscribed neuronal subpopulations, along with limited numbers of oligodendrocytes, in the adult CNS. This switch in 1B236 gene expression from oligodendrocytes to neurons during development suggests that its products may act similarly in myelinogenesis and mature neuronal function.

CONCLUSION

In summary, in situ hybridization can be used to discriminate sites of gene expression in the developing CNS. Examples have been provided illustrating its application both as a cell-type marker to follow specific neuronal subpopulations during development, and to examine a developmental shift in the cellular expression of a novel mRNA. When combined with immunocytochemistry using antisera against synthetic peptides deduced from the nucleotide sequences of mRNAs, these techniques provide a complete neuroanatomical complement to cDNA cloning and gene isolation methods for the study of brain development.

REFERENCES

(1) Branks, P. and Wilson, M.C. (1986) Mol. Brain Res. 1, 1-16.
(2) Milner, R.J. and Sutcliffe, J.G. (1983) Nucleic Acids Res. 11, 5497-5520.
(3) Lenoir, D., Battenberg, M., Kiel, M, Bloom, F.E. and Milner, R.J. (1986) J. Neurosci. 6, 522-530.
(4) Bloom, F.E., Battenberg, E.L.F., Milner, R.J. and Sutcliffe, J.G. (1985) J. Neurosci. 5, 1781-1802.
(5) Higgins, G.A. and Wilson, M.C. (1987) in In situ Hybridization: Applications to Neurobiology (Valentino, K., Eberwine, J. and Barchas, J., eds.) pp. 279-307, Oxford University Press, Oxford, U.K.

DEVELOPMENTAL REGULATION OF ARGININE KINASE IN DROSOPHILA IMAGINAL DISCS

J. M. James and G. E. Collier, Department of Biological Sciences, Illinois State University, Normal Illinois 61761

The insect hormone ecdysone is responsible for changes in gene expression during developmental stages of Drosophila melanogaster, as reflected in puffing patterns of salivary gland polytene chromosomes. Three sequential classes of puffs have been described. Intermolt puffs of third instar larvae regress as ecdysone levels rise at the end of larval development, whereas a few early puffs appear rapidly under influence of the hormone. Numerous late puffs appear at least three hours later as a consequence of the inductive influence of early puff gene products (1). The identity and function of most of the late gene products are unknown.

Arginine kinase (AK) catalyzes the reversible formation of ATP and L-arginine from ADP and arginine phosphate and is particularly abundant in indirect flight muscles of insects. The gene for AK (Argk) has been localized to cytological region 66F of chromosome 3L (2). The present study traces the expression of Argk during morphogenesis and examines the role of ecdysone in its regulation.

DEVELOPMENTAL PROFILE AND TISSUE SPECIFICITY

Animals grown at 25°C were staged according to the method of Bainbridge and Bownes (3), except for the designation of postfeeding larvae as PL. AK activity levels were measured spectophotometrically (2). The developmental profile (Fig. 1) reveals two peaks of AK activity: one during the prepupal period and another at the time of eclosion. The second peak corresponds with adult muscle development, whereas the first occurs six hours after puparium formation, at stage P3.

Fig. 1. Developmental profile of arginine kinase activity in Drosophila melanogaster.

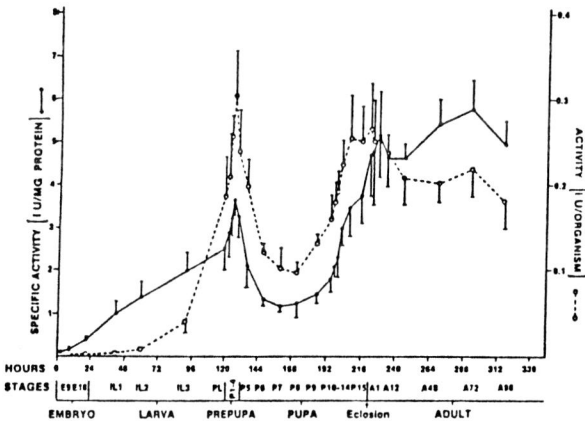

The tissue specificity of the prepupal peak was determined by dissection and homogenization of parts of late third instar larvae, postfeeding larvae, and stage P3 prepupae. A 5-fold increase in specific activity from the 108 hour old third instar larvae to stage P3 prepupae was observed for imaginal discs. Other tissues (muscle, digestive tract, brain and ganglion, and salivary glands) showed declines, or increases of less than 2-fold.

EFFECTS OF ECDYSONE ON ARGININE KINASE EXPRESSION

The prepupal peak does not appear when the temperature-sensitive mutant, ecd-1 (4), is shifted from 20°C to 29°C at the mid third instar stage (Fig. 2). It reappears when ecd-1 larvae, kept at 29°C for 60 hours are shifted back to 20°C. Feeding experiments support the inference of hormonal control of the prepupal peak. ecd-1 larvae, which had been at 29°C for 36 hr, were fed β-ecdysone in vials containing equal amounts instant Drosophila

medium (Carolina Biological) and 0.5mg β-ecdysone/ml H_2O. Pupariation occurred after 18 hr of feeding at 29°, and the stage P3 peak of AK activity was seen at 26 hr (Figure 3). No change in AK activity was seen in control larvae not fed β-ecdysone. Dissected imaginal discs from β-ecdysone fed larvae showed a four fold increase in AK activity betweeen stages PL and P3.

Fig. 2. Developmental profile of AK in ecd-1. Lower legend - control stages and concurrently sampled ecd-1 larvae after shift to 29°C at mid-third instar. Upper legend - ecd-1 larvae returned to 20°C after 60h at 29°.

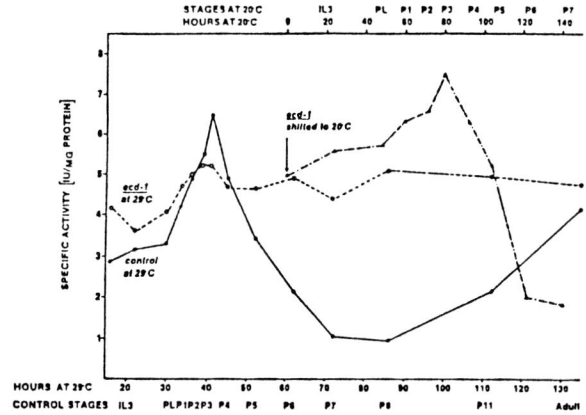

Fig. 3. Profile of AK activity after feeding β-ecdysone to ecd-1 larvae at 29°. Controls lacked β-ecdysone.

Fig. 4. Profile of AK activity for genotypes deficient, duplicated, or euploid for the 74A -75D region.

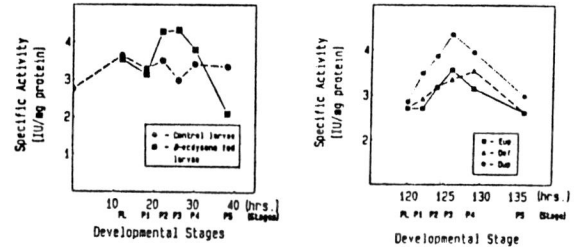

EFFECT OF EARLY GENE DOSAGE ON AK EXPRESSION

Post-feeding larvae through P5 prepupae of genotypes (5) deficient, duplicate, or euploid for 74A - 75D (which contains both the 74EF and 75B early puffs) were assayed for AK activity. Higher activity is seen when this region is in three doses, whereas a slower and lower response is seen in the flies carrying only one dose (Fig. 4). The corresponding aneuploids for the 2B5 early gene region showed no dosage effect on the prepupal peak of AK activity. Analysis of separate deficiencies of 75B and 74EF suggest that the 75B region is responsible for the dosage effect.

We conclude that in imaginal discs Argk is a target gene regulated by the product of the 75B ecdysone-regulated "early" gene.

REFERENCES

(1) Ashburner, M. and Berendes, H. D. (1978) In The Genetics and Biology of Drosophila, vol. 2b (Ashburner, M. and Wright, T. R. F., eds.) pp. 315-395, Academic Press, NY.

(2) Fu, L. and Collier, G. E. (1983) Bull. Inst. of Zool., Acad. Sinica 22, 25-35.

(3) Bainbridge, S. P. and Bownes, M. (1981) J. Embryol. exp. Morph. 66, 57-80.

(4) Garen, A., Kauvar, L. and Lepesant, J. (1977) Proc. Natl. Acad. Sci. USA 74, 5099-5103.

(5) Walker, V. K. and Ashburner, M. (1981) Cell 26, 259-277.

c-Ha-ras AND MYOGENIC DIFFERENTIATION

D.J. Kelvin, K.A. Zito and J.A. Connolly

Department of Anatomy, University of Toronto, Toronto, Canada M5S 1A8

INTRODUCTION: Myogenic differentiation is characterized by the dramatically increased expression of a family of muscle specific proteins, for example the acetylcholine receptor (AChR), creatine phosphokinase (CK), and myosins. The BC3H1 tissue culture cell line serves as an excellent model for myoblast differentiation (1). If kept in the logarithmic growth phase in the presence of 20% FBS, these cells do not differentiate (as judged by the turnon of these muscle proteins). However, when the FBS concentration is reduced to 1% these cells stop growing and within 48 to 72 hours begin expressing several muscle specific proteins including AChRs and CK. The control process is apparently FGF dependent (2).

Recently, the viral oncogene v-Ha-ras as well as the proto-oncogene c-ras have been shown to have an effect on the differentiation of the neuronal cell line PC-12 (3,4). Further experiments indicate that ras may transduce growth factor signals in the PC-12 system. In an attempt to understand the transduction of signals during myogenic differentiation we explored the effects of the introduction of c-Ha-ras and activated c-Ha-ras on BC3H1 differentiation.

MATERIALS AND METHODS: BC3H1 cells were transfected with either pHO6T1 or pHO6N1 plasmid DNA. The pHO6T1 construct contains the activated T24 Ha-ras gene and the aph gene. The pHO6N1 contains the c-Ha-ras gene along with the aph gene (5). Cell lines were selected for by treatment with G418. BC3H1, BCN1 (cells transfected with pHO6N1), and BCT1 (cells transfected with pHO6T1) cells were grown to near confluency and split at a ratio of 1:5. Twentyfour hours following the split, culture medium was replaced with fresh media containing either 20% or 1% FBS for the purpose of inducing differentiation. CK activity and AChR expression was determined 48 and 72 hours after the media change.

RESULTS: Table I shows CK activity in cell homogenates prepared from BC3H1, BCN1, and BCT1 cultured with either 20% or 1% FBS. BC3H1 and BCT1 grown in 20% FBS (normally a "non-differentiation condition") showed little CK activity on day 2 with a slight increase on day 3. BCN1 differed dramatically from BC3H1 and BCT1, having higher levels of CK activity that increased 7 fold from day 2 to day 3.

TABLE I
CK EXPRESSION IN MUSCLE CELL LINES

Cell Line	Percent FBS	CK activity/ug prot x 10^{-4}	
		Day 2	Day 3
BC3H1	20%	2.1	4
BCN1	20%	9.2	61
BCT1	20%	0.7	5
BC3H1	1%	11	20
BCN1	1%	13	70
BCT1	1%	0.9	0.4

When induced to differentiate by reducing the FBS concentration to 1%, BC3H1 cells respond by increased CK activity when compared to cells grown in 20%. BCN1 cells grown in 1% FBS also have increased levels of CK activity when compared to BCN1 cells grown in 20% FBS. Unlike BC3H1 and BCN1, BCT1 cells do not respond to the the inductive stimulus and in fact have lower CK levels than BCT1 cells grown in 20% FBS. This phenotype was stable and after several months these cells still cannot differentiate.

Table II shows the number of AChRs expressed on BC3H1, BCN1, and BCT1 grown in either 20% or 1% FBS. BC3H1 and BCT1 behave similarly when grown in 20% FBS, having low numbers of receptors on day 2 but increasing on day 3. BCN1 on day 2 have numbers of receptors comparable to the numbers expressed on BC3H1 and BCT1 on day 3. On day 3 BCN1 grown at 20% express an 8 fold increase in receptor numbers over day 2.

Under the inductive signal of 1% FBS BC3H1 express high levels of receptors on day 2 followed by a dramatic increase of receptor numbers on day 3. BCN1 responds to the inductive signal by expressing extremely high levels of receptors on day 2 with a slight increase in receptor number on day 3. BCT1 respond poorly to the reduction in FBS with only slightly elevated levels of receptor expression.

TABLE II
EXPRESSION OF AChRs IN MUSCLE CELL LINES

Cell Line	Percent FBS	FMoles [125I] α-BGT bound/ug protein	
		Day 2	Day 3
BC3H1	20%	0.26	1.97
BCN1	20%	1.81	16.23
BCT1	20%	0.61	1.43
BC3H1	1%	9.95	22.5
BCN1	1%	31.44	40.4
BCT1	1%	2.37	7.18

DISCUSSION: These data contain two very interesting points which lead to the conclusion that ras expression influences the differentiation of muscle cells. First, the introduction of an activated ras gene into BC3H1 muscle cells inhibits either the expression of differentiation specific proteins or blocks the differentiation process. Second, the introduction of a normal ras gene into these muscle cells induces them to express more differentiation specific proteins when compared to the parental BC3H1 strain. Even in the absence of an inductive signal BCN1 still behave as differentiating cells by expressing higher levels of muscle specific proteins.

Similar results were obtained with a second transfection. Further, transfection within the plasmid minus either ras gene induces G418 resistance but has no effect on differentiation. Individual clones derived from BCT1 showed no ability to differentiate; clones derived from BCN1 showed marked differences in this ability and we are presently analyzing copy number and expression of the introduced ras gene.

We hypothesize that ras plays a role in the transduction of growth and differentiation signals in differentiating muscle cells.

REFERENCES:
(1) Munson, R. et al. (1982). J. Cell Biol. 92:350-356.
(2) Lathrop, B. et al. (1985). J. Cell Biol. 100: 1540-1547.
(3) Noda, M. et al. (1985) Nature 318: 73-75.
(4) Hagag, N. et al. (1986) Nature 319: 680-682.
(5) Spandidos, D.A. and N.M. Wilkie. (1984) Nature 310: 469-475.

THE ROLE OF C-MYC IN MYOBLAST DIFFERENTIATION: ANALYSIS OF TRANSFECTED AND FUSION DEFICIENT LINES

J. D. Kohtz, D. S. Kohtz, and E. M. Johnson

Brookdale Center for Molecular Biology and
Department of Pathology, Mount Sinai School of
Medicine, New York, New York 10029

INTRODUCTION:

Mouse C2 myoblasts (C2M) can be induced to undergo fusion to myotubes by reducing serum and growth factor concentration (1). Within 24 hours of induction, cells cease to proliferate, and within 48-72 hours they begin to express several muscle gene products. A role for c-myc in the induction of cell proliferation is suggested by its response to PDGF stimulation of fibroblasts (2). In addition, virally-promoted c-myc expression can block mouse erythroleukemia cell differentiation (3). Analysis of teratoma-derived myoblasts (4) and rat L6 myoblasts (5) show that c-myc mRNA levels decline upon induction of differentiation. Is this decrease obligatory for fusion? We addressed this question in two ways: first, by examining c-myc mRNA in cloned, fusion-deficient C2 lines, and second by examining fusion in transfected C2 lines expressing high levels of c-myc.

RESULTS:

We cloned C2 myoblasts resistant to fusion upon changes in serum concentration. One such line (FD-2) expresses certain differentiation markers, including myosin heavy chain mRNA, but does not display the morphological changes characteristic of fusing C2 cells. Fusion-resistant rat myoblasts that express biochemical markers have previously been described (6). Northern blot analysis (Figure 1) was performed on C2M and FD-2 induced to differentiate by a changing their medium from 20% fetal calf serum + 0.5% chick embryo extract to 4% horse serum. RNA was extracted from undifferentiated cells (lane 0) and at 3, 8, 30, and 52 hours post-induction. The c-myc mRNA is virtually eliminated from fusing C2M, but remains high relative to histone H3 mRNA in FD-2. (Compare c-myc with H3 mRNAs in the 30 hr lane of each cell type.) H3, which is coupled to DNA synthesis, declines in both cell lines but at a slower rate in FD-2. Our results suggest that FD-2 is capable of expressing certain differentiation markers without reducing its levels of c-myc mRNA. To see whether altered c-myc expression is coupled to fusion resistance, we studied C2M co-transfected with plasmids pSVmyc-1 and pSV2neo. Several G418 resistant cell lines were cloned. Controls were transfected with pSV2neo alone. pSVmyc-1 allows expression of c-myc under control of the SV40 promoter. Southern blot analysis (Figure 2A) of Hind III digested genomic DNA from transfected lines 27, 19, and 16, and C2M demonstrates the presence of multiple copies of pSVmyc-1 integrated in the genome of the three transfected lines. The positions of lamda-Hind III bands are shown on the left. The position of the endogenous c-myc fragment (4.6 kb), barely visible in C2M, is marked E on the right. Figure 2B is a Northern blot comparing relative c-myc mRNA levels in the various transfected cell lines. The blot was over-developed to visualize the endogenous C2 c-myc transcript marked E. The shorter transcript from the transfected plasmid (T) lacks exon 1. It is present at highest levels in the transfected lines 27 and 19. When induced to differentiate, these two lines fail to fuse even after extended periods, although changes in H3 expression associated with growth arrest are still seen. Suprisingly, a decrease in levels of mRNA encoded by the transfected gene is seen paralleling that of the endogenous gene even though G418 resistance continues. Fusion is inversely proportional to the initial levels of c-myc mRNA in the transfected lines. Our results suggest that the decrease of c-myc observed in myoblasts is obligatory for an aspect of C2M terminal differentiation, but that certain genes remain unaffected. They further suggest that the transfected c-myc gene encodes signals for regulation of its mRNA levels upon induction of C2M differentiation.

Figure 2. High level c-myc expression by transfected lines.

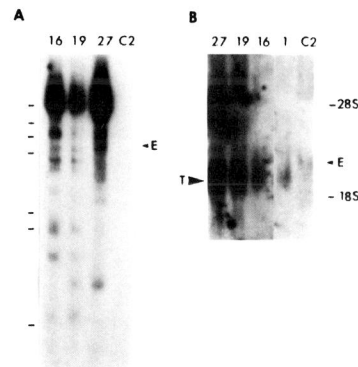

A. 15 µg of genomic DNA probed with a Hind III fragment of pMc-myc54. B. 20 µg RNA probed as in (A).

REFERENCES
(1) Blau, H.M. Chiu, C.P. Webster, C. (1983) Cell 32, 1170-1180.
(2) Kelly, K. Cohran, B.H., Stiles, C.D., Leder, P. (1983) Cell 35, 603-610.
(3) Coppola, J.A., Cole, M.D. (1986) Nature 320, 760-763.
(4) Sejersen, T. Sumegi, J., Ringertz, N.R. (1985) Exp. Cell. Res. 160, 19-30.
(5) Endo, T., Nadal-Ginard, B. (1986) Mol. Cell. Biol. 6, 1412-1421.
(6) Nguyen, H.T., Medford, R.M., Nadal-Ginard, B. (1983) Cell 34, 281-293.
(7) Stanton, L.W., Watt, R., Marcu, K.B. (1983) Nature 303, 401-406.
(8) Zong, R., Roeder, R.G., Heintz, N. (1983) Nucl. Acid. Res. 11, 7409-7425.

Figure 1. c-myc levels decrease in C2M induced to differentiate, but not in FD-2.

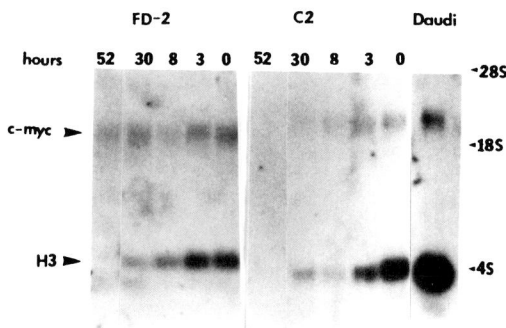

20 µg total RNA/lane was probed with a HindIII fragment of pMc-myc54 (7) and an EcoRI fragment of pHh5B (8).

HEAT SHOCK GENE EXPRESSION IN MOUSE EMBRYOGENESIS

R.K. Kothary*, M.D. Perry+, S. Clapoff*,
V. Maltby*, L.A. Moran+, and J. Rossant*.

*Division of Molecular & Developmental Biology,
Mt. Sinai Hospital Research Institute, and
+the Department of Biochemistry, U. of Toronto,
Toronto, Ontario, CANADA M5G 1X5

The heat shock response is a universally occurring phenomenon. Organisms from all branches of the phylogenetic tree respond to heat and other stresses by synthesizing a novel set of polypeptides (1). The major heat shock protein (hsp) in all organisms has a molecular weight of approximately 70Kd. In the mouse, the hsp70 gene family can be distinguished into two types. One is the cognate gene group which appears to produce the constitutively expressed hsc70 and hsc74. The second is the inducible gene, hsp68, which is usually only expressed in stressed cells. However, there are some indications that hsp68 may be developmentally regulated during embryogenesis. Constitutive expression of hsp68 has been detected in 2-cell mouse embryos using 2-D protein gel electrophoresis (2,3).

We have used a mouse hsp68 cDNA clone (4) to investigate the expression of hsp68 and hsc70 in various embryonic and extra-embryonic tissues from 8.5 day to 18.5 day old mouse embryos. In addition, we present data on the use of a fusion gene, consisting of the E. coli lac Z gene under the control of the promoter from a mouse hsp68 gene, to assay the activity of hsp68 in earlier stages of development.

Hsp68 expression in adult tissues.
RNA samples from various adult tissues were analysed on a Northern blot. Although hsp68 behaves as a true heat-inducible gene in cultured cells, constitutive expression was observed in certain tissues. Tissue specific expression of hsp68 was detected in the kidney. The same level of hsp68 expression was observed in kidneys from several different mice. This tissue specific expression may be a result of a separate regulatory mechanism from the overall heat-inducibility of the gene. Testes RNA contained an abundant species that cross-hybridized to the hsp68 cDNA probe. This transcript was slightly smaller and was probably derived from one of the other hsp68 genes known to be present in mouse cells (4).

Hsp68 expression during mouse embryogenesis.
Analysis of hsp68 RNA transcription was carried out in tissues from embryos between 8.5 day and 18.5 day of age. Northern blot analysis showed constitutive expression of the hsp68 gene specifically in the extra-embryonic tissues at several stages examined. The ectoplacental cone (or placenta at later stages) had the most significant amount of hsp68 mRNA. The yolk-sac tissue contained detectable levels of hsp68 message after about day 11.5 of development; expression was confined to visceral endoderm of the yolk-sac and absent from yolk-sac mesoderm. The pattern of hsp68 expression thus appears to be lineage specific. Of the three early lineages, hsp68 appeared to be constitutively expressed first in the trophectoderm and primitive endoderm lineages. Expression of hsp68

in the third lineage, primitive ectoderm, was not detected until late fetal stages. It is not known what tissues of the fetus express this gene. One possibility is the fetal liver, which is known to take over some of the functions of the visceral endoderm.

Transient expression of hsp68-lac Z fusion gene in EC cells and mouse embryos.
It is difficult to obtain enough pre-implantation stage embryos to determine whether hsp68 mRNA is transcribed very early in development. We are currently taking a different approach to examining expression in early mouse embryos by utilizing marker enzymes driven off the mouse hsp68 promoter. To study hsp68 expression in pre-implantation stage embryos, a highly sensitive assay is required. Such assays exist for B-galactosidase activity in single one-cell embryos. Thus, a mouse hsp68-E. coli lac Z fusion gene was constructed.

The plasmid containing the fusion gene was transfected into P19 cells. Transfected cells were either treated as controls (kept at 37°C) or induced with a heat shock (42°C) for increasing lengths of time prior to a standard recovery of one hour at 37°C. Cells were then assayed for B-galactosidase activity. Control cells had no expression of the fusion gene whereas heat-induced cells showed increasing B-gal activity with time of heat shock. Thus the hsp68-lac Z fusion gene was functional and was heat-inducible in P19 cells.

Purified hsp68-lac Z DNA (8 ng/uL) was injected into the pro-nuclei of one-cell mouse embryos. These embryos were either cultured to the two-cell stage or re-transferred to foster mothers and recovered at later stages of development. Embryos were subsequently fixed and stained for B-gal activity. The results are summarised in table I.

Table I. Transient expression of hsp68-lac Z in embryos.

EMBRYO STAGE	NO. OF EGGS INJECTED	NO. OF BLUE EMBRYOS
1-cell	35	15
2-cell	65	18
8-cell	37	0

Hsp68-lac Z was active in both one- and two-cell mouse embryos, but silent in 8-cell stage embryos. Thus it seems that the hsp68 gene may be developmentally regulated very early in embryogenesis in addition to the lineage specific expression we see in later stages.

References
1. Schlesinger, M.J., et al., eds. (1982). Heat Shock: From Bacteria to Man. CSH Lab.
2. Bensaude, O., et al. (1983). Nature 305, 331-333.
3. Morange, M., et al. (1984). Mol. Cell. Biol. 4, 730-735.
4. Lowe, D., & L.A. Moran. (1986). J. Biol. Chem. 261, 2102-2112.

INDUCED SYNTHESIS OF CELLULAR TRANSGLUTAMINASE BY SODIUM BUTYRATE IN TRANSFORMED HUMAN FIBROBLASTS

K.N. Lee, P.J. Birckbichler, and M.K. Patterson, Jr.

The Noble Foundation, Box 2180, Ardmore, Oklahoma 73402

Transglutaminase is a protein-modifying enzyme which catalyzes the exchange of primary amines for ammonia at the γ-carboxamide groups of protein-bound glutamine residues (1). Our previous studies demonstrate that transglutaminase activity in simian virus-transformed human embryonic lung fibroblasts (WI-38 VA13A) could be increased by sodium butyrate treatment (2). In this report we present evidence that the enhanced enzyme activity is due to an accumulation of enzyme and that sodium butyrate stimulates the de novo synthesis of transglutaminase mRNA, resulting in the increase of enzyme synthesis and a corresponding accumulation of the enzyme.

(I) INDUCTION OF TRANSGLUTAMINASE BY SODIUM BUTYRATE. The kinetics for the induction of transglutaminase activity in cells exposed to 1 mM sodium butyrate were examined (Fig. 1A). The level of enzyme activity approached a maximum by 6 days; 9 to 11 fold higher in the presence of sodium butyrate than in its absence. Immunoreactive transglutaminase protein measured by inhibition ELISA assay (3) also increased in a manner similar to the increase in enzyme activity (Table I). The disappearance of transglutaminase following inhibition of protein synthesis by addition of cycloheximide was also measured; the level of enzyme of both control and butyrate-treated cells were degraded exponentially with a half-life of 24 h. (Fig. 1B). Since the rate of degradation of enzyme was the same, the increased amount of enzyme might be due to an increased rate of enzyme synthesis. To examine this possibility, cell cultures were pulsed with L-[^{35}S] methionine for 60 min prior to harvest and labeled transglutaminase immunoprecipitated and quantitated. The relative rate of enzyme synthesis increased 9-11 fold in cells treated for 6 days with sodium butyrate (Table I); longer treatment did not appreciably affect the rate of enzyme synthesis.

(II) EFFECT OF SODIUM BUTYRATE ON TRANSGLUTAMINASE mRNA ACTIVITY. In order to estimate the effect of sodium butyrate on transglutaminase mRNA levels, total RNA was isolated from both control and butyrate-treated cells (4) and translated in a nuclease-treated rabbit reticulocyte lysate with L-[^{35}S] methionine. Anti-transglutaminase precipitable proteins in the cell-free translation products were isolated by immunoprecipitation. Protein synthesis was quantitated by precipitation of aliquots of the radioactive translation products with trichloroacetic acid. There was no significant difference in the total protein synthesis translated by RNA from either control or butyrate-treated cells, but translatable transglutaminase mRNA activity markedly increased in the latter (Table I).

Fig. 1. Effects of sodium butyrate on transglutaminase activity and disappearance of transglutaminase following culture of cells in cycloheximide. Cells were cultured in the absence (0) or presence (●) of sodium butyrate (1 mM), homogenized, and assayed for enzyme activity (A). For measuring degradation of transglutaminase (B), cells which had been grown for 4 days in the absence (0) or presence (●) of sodium butyrate (1 mM), were replenished with fresh medium containing 18 μM cycloheximide and 1 mM sodium butyrate. At various times following culture in cycloheximide media, cells were homogenized and the levels of enzyme were determined by inhibition ELISA assay.

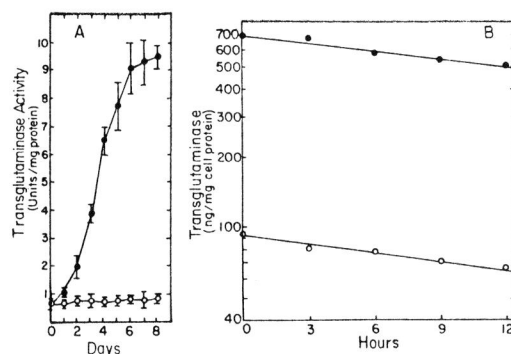

Table I. Effects of sodium butyrate on enzyme amount, relative synthetic rate, and mRNA activity of transglutaminase. Relative rate of synthesis and relative mRNA activity of transglutaminase was calculated as (immunoprecipitated radioactivity in transglutaminase/total trichloroacetic acid precipitable radioactivity). The values presented are the mean ± S.E. of four determinations.

Transglutaminase Parameter	Control	Sodium Butyrate
Enzyme amount (μg/mg protein x 10^2)	8.0±0.7	83.2±7.6
Relative rate of synthesis (x 10^5)	6.1±0.8	66.8±4.2
Relative mRNA activity (x 10^5)	5.3±0.6	64.6±7.6

REFERENCES

(1) Folk, J.E. and Finlayson, J.S. (1977) Adv. Protein Chem. 31, 1-133.
(2) Birckbichler, P.J., Orr, G.R., Patterson, M.K., Jr., Conway, E., Carter, H.A. and Maxwell, M.D. (1983) Biochim. Biophys. Acta 763, 27-34.
(3) Birckbichler, P.J., Upchurch, H.F., Patterson, M.K., Jr. and Conway, E. (1985) In Vitro 21, 27A.
(4) Deeley, R.G., Gordon, J.I., Burns, A.T.H., Mullinix, K.P., Binastein, M. and Goldberger, R.F. (1977) J. Biol. Chem. 252, 8310-8319.

MOLECULAR CLONING OF FACTORS EXPRESSED DURING THE DECISION-TO-DIFFERENTIATE IN MYOGENIC CELLS

Victor K. Lin, Massoud Mahmoudi, and Woodring E. Wright

Department of Cell Biology and Anatomy, University of Texas Southwestern Medical School, Dallas, TX 75235

We are attempting to isolate the factor(s) regulating the decision to differentiate using a subtraction hybridization approach (1). By selecting for clones that were able to form myotubes in the presence of increasing concentrations of BUdR during multiple cycles of selection, we have produced a variant of L6 rat myoblasts that incorporates 5-bromodeoxyuridine (BUdR) but is nonetheless able to differentiate in the presence of 32 µM BUdR (BUdif$^+$). Cesium chloride gradient analysis of the change in buoyant density of the DNA indicates that under these conditions 70% of the thymidine residues were replaced by BUdR, thus the cells are not transport or incorporation mutants. These cells essentially are dependent on the continued presence of BUdR to prevent them from differentiating under normal culture conditions. Cell hybrids between the BUdif$^+$ cells and differentiation defective cells are able to differentiate at a much higher frequency than comparable hybrids between the parental L6 line and the differentiation defective myoblasts. The resistance to BUdR thus probably results from overproducing the factor(s) regulating the decision to differentiate (2). Poly(A)$^+$ RNA was isolated from BUdR-resistant cells 24 hours after being stimulated to differentiate, a time just prior to overt myogenesis, when the molecules regulating the decision to differentiate should be at maximal levels. A cDNA against this "mid-decision" RNA was exhaustively hybridized to mRNA isolated from differentiation defective myoblasts. Sequences common to both cell types were removed by selective binding of double-stranded nucleic acids to hydroxylapetite, leaving a cDNA enriched for sequences unique to the "mid-decision" phase. A partially enriched "mid-decision" cDNA was cloned into λgt10 and screened with a fully enriched "mid-decision" probe. The filters were then stripped and reprobed with a cDNA against parental L6 mRNA from which differentiation defective sequences had been subtracted. Plaques exhibiting a differential signal were then purified and analyzed. The factor(s) regulating the decision to differentiate should be absent (or at very low levels) in undifferentiated cells, be at maximal levels during the "mid-decision" period, remain constant or decline in fully differentiated myotubes (versus contractile/structural proteins which should increase), and be expressed at amplified levels in BUdif$^+$ as compared to normal myoblasts. Most of the plaques from this initial screening contained sequenes that were overexpressed at all stages of differentiation in the BUdif$^+$ cells and which were at very low levels in the parental L6 myoblasts. Many of these cDNA inserts probably code for regulatory factors controlling household functions that are overexpressed as an adaptation to growth in high concentrations of BUdR. A differentiation-defective variant of the BUdif$^+$ cells was then derived by five cycles of clonal selection for cells that failed to form myotubes in the presence of BUdR. These cells (BD5) maintained the ability to grow in 32 µM BUdR and should thus express all of the adaptations to BUdR except for the overproduction of the factor that regulates the decision to differentiate. Enriched probes were then prepared by subtracting differentiation defective sequences

(from the original BUdR sensitive defective cells) from cDNAs prepared from either "mid-decision" BUdif$^+$ myoblasts or their differentiation defective BD5 derivative. Plaques exhibiting a differential signal were then again purified and analyzed. One clone, pBU65, contains a cDNA that is specific for the "mid-decision" stage of myogenesis and thus may be involved in regulating myogenesis. Northern analysis indicates that pBU65 hybridizes to a single band in "mid-decision" cells of approximately 2kb in length. Figure 1 compares the relative abundance of the pBU65 insert in undifferentiated, "mid-decision" and differentiated myotubes in the parental L6 cells, the overexpressing BUdif$^+$ cells, and the differentiation defective derivative (BD5) of the BUdif$^+$ cells. The intensity of hybridization to 0.25 µg of poly(A)$^+$ RNA isolated from the various cell types is compared to that of a constitutive probe, p85. Although not conclusive, the roughly equivalent expression in the parental and BUdif$^+$ cells argues against it being a DNA-binding protein whose overexpression results in the ability of the BUdif$^+$ cells to differentiate in the presence of BUdR. Antibodies to the protein coded for by pBU65 are being generated in order to characterize its cytochemical distribution during differentiation and contribute to identifying its functional role. Additional clones from the library that exhibit some of the desired characteristics are also being characterized.

Fig. 1. Relative expression of pBU65 and a constitutive marker (p85) during differentiation of parental L6, BUdif$^+$ (BU36) and a differentiation defective derivative of BUdif$^+$ (BD5) cells.

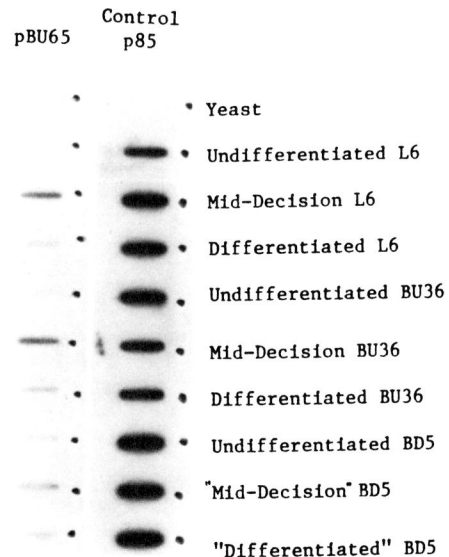

REFERENCES

(1) Wright, W.E. (1985) BioEssays 3, 245-248.
(2) Wright, W.E. (1985) J. Cell Biol. 100, 311-316.

DETECTION OF THYROID HORMONE (TH)-ENHANCED AND -INHIBITED HEPATIC mRNA SPECIES IN R.CATESBEIANA DURING TH-INDUCED AND SPONTANEOUS METAMORPHOSIS.

D. Lyman and B.A. White, Anatomy Department,
UConn.Health Ctr., Farmington, CT. 06032, USA

In R. catesbeiana, the larval tadpole is transformed into an adult by a developmental process termed metamorphosis. In the liver, significant TH-induced (1) changes occur in the profile of hepatocyte proteins (2) during metamorphosis. Studies with adult rat tissues (3-5) support a pre-translational action of TH, which is consistent with the presence of nuclear TH-binding sites in tadpole liver cells (6). However, the mechanisms by which TH induce stable changes in specific liver gene expression during metamorphosis remain largely unstudied.

We have initiated a study of TH regulation of liver gene expression during metamorphosis by employing differential screening of a cDNA library made from T_4-treated bullfrog tadpole liver poly (A)+RNA in order to detect clones which contain sequences corresponding to TH-responsive liver genes. From an initial repeated screening of 6000 transformant colonies, recombinant plasmids of 13 positive and 2 nonpositive clones were purified by the cleared lysate method and cesium chloride gradient centrifugation. The TH-responsive nature of the cognate mRNAs of the plasmids was determined by Northern and dot blot analysis using ^{32}P-labeled, nick translated plasmids. For Northern analysis, equivalent amounts of untreated premetamorphic tadpole, T_4-treated premetamorphic tadpole and adult bullfrog liver RNA were used. For dot blot analysis, total cell RNA was used from livers of individual tadpoles at selected developmental stages and adult bullfrogs and kidneys of adult bullfrogs and premetamorphic tadpoles.

We obtained 3 TH-enhanced, 2 TH-nonresponsive and 2 TH-inhibited independent cDNA plasmids. Dot blot analysis of spontaneous metamorphic samples confirmed that the 3 TH-enhanced cDNAs represent developmentally regulated, TH-enhanced, liver genes. Five of the 13 potentially positive plasmids from the differential screening in fact detected no change in mRNA levels between samples in Northern analysis. The 2 TH-inhibited plasmids detected an inhibition in the steady state levels of their cognate mRNAs due to T_4 treatment. Table I summarizes various properties of these plasmids.

Importantly, dot blot analysis of RNA samples with the cloned cDNAs demonstrated their induction profiles during spontaneous metamorphosis. The TH-enhanced plasmids (pD1-12,pD1-30) may be liver-specific (Table I). Transcripts are detected in the earliest premetamorphic tadpole liver samples and then increase through metamorphosis, when T_3/T_4 plasma levels have been reported to increase (e.g.1). Significantly, the steady state mRNA levels remain elevated in adult livers despite the reported decline in plasma T_3/T_4 levels. In contrast, transcripts of pD1-37(TH-inhibited) are present only in early premetamorphic livers until mid-prometamorphosis (stage XV). pD1-1 (TH-inhibited) is not liver-specific (Table I) and transcripts are equally present in liver and kidney at all stages. However, exogenous T_4 treatment of premetamorphic tadpoles consistently decreased the level of pD1-1 mRNA. pD1-1 may correspond to a general cell constitutive gene under multihormonal regulation during metamorphosis. pD 1-18 (TH-nonresponsive) appears to be liver-specific but is not induced by exogenous T_4, or during spontaneous metamorphosis. It is possibly a liver-specific constitutive gene.

Thus, we have cloned several cDNAs of TH-responsive, developmentally regulated, genes in the liver of R. catesbeiana. Although we have not yet demonstrated a direct effect of T_4 on the responsiveness of these transcripts in cultured tadpole hepatocytes, the demonstration of specific nuclear TH receptors in these cells (6) is consistent with such an action of T_4 in premetamorphic tadpoles. Moreover, our observations of positive and negative pre-translational T_4 effects are notably similar to the demonstrated direct pre-translational effects of TH on the expression of several genes in differentiated rat liver cells (3-5).

Interestingly, we have not yet obtained any recombinants which detect TH-induced transcripts that are completely absent in larval hepatocytes and then become expressed under the influence of TH. We continue to search for these types. Apparently, many TH-regulated liver transcripts are present in the larva and attain their mature level of expression in the adult through the mediation of TH. Perhaps most interestingly, the maintained steady state transcript levels of both induced (pD1-12,pD1-30,pD1-34) and inhibited (pD1-37) mRNAs in the adult liver, when TH plasma levels have declined, suggests a possible stable and persistent aspect in the mechanism of TH action in the developmental regulation of expression of the cognate genes.

TABLE I

Clone	mRNA(kb)	cDNA(Kb)	Liver T4	SM	Kidney Larva	Adult
pD1-12	2.6	2.0	↑	↑	ND	ND
pD1-30	0.7	0.6	↑	↑	ND	ND
pD1-34	1.0(major) 3.3(minor)	0.97	↑			
pD1-1	0.93	1.0	↓	C	C	C
pD1-37	0.74	0.5	↓	↓	ND	ND
pD1-8	0.93	0.45	C			
pD1-18	6.5	0.35	C	C	ND	ND

SM-Spontaneous Metamorphosis; ND-Not Detected
C-Constant

REFERENCES

(1) White, B.A. and Nicoll, C.S. (1981) in Metamorphosis. A Problem in Developmental Biology. 2nd ed. (L.I. Gilbert and E. Frieden, eds.) pp. 363-396. Plenum Press, New York.

(2) Cohen, P.P. et al. (1978) in Hormonal Proteins and Peptides. Vol.VI (C.H. Li, ed.) pp. 273-381. Academic Press, New York.

(3) Scarpulla, R.C. et al. (1986) J. Biol. Chem. 261, 4660-4662.

(4) Wong, N.C.W. and Oppenheimer J.H. (1986) J. Biol. Chem. 261,10387-10393.

(5) Narayan, P. and Towle, H.C. (1985) Mol. Cell. Biol. 5, 2642-2646.

(6) Galton V.A. and St. Germain, D. (1985) Endocrinology. 117, 912-916.

AN EARLY DEVELOPMENTAL PHASE OF SRC- EXPRESSION IN THE NEURAL ECTODERM

Patricia F. Maness

Department of Biochemistry, University of North Carolina, School of Medicine, Chapel Hill, North Carolina 27514

The study of oncogenes has, provided an unexpected probe for early development. The normal cellular oncogene c-src encodes and expresses a tyrosine-specific protein kinase, $pp60^{c-src}$ that is highly homologous to the transforming protein of Rous sarcoma virus (1). $pp60^{c-src}$ is expressed at elevated levels in developing neural tissues (2,3), however, its function is unknown. It was shown in this laboratory that $pp60^{c-src}$ is a product of neurons, appearing at the onset of terminal differentiation in the developing chick neural retina when proliferation ceases (4). Neurons of the cerebellum also express $pp60^{c-src}$, and here too expression correlated with cessation of DNA synthesis (5).

To investigate the expression of $pp60^{c-src}$ during earlier stages of embryo genesis, we analyzed chick embryos for expression of $pp60^{c-src}$ at stages 4-12 (18- to 49-hr incubation) by immunocytochemical staining. The antibodies used for localization (6) were raised to bacterially expressed $pp60^{v-src}$, the product of the Rous sarcoma virus v-src gene. Due to the close sequence homology demonstrated between the viral and cellular forms of $pp60^{src}$, these antibodies cross react with $pp60^{c-src}$. We show here that an early phase of src- expression in the chick embryo occurs transiently during gastrulation and neural tube formation. During this phase of expression, $pp60^{c-src}$ immunoreactivity (IR) was localized primarily within the neural ectoderm, but after neural tube closure was complete, IR was not observed. Taken together with our previous studies on $pp60^{c-src}$ expression during neuronal differentiation, the present results show that there are two developmental phases of src-immunoreactivity in the neural lineage (8).

RESULTS AND DISCUSSION

Chick embryos engaged in gastrulation and neural tube formation were examined for expression of $pp60^{c-src}$ immunoreactivity (IR) by immunoperoxidase staining using antibodies raised against $pp60^{v-src}$ purified from Escherichia coli expressing a cloned v-src gene (6). $pp60^{c-src}$ IR was found to be present in the chick embryo at the definitive primitive streak stage (stage 4, 18-19 hr incubation time). At this stage $pp60^{c-src}$ IR was localized primarily within the cells of the developing ectodermal germ layer. Cells of the developing endodermal and mesodermal germ layers showed much lower levels of IR staining. Staining observed in the ectoderm was completely blocked by preincubating the antibody with low concentrations (20-25 µg/ml) of purified antigen.

A few hours later, at the head process stage (stage 6, 23-24 hr) the neural ectoderm exhibited prominent IR staining. In contrast, neither the notochordal mesoderm nor the endoderm exhibited significant amounts of IR. IR in the neural ectoderm terminated at a location peripheral to the

notochord. The peripherally located ectoderm in this region does not develop into neural derivatives but instead into epidermis and extraembryonic ectoderm later on. High levels of $pp60^{c-src}$ IR were also not observed in the nonneural ectoderm located anterior to the head process.

During neural tube closure $pp60^{c-src}$ IR began to disappear from the wall of the neural tube in the stage 9 (29-33 hr) chick embryo. IR was still observed in the neural crest region and in the ventral floor of the neural tube overlying the notochord.

At stage 12 (45-49 hr) the neural tube of the chick embryo is completely closed and consists entirely of proliferating neuroepithelial cells which have not yet differentiated along neuronal or glial lines. At this stage, specific IR staining was not observed at any level of the neural tube examined--the telencephalon, metencephalon, optic vesicles, optic tectum, or spinal cord region.

It is possible that the IR in the neuroectoderm corresponds to the expression of a member of the src family of tyrosine kinases closely related to $pp60^{c-src}$. However, the antiserum used here does not recognize the products of the v-yes or v-fgr genes. It will therefore be important to confirm these results by in situ hybridization with oncogene specific DNA probes. It is of interest that an analogous pattern of c-src expression in Drosophila has been shown by in situ hybridization (7). c-src RNA is abundant in early Drosophila embryos during gastrulation, low in larvae, and high in neural tissue and smooth muscle of pupae at later stages.

To recapitulate, $pp60^{c-src}$ IR was present during the primitive streak and head process stages of chick embryogenesis, primarily in the neural ectoderm. $pp60^{c-src}$ IR was either absent or much less prominent in the nonneural ectoderm, mesoderm, and endoderm. During neural tube closure, $pp60^{c-src}$ IR began to disappear from the wall of the neural tube, but remained in the neural crest region, and ventral floor of the neural tube. After neural tube closure was complete, specific $pp60^{c-src}$ IR in the neural tube was not observed until terminal neuronal differentiation later on. There studies demonstrate two phases of expression of $pp60^{c-src}$, or an immunologically related protein in the neuronal lineage, and lend support to the notion that tyrosine kinases function in differentiation signalling.

REFERENCES

(1) Collett, M.S., Erikson, E., Purchio, A.F., Brugge, J.S., and Erikson, R.L. (1979). Proc. Natl. Acad. Sci. 76:3159-3163.
(2) Cotton, P.C., and Brugge, J.S. (1983). Mol. Cell. Biol. 3:1157-1162.
(3) Levy, B.T., Sorge, L.K., Meymandi, A., and Maness, P.F. (1983). Dev. Biol. 104:9-17.
(4) Sorge, L.K., Levy, B.T., and Maness, P.F. (1984b). Cell 46:249-257.
(5) Fults, D.W., Towle, A.C., Lauder, J.M., ad Maness, P.F. (1985). Mol. Cell. Biol. 5:5-27.
(6) Gilmer, T.M., and Erikson, R.L. (1983). J. Virol. 45:462-465.
(7) Simon, M.A., Drees, B., Kornberg, T.A., and Bishop, J.M. (1985). Cell 42:831-840.
(8) Maness, P.F., Sorge, L.K., and Fults, D.W. (1986). Devel. Biol. 117:83-89.

THE INDUCTION OF KERATIN GENE EXPRESSION DURING XENOPUS DEVELOPMENT

P.M. Mathisen, E. Touma, A. Jurkowski, and L. Miller

Laboratory for Cell, Molecular, and Developmental Biology, Department of Biological Sciences, University of Illinois at Chicago, Box 4348, Chicago, IL 60680.

We are currently studying the role of the thyroid hormone, T_3 in the induction of keratin gene expression in the epidermis of Xenopus during development. In Xenopus laevis the epidermis is an unkeratinized, two-cell layered epithelium from gastrulation until metamorphosis. The bilayered larval epidermis is then converted into a stratified, keratinized epithelium, under the control of T_3, and is characterized by the appearance of an adult specific 63kd keratin. In the adult epidermis the 63kd keratin is the most abundant protein. Thus the developmental activation of the 63kd keratin is associated with the terminal differentiation of the epidermis.

To determine when the synthesis of the 63kd keratin begins during development, we isolated head skin of embryos at different developmental stages, incubated with ^{35}S-met, extracted the keratins and analyzed by PAGE. We found that there is a very low level of the 63kd keratin synthesized as early as Stage 48. The relative rate of synthesis of the 63kd remains low until Stage 54 and then increases progressively to become the most abundant keratin synthesized during and after metamorphosis. Thus during development the 63kd keratins are expressed at two levels; initially at a very low basal level, and later at a high level of synthesis.

In order to examine the effect of T_3 on the skin in more detail, we are using an in vitro explant culture technique. Skin was dissected from the dorsal surface of the head of embryos at different stages and placed on a matri-gel coated surface of a culture dish. The epithelial cells migrate out of this explant and form a monolayer, but cell division does not occur. These cultures are then treated with T_3 to induce the 63kd keratin. Skin explants from embryos not expressing the 63kd keratin (prior to Stage 48) maintain the in vivo pattern for 3-4 days with or without T_3. After 4-5 days with T_3, a large increase in the synthesis of the 63kd keratin occurs. At the same time, there is also a very low basal level of expression in cultures without T_3. Thus the in vitro skin explants reproduce the two levels of 63kd keratin gene expression seen in vivo. Furthermore, our in vitro experiments have demonstrated that the basal level of 63kd keratin synthesis proceeds independently of T_3 and is not inhibited by prolactin or retinoic acid. Conversely, high level synthesis of the 63kd keratins requires T_3 in the culture media and is inhibited by prolactin and retinoic acid. One possible explanation for these results is that different 63kd keratin genes are activated at different developmental stages, one independently of T_3, at stage 48, and the remaining genes, whose activation is dependent on T_3 at stage 54. Alternatively, all three 63kd keratin genes are activated at the same developmental stage and their expression remains low until stage 54 when high concentrations of T_3 are present. In this case, all 63kd keratin genes would be activated independently of T_3. RIA measurements by Leloup and Buscaglia (2) indicate that T_3 and T_4 cannot be detected in Xenopus blood until stage 54 and thereafter rise reaching the maximum concentration at stage 60-62.

We have obtained three different cDNAs for the 63kd keratin (1). By using these cDNAs we are determining if the appearance of the 63kd keratin protein is controlled at the mRNA level. We have subcloned the 3' end of each of the cDNAs in order to study the appearance of each of the mRNAs. Using both slot blot and Northern analysis, we have shown that two of these mRNAs, pUF164 and pUF23, appear at about stage 55, which is concomitant with a high rise of T_3. We are currently using more sensitive detection techniques (SP6 generated RNA probes) to ensure that we will detect even very low levels of these RNA transcripts.

By measuring both protein and mRNA synthesis we are beginning to determine the developmental profile of the first-time activation of a developmentally regulated gene which is induced by both in vivo and in vitro by the T_3.

References

(1) Hoffman et al. (1985) J. Mol. Biol. 184:713-724.
(2) Leloup, J. & Buscaglia, M. (1977) Endocrinologie 284:2261-2262.

90

DEVELOPMENTALLY-REGULATED RAT BRAIN mRNAS: MOLECULAR AND ANATOMICAL CHARACTERIZATION

F.D. Miller, C.C.G. Naus, G.A. Higgins,
F.E. Bloom and R.J. Milner

Division of Preclinical Neuroscience and
Endocrinology, Scripps Clinic and Research
Institute, La Jolla, CA

ISOLATION OF EMBRYO-ENRICHED RAT BRAIN mRNAS.

In order to identify markers for developing neural cell populations and gain molecular insights into the processes of neural development and differentiation, we have selected cDNA clones of rat brain mRNAs that are expressed in brain at embryonic day 16 (E16) at least tenfold greater abundance than in adult brain. Eleven such clones were obtained from a cDNA library of E16 brain poly (A)+ RNA using a combination of differential and subtractive hybridization screens (Figure 1). The temporal and spatial patterns of expression of the mRNAs corresponding to these clones were characterized by Northern blotting and by in situ hybridization. Although all the mRNAs are enriched in embryonic brain, different mRNAs demonstrate their maximum abundance at different times in late embryogenesis and can be grouped into three classes on the basis of their time courses of expression. These mRNAs also display different patterns of spatial expression in the embryo: some are highly enriched in brain, while others are expressed in a wide variety of embryonic tissues. Nucleotide sequence analysis revealed that the majority of the selected cDNA clones represent novel mRNA species, while one clone encodes the brain-enriched $T\alpha 1$ isotype of α-tubulin (1). A number of the cDNA clones were selected for further detailed analysis on the basis of the temporal and spatial patterns of expression of the corresponding mRNAs as revealed by Northern blot analysis and in situ hybridization.

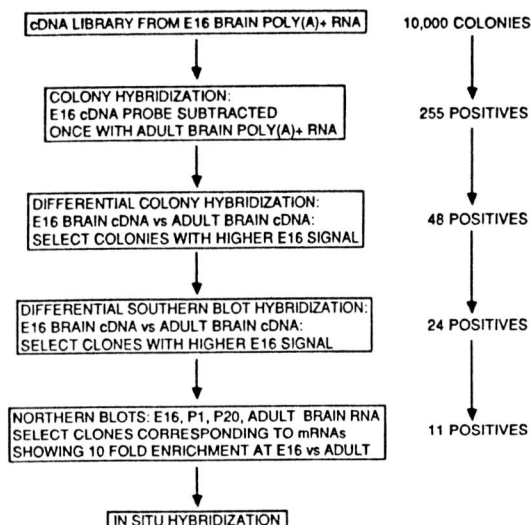

Figure 1. Strategy for selection of cDNA clones.

TUBULIN ISOTYPES ARE DIFFERENTIALLY-REGULATED DURING NEURONAL MATURATION

Detailed in situ hybridization analysis with a probe specific for the 3' untranslated region of the $T\alpha 1$-tubulin mRNA demonstrated that this mRNA is highly enriched in, and possibly specific to, the developing nervous system during embryogenesis and the early postnatal time period. Further, this mRNA is expressed at the highest levels in those neuronal populations that have completed migration and are extending neuritic processes, such as the neurons of the developing cortical plate. By P23, when process extension is complete in the developing brain, $T\alpha 1$ mRNA is present at very low levels, and is predominantly expressed in those neurons that extend long axonal projections. The temporal correlation between expression of $T\alpha 1$ mRNA and neurite extension was confirmed by regional Northern blot analysis, and the specificity of induction of this particular tubulin isotype was confirmed by RNase-protection experiments. To extend the analysis, expression of the T26 α-tubulin isotype, which is also present in the developing embryonic nervous system (2,3), was analysed by in situ hybridization and Northern blot analysis. In contrast to $T\alpha 1$, the T26 mRNA is expressed throughout the developing embryo and nervous system. Within the developing nervous system, T26 mRNA is expressed in a wide variety of cell types, with some enrichment in the proliferative zones of the embryonic and early postnatal brain. Levels of this mRNA in the brain remain approximately the same from E16 through to adulthood. Thus, expression of the $T\alpha 1$ α-tubulin isotype mRNA is both temporally and spatially correlated with neurite extension in the developing nervous system, while the T26 mRNA appears to be constitutive, with some degree of enrichment in cell populations that are undergoing mitosis. This differential regulation of expression of these two closely-related α-tubulin isotypes has been confirmed in primary explant cultures of neuroblasts, and in PC12 cells that have been induced to undergo differentiation with nerve growth factor.

DEVO 2 mRNA: EXPRESSION IN GERMINAL ZONES OF THE DEVELOPING NERVOUS SYSTEM

Several of the cDNA clones selected in the initial screening analysis represent mRNA species that are expressed in the germinal zones of the developing nervous system. One of these cDNA clones, pDEVO 2, represents a moderately abundant, 1.0 kb novel mRNA that is primarily translated on membrane-bound polyribosomes. Devo 2 mRNA is expressed as early as E10, increases to maximum abundance at birth, declines approximately 10-to-20-fold by P23, and is expressed at low levels throughout adulthood. In situ hybridization analysis of this mRNA revealed that it is expressed in a wide variety of embryonic tissues. Within the developing nervous system, however, Devo 2 is primarily expressed in the ventricular germinal zones prenatally, and in the external germinal layer of the cerebellum postnatally. Detailed analysis of this mRNA in the developing cerebellum revealed that there is a close temporal and spatial correlation between the expression of Devo 2 mRNA and neurogenesis in this region of the brain.

REFERENCES
1. Lemischka, I.R., Farmer, S., Racaniello, V.R., and Sharp, P.A. (1981), J. Molec. Biol. 150, 101-120.
2. Ginzburg, J., Behar, L., Givol D., Littauer U.Z. (1981) Nuc. Acids Res. 9, 2691-2697.
3. Lewis S.A., Lee M.G.-S., and Cowan N.J. (1985) J. Cell Biol. 101, 852-861.

THE HEAT SHOCK-ENHANCED SYNTHESIS OF GAPDH IN XENOPUS EMBRYOS IS INVERSELY CORRELATED WITH THE CONSTITUTIVE LEVELS OF ENZYME ACTIVITY

R. W. Nickells, T. I. Wang and L. W. Browder

Dept. of Biology, Univ. of Calgary, Calgary, Alberta, Canada T2N 1N4

In many cell types, heat shock stimulates changes in the cellular metabolism such as loss of cellular ATP (1) and increases in lactic acid (2), as well as altering the structure and possibly the function of mitochondria (3). These results imply that energy requirements of a cell during heat shock are dependent on increases in glycolysis. This correlates with recent observations that two glycolytic enzymes in yeast (enolase [4] and glyceraldehyde-3-phosphate dehydrogenase [GAPDH, 5]) are heat shock proteins (hsps).

Our preliminary results indicate that Xenopus embryos also respond to heat shock by accumulating large amounts of lactic acid. We have investigated the possibility that Xenopus embryos respond to an increased demand on glycolysis by synthesizing more of some glycolytic enzymes. By immunoprecipitation and peptide mapping, we have identified a 35 kilodalton Xenopus hsp (hsp35) as being an isozyme of GAPDH (data not shown). Hsp35 synthesis is variable, however, and is sometimes not detected. In this report, we present evidence that GAPDH accumulates during heat shock, resulting in an increase in GAPDH specific activity. We also report that heat shock-induced increases in GAPDH specific activity are inversely correlated with the constitutive level of enzyme activity. This inverse correlation may account for the variable synthesis of hsp35.

(I) GAPDH ACCUMULATES DURING HEAT SHOCK.

Figure 1A shows an immunoblot of equal amounts of protein from control and heat-shocked neurulae, which has been probed with rabbit antiserum prepared against yeast GAPDH. The band most strongly recognized by the antiserum comigrates with hsp35 and GAPDH standard. The immunoblot shows that there is accumulation of this antigen during heat shock. The apparent increase in GAPDH during heat shock results in significant increases in GAPDH specific activity in heat-shocked embryos (Figure 1B).

Figure 1. (A) The 35 kDa region of an immunoblot of control (C) and heat-shocked (HS) neurula proteins. The relative position of GAPDH standard is indicated. (B) A histogram of GAPDH specific activity in control (C) and heat-shocked (HS) neurulae. The heat-shocked sample shows a significant increase in specific activity (Mann-Whitney U-test, p<0.025). For these experiments, the neurulae were heat-shocked for 20 min at 37°C.

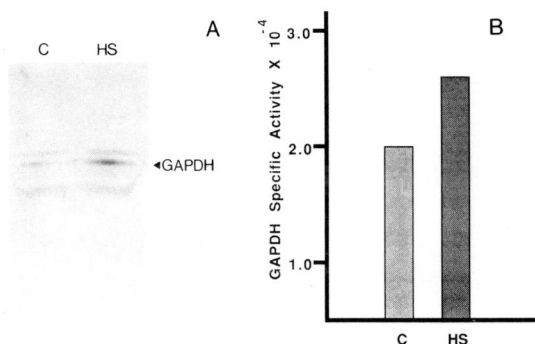

This increase in specific activity, like the synthesis of hsp35, is often variable. Hsp35 synthesis, however, is only detected in embryos showing relatively large increases in specific activity (data not shown). Our further experiments have indicated that heat-shocked embryos have enhanced synthesis of a specific isozyme of GAPDH (data not shown).

(II) THE VARIABLE INCREASES IN GAPDH SPECIFIC ACTIVITY ARE INVERSELY CORRELATED WITH CONSTITUTIVE LEVELS OF ENZYME ACTIVITY.

Heat shock-induced increases in GAPDH specific activity are variable. We have detected an inverse correlation between the specific activity of embryos before and after heat shock, which we believe explains this variability. Figure 2 shows the plot of constitutive GAPDH specific activity from dorsal lip gastrulae and neurulae versus the % change in activity observed after a 20 min heat shock at 37°C. Control embryos with high levels of GAPDH activity have little or no increase in activity after heat shock. Embryos with low levels of GAPDH activity, however, show predictably larger increases in activity after heat shock.

Figure 2. Plot of seven experiments showing the inverse correlation between constitutive GAPDH activity and the subsequent heat shock-induced change. The best fit line is shown. The correlation coefficient for these data is 0.987.

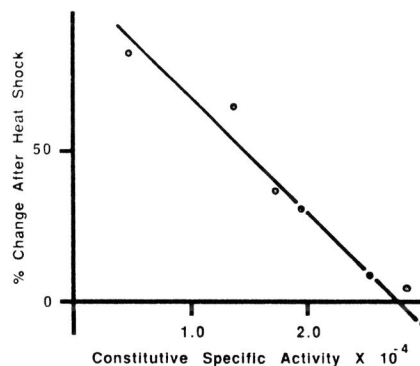

DISCUSSION

We interpret these data as indicating that heat shocked embryos require a threshold of GAPDH activity as part of their response to a thermal stress. Embryos with low constitutive levels of GAPDH activity may meet this threshold by synthesizing more GAPDH, which is observed as the previously described hsp35. We have observed that embryos from different females have different basal levels of GAPDH activity. This inherent variability may account for the variable synthesis of hsp35.

REFERENCES

1. Findly, R.C., Gillies, R.J. and Schulman, R.G. (1983). Science. 219, 1223-1225.
2. Hammond, G.L., Lai, Y.-K. and Markert, C.L. (1982). Proc. Natl. Acad. Sci. USA. 79, 3485-3488.
3. Lepock, J.R., Cheng, K.-H., Al-Qysi, H. and Kruuv, J. (1983). Can. J. Biochem. Cell Biol. 61, 421-427.
4. Iida, H. and Yahara, I. (1985). Nature. 315, 688-690.
5. Lindquist, S. (1986). Ann. Rev. Biochem. 55, 1151-1191.

cDNA CLONING OF ALKALINE PHOSPHATASE FROM RAT
OSTEOSARCOMA CELLS

M. Noda, M. Thiede, R. Buenaga, K. Yoon, M.
Weiss,* P. Henthorn,* H. Harris* and G.A. Rodan

Department of Bone Biology and Osteoporosis
Research, Merck Sharp & Dohme Research
Laboratories, West Point, PA 19486, U.S.A. and
*Department of Human Genetics, University of
Pennsylvania School of Medicine, Philadelphia, PA
19104

Alkaline phosphatases are ubiquitous enzymes
found in most organisms from bacteria to man.
Vertebrate alkaline phosphatases are dimeric,
plasma membrane-associated glycoproteins. In man
they comprise a family of tissue-specific,
electrophoretically distinguishable isoenzymes
which are the products of at least three gene loci
(placental, intestinal, and liver/bone/kidney) and
are distinguishable by a variety of structural,
biochemical and immunological methods (1,2).
Recently several allelic forms of placental
alkaline phosphatases were cloned and their full
length cDNA sequences were reported (3,4,5). The
function of alkaline phosphatases in vertebrate
tissues is not well understood. This is
especially evident in the case of bone, where
alkaline phosphatase has been implicated in the
process of tissue mineralization (6). Alkaline
phosphatase is known as a phenotypic marker for
osteoblasts and is expressed in a characteristic
fashion during osteoblast differentiation in vivo
and in vitro. To obtain a molecular probe for
osteoblast differentiation, we isolated the cDNA
of rat alkaline phosphatase.

(I.) ISOLATION AND CHARACTERIZATION OF
ALKALINE PHOSPHATASE. An expression library (λ
gt11) constructed from mRNA of a rat osteosarcoma
cell line (ROS 17/2.8) was screened with a
nick-translated full length human bone/liver/
kidney alkaline phosphatase insert (7).
Relatively nonstringent conditions were used for
screening, to overcome sequence diversity between
species: hybridization was carried out with 6xSSC,
0.1% SDS at 42°C and filters were washed at 55°C
in 2xSSC, 0.1% SDS. Two independent clones
designated RAP I (600 bp) and RAP II (500 bp) were
isolated after screening 280,000 plaques, using
the nick-translated human alkaline phosphatase
probe. Sequence analysis showed that both cDNAs
contain poly A tails. The longer clone (RAP I)
was further characterized and its restriction map
and corresponding region in the human bone/liver/
kidney alkaline phosphatase is shown in Fig. 1.
Arrows indicate the direction of dideoxy
sequencing. DNA sequencing data showed that RAP I
corresponds to the 3' untranslated region of human
cDNA and has approximately 55% homology to human
alkaline phosphatase cDNA. Approximately 100 base
pairs at the poly adenylation site of the human
alkaline phosphatase are absent in RAP I as
indicated in Fig. 1.

Fig. 1.

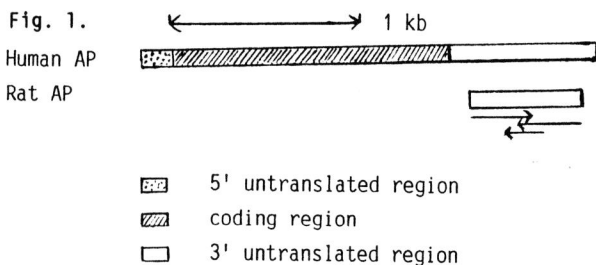

Human AP
Rat AP

▨ 5' untranslated region
▨ coding region
▢ 3' untranslated region

II. MEASUREMENT OF STEADY STATE mRNA LEVELS
USING RAT AP PROBE. Expression of alkaline
phosphatase in osteoblasts increases during
differentiation and can be modulated by several
agents such as glucocorticoids, retinoic acid and
1,25 vitamin D. Rat AP cDNA was used to measure
the steady state level of mRNA in ROS 17/2.8 cells
stimulated by dexamethasone and in two other
osteoblastic cell lines, ROS 2/3 and UMR 106. For
Northern analysis total RNA was isolated from
tissue culture cells by the proteinase K method in
the presence of vanadyl complex (8); 10 μg of
total RNA were separated on formaldehyde
denaturing gel; transferred on to nylon membranes;
and hybridized with oligo-labelled insert from RAP
I. As shown in Fig. 2, RNAs from osteoblastic
cells, ROS 17/2.8, ROS 2/3 and UMR 106 hybridize
to the alkaline phosphatase mRNA (2.4 kb), while
RNA from the non-osteoblastic cells ROS 25/1 does
not. This band was seen also in poly A RNA
obtained from rat liver where alkaline phosphatase
is more abundant than in other tissues.
Dexamethasone induction of alkaline phosphatase
was accompanied by a rise in message level in a
dose-dependent manner and can be attributed to
enhanced transcription (Fig. 2). Among these
samples the specific activity of the alkaline
phosphatase enzyme showed a close correlation with
the mRNA level measured by Northern blots.

Fig. 2.

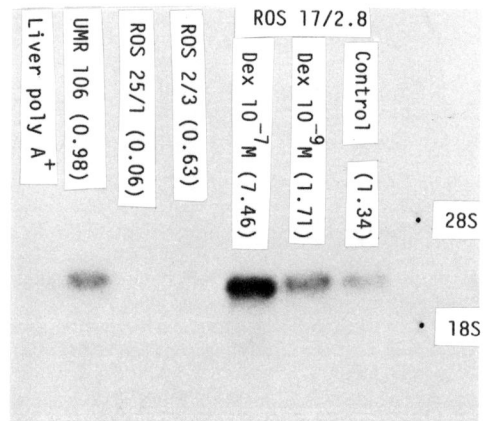

Numbers in parentheses indicate AP activity
(μmole/min/mg protein).

REFERENCES

1. Goldstein, D., Rogers, C.E. and Harris, H.
   (1980) Proc. Natl. Acad. Sci. 77, 2857-2860.
2. Slaughter, C.A., Cosco, M.C., Cancro, M.P. and
   Harris, H. (1981) Proc. Natl. Acad. Sci. 78,
   1124-1128.
3. Henthorn, P., Knoll, B., Raducha, M.,
   Rothblum, K.N., Slaughter, C., Weiss, M.,
   Lafferty, M.A., Fisher, T. and Harris, H.
   (1986) Proc. Natl. Acad. Sci. 83, 5597-5601.
4. Kam, W., Clauser, E., Kim, Y.S., Kan, Y.W. and
   Rutter, W.J. (1986) Proc. Natl. Acad. Sci. 82,
   8715-8719.
5. Millan, J.-L. (1986) J. Biol. Chem. 261,
   3112-3115.
6. Wuthier, R.G. and Register, T.C. (1985) The
   Chemistry and Biology of Mineralized Tissues
   (W.T. Butler, ed.) pp. 114-124, EBSCO.
7. λgt11 library was kindly provided by Dr.
   Heinegard.
8. Greenberg, M.E. and Ziff, E.B. (1984) Nature
   311, 433-437.

CHARACTERIZATION OF THE ets-2 GENE IN DROSOPHILA MELANOGASTER

L.J. Pribyl[1,3], D.K. Watson[1], M.J. McWilliams[2], R. Ascione[1], and T. Papas[1].
[1]NCI-FCRF, LMO; [2]PRI,FCRF; Frederick, MD 21701 and [3]Georgetown University, Department of Biochemistry, Washington, D.C. 20005.

Drosophila melanogaster provides a unique system for the study of genes due to the extensive biochemical, genetic, and developmental body of knowledge that has accumulated over the years. Many genes have been positioned genetically and some have been localized molecularly on chromosomes. These factors make this an attractive system for the study of cellular sequences related to viral oncogenes, based on the idea that at least some of these genes might have been conserved due to the key nature of their function.

Avian and mammalian ets genes.

Avian and mammalian organisms have been found to contain cellular sequences that are homologous to the ets-region from the avian retrovirus, E26. This virus is a replication-defective leukemia virus that contains two cell derived sequences, myb and ets, within its transforming protein, p135 (1). The ets-region consists of at least two domains, ets-1 and ets-2. These domains are separated and have been shown to be on different chromosomes in human, mouse, and cat and both loci direct the synthesis of unique mRNAs (2,3). In contrast to the mammalian ets genes, the ets-1 and ets-2 domains are contiguous as a single locus in the avian gene, which encodes for a single predominant message (4). Thus, to help understand the evolutionary development of these domains, this study has been undertaken to search for and characterize the sequences related to the ets domains in Drosophila.

Identification of a Drosophila ets gene.

Sequences related to ets have been detected by Southern blot analysis after digestion of genomic DNA from Schneider cell line #2 with various restriction enzymes, and hybridized with a v-ets probe.

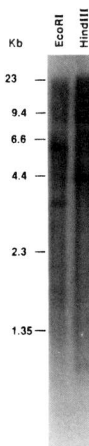

Fig. 1. DNA was isolated from Schneider Cell Line #2 and digested to completion with the indicated enzyme, separated on 1% agarose, and the DNA immobilized on nitrocellulose, and hybridized under low stringency (35% formamide, 5X SSC, 32°C) to the v-ets probe, E1.28. Under these conditions a predominant HindIII bank fo 4.4kb was detected.

To enrich for the presumptive Drosophila ets gene, Hind III resistant fragments of 4-6kb were isolated and used to construct a partial library in λ Charon 28. An ets positive phage (λ C21) containing 4.4kb of DNA was isolated and has been characterized by restriction endonuclease mapping. Partial sequence analysis confirms that the homology initially identified by Southern analysis is related to v-ets. The two regions of homology present in this clone are separated by a presumptive intron, having the proper consensus splice donor and acceptor sequences, maintaining the viral homologous open reading frame. This region corresponds to ets-2, which has been defined as the 3' (carboxy) end of the v-ets gene. A structural comparison of v-ets, with avian, human, and Drosophila c-ets genes is summarized in the diagram below.

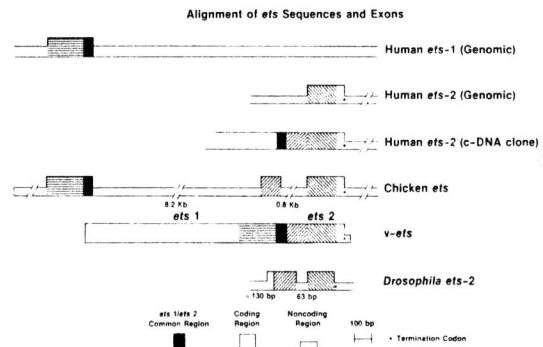

There is greater than 90% conservation of the predicted amino acid sequence, when the Drosophila ets-2 gene is compared with either v-ets or human ets. This degree of homology is greater than that of any other characterized Drosophila oncogene-related sequence.

Expression of this ets-related sequence has been detected in Schneider cells, and is being searched for in the major developmental stages of an Oregon R strain. In situ chromosomal mapping studies are also in progress.

REFERENCES

1. Nunn, M.F., Seeburg, P.H., Moscovici, C., and Duesberg, P.H. (1983) Nature 306, 391-395.

2. Watson, D.K., McWilliams-Smith, M.J., Kozak, C., Reeves, R., Gearhart, J., Nunn, M.F., Nash, W., Fowle III, J.R., Duesberg, P.H., Papas, T.S., and O'Brien, S.J. (1986) Proc. Natl. Acad. Sci. U.S.A. 83, 1792-1796.

3. Watson, D.K., McWilliams-Smith, M.J., Nunn, M.F., Duesberg, P.H. O'Brien, S.J., and Papas, T.S. (1985) Proc. Natl. Acad. Sci. U.S.A. 82, 7294-7298.

4. Watson, D.K., McWilliams-Smith, M.J., Flordellis, C.S., and Papas, T.S. (1985) Advances in Gene Technology: Molecular Biology of the Endocrine System. 158-159.

ISOLATION AND EXPRESSION OF GENES WITH HOMOLOGY TO THE TYROSINE KINASE DOMAIN OF VIRAL ONCOGENES DURING SEA URCHIN EMBRYOGENESIS

E. William Radany

Marine Research Laboratory Battelle, Pacific Northwest Laboratories, Sequim, WA 98362

(I) INTRODUCTION. Phosphorylation of tyrosine residues by tyrosine-specific protein kinases has been recently implicated as being a key regulatory step in the regulation of cellular proliferation. Increased tyrosine protein kinase activity has been associated with cell transformation induced by several retroviruses (1), early embryogenesis (2) and with the action of several hormones or growth factors (3).

Previous work has demonstrated that tyrosine protein kinase activity of sea urchin eggs increases following fertilization and continues to increase during early embryonic development (4-6). Antibodies raised against a synthetic peptide based upon the phosphorylation site of SRC interact specifically with a M_r 58 K dalton protein from sea urchin embryo's (7) suggesting that proteins with homology to viral oncogenes play a role during sea urchin embryogenesis.

The ability to immunologically identify tyrosine protein kinases suggests that tyrosine kinase genes could be detected by hydridization with DNA probes isolated from the conserved kinase region. Using this approach, we have isolated cDNA and genomic sequences that likely represent additional members of the tyrosine kinase gene family.

(II) METHODS. Construction of cDNA and Genomic Libraries. Total RNA from various stage sea urchin (Strongyloceatrotus purpuratus) embryo's was isolated by the guanidinium thiocyanate-çesium chloride cushion method (8) and poly (A)$^+$ RNA purified by two passages over an oligo (dt) column. Double-stranded cDNA was sized on a 5 - 20% sucrose density gradient and fractions containing cDNA greater than 1.0 Kb were combined and used to construct a λgt11 cDNA library.

Sea urchin (Strongylocentrotus purpuratus) sperm high molecular weight DNA (>200Kb) was isolated as previously described (9). Partial Sau 3A1 digest of high molecular weight DNA was sized on a 10-40% sucrose density gradient and fragments between 15-20 Kb combined and cloned into the BamHI site of λL47.

Screening of cDNA and Genomic Libraries. A portion of the V-fps gene (pBR-F04) (10) which shares homology with several viral tyrosine kinase genes was used as a probe to screen cDNA and genomic libraries using the method of Benton and Davis (11).

(III) RESULTS AND DISCUSSION. Initial screening of the cDNA library at moderate stringency resulted in the isolation of three clones. One clone found positive was rescreened and plaque purified. From the plaque purified clone, a 1.8 Kb cDNA insert was isolated and subcloned into the plasmid pUC8 (pSUTK-1). A restriction map and partial DNA sequence analysis were generated. The 1.8 Kb fragment revealed an open reading frame coding for amino acids with homology to several cellular tyrosine kinases.

The plasmid pSUTK-1 was then used as a probe for the screening of RNA from various stage sea urchin embryo's. Northern blots revealed a hybridizing band at 2.3 Kb. The presence of this transcript was developmentally regulated with maximal intensity of the band occurring at the late gastrula stage.

A Strongylocentrotus purpuratus genomic library was screened at high stringency using pSUTK-1 as a probe. Five separate clones were isolated after an initial screening, plaque purified and subcloned.

Three of the clones appear to be overlapping based on restriction mapping and hybridization experiments.

Based on these observations, we conclude that sea urchins express genes related to cellular and viral tyrosine kinase genes and the expression of these genes appear to be developmentally regulated.

REFERENCES

(1) Hunter, T. and Sefton, B.M. (1980) Proc. Natl. Acad. Sci. U.S.A. 77,1311-1315.
(2) Dasgupta, J.D., and Garbers, D.L. (1983) J. Biol. Chem. 258,6174-6178.
(3) Ushiro, H. and Cohen, S. (1980) J. Bio. Chem. 255,8363-8365.
(4) Kamel, C., Veno, P.A. and Kinsey, W.H. (1986) Bioch. Biophys. Res. Comm. 148, 349-355.
(5) Satoh, N. and Garbers, D.L. (1985) Dev. Bio. 111, 515-519.
(6) Ribot, H.D., Eisenman, E.A. and Kinsey, W.H. (1984) J. Biol. Chem. 259, 5333-5338.
(7) Radany, E.W. (1986) Submitted for publication.
(8) Glisin, J., Crkvenjakov, R. and Byus, C. (1974) Biochem. 12, 2633-2637.
(9) Cohn, R.H., Lowry, J.C. and Kedes, L.H. (1976) Cell 9, 147-161.
(10) Shibuya, M., Wang, L.-H. and Hanafusa, L.H. (1976) J.Virol. 42,1007-1016.
(11) Benton, W.D. and Davis, R.W. (1977) Science 196,180-182.

GENE EXPRESSION DURING EMBRYOGENESIS AND TUMOR FORMATION IN XIPHOPHORINE FISH

S.M. Robertson[1], F. Raulf[1], W. Maeueler[1], G. Raivich[2], and M. Schartl[1]

1. Molec. Embryol. Lab., Genecenter, MPI for Biochem. , 8033 Martinsried, F.R.G. 2. Dept. of Neuromorphology, M.P.I. for Psychiatry, 8033 Martinsried, F.R.G.

The live-bearing Xiphophorine fish, the platyfish and the swordtail, offer unique advantages as a vertebrate system for studying embryogenesis, as well as tumorigenesis. Firstly, a particular cell lineage, the melanocyte-responsible for population-specific spot patterns- is readily observable macroscopically in the skin of the fish. Secondly, embryos from an early stage (i.e. early organogenesis) can be cultured *in vitro* for observation and/or manipulation for the duration of embryogenesis and thereafter raised to adulthood. Thirdly, crosses between particular platyfish and swordtail populations results, in a Mendelian manner, in the appearance of a spontaneous malignant melanoma, demonstrating the participation of a particular genetic locus, termed Tu, in this process (reviewed in ref. 1). Here we briefly present our results pertaining to three aspects of this system.

(1) C-ONC EXPRESSION DURING EMBRYOGENESIS AND TUMOR FORMATION

c-src expression, as measured by either tyrosine-kinase activity or by Northern blot analysis, peaks during the organogenesis stages of embryonic development and thereafter decreases to the low level found in adult fish (2). c-src tyrosine-kinase activity also displays a tissue-specificity, with high levels in embryonic brain and muscle, whereas in the adult activity remains high in the brain but is barely detectable in muscle. *In-situ* hybridizations, utilizing a Xiphophorus c-src genomic clone, confirms the localization of c-src mRNA to neural tissue in adult fish (grey matter of the brain and the neuronal layers of the retina), and *in-situ* hybridizations are presently underway to determine tissue and temporal distribution of c-src mRNA in the developing embryo.

c-src expression, as measured by immunoreactive tyrosine-kinase activity, is coordinately elevated in the melanomas, depending upon tumor severity (3). This result has been confirmed at the mRNA level, and we are presently persuing this coordinated regulation between tumor severity (i.e. degree of Tu activation- see ref. 1) and c-src expression by utilizing fish strains carrying chromosomal abnormalities (deletions/translocations) known to affect the Tu locus. These analyses to date have shown that Tu and c-src are not one and the same locus, which therefore implies a more complex interaction between the two genes.

c-sis mRNA expression as measured by Northern blot analyses demonstrates a dramatic increase in 5 day-old neonates (suggesting a possible function in maturation of the immune system). As well, a complex pattern of differential expression during embryogenesis of the three mRNA species detected by a v-Ha-ras probe has been detected.

These data, therefore, include the first report of c-onc mRNA detection by *in-situ* hybridization in fish, as well as the first localization of c-src mRNA by *in-situ* hybridization in adult neural tissue. For *in-situ* hybridizations, the fish embryo presents certain advantages over other vertebrate embryos in that the embryo is culturable in vitro to term, and, although the embryo contains a large amount of yolk, the embryo proper develops above this bulk of yolk, and is physically separable from it, avoiding one of the major obstacles to efficient *in-situ* hybridization.

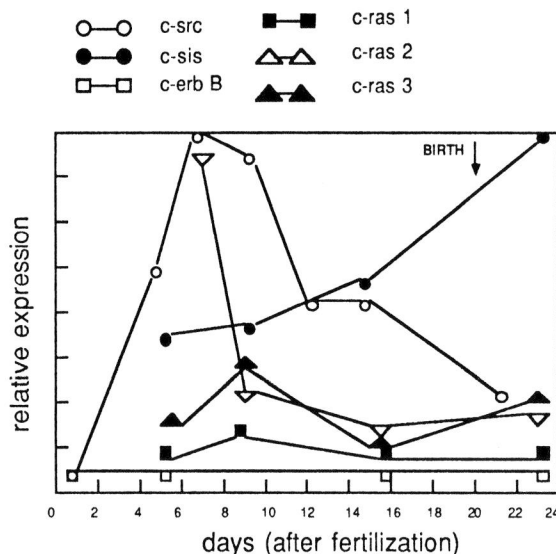

Fig. 1. PROTO-ONCOGENE EXPRESSION DURING EMBRYONIC AND NEONATAL DEVELOPMENT OF *XIPHOPHORUS HELLERI*

o—o c-src   ■—■ c-ras 1
●—● c-sis   △—△ c-ras 2
□—□ c-erb B ▲—▲ c-ras 3

(2) EXPRESSION OF HOMEOBOX-CONTAINING GENES IN XIPHOPHORUS

A number of homeobox-containing genes have been isolated from a Xiphophorus genomic library. These genes demonstrate several characteristics of homeobox-containing genes of other systems in that they appear to be clustered in the genome (eg. as in Drosophila, Xenopus, mouse, man). Three Xiphophorus homeobox sequences have been detected within two overlapping clones encompassing at least 18 kb of genomic DNA. For the reasons outlined above, the Xiphophorus embryo is an excellent system for the study of the expression of these genes. Northern blot and *in-situ* hybridization analyses are presently underway in an attempt to define the role(s) of these genes in vertebrate embryogenesis.

(3) APPROACHES TO CLONING TU

A genetic model involving negative regulation of Tu has been proposed (1)- i.e. activation as an oncogene through deregulation- and therefore the cloning of such a gene and the analysis of its mode of regulation (and deregulation) are of particular importance.

The following bio-assay exists for Tu : when genomic DNA from a particular platyfish population is microinjected into the neural-crest region of swordtail embryos, a small percentage of these embryos later present a number of dermal pigmented cells of a type characteristic of the melanoma (4). We have prepared a cosmid gene library from such a platyfish population and are initiating embryo injections. The library will be sub-divided and sub-divisions assayed by way of this micro-injection protocol ('sib-selection'; 'zoom-cloning'). This would represent a novel approach towards cloning a gene involved in the determination/differentiation of the pigment cell lineage, as well as being involved in realization of the neoplastic potential within this cell lineage.

REFERENCES

(1) Anders, F., Schartl, M., Barnekow, A., and Anders, A. (1984) Adv. Cancer Res. 42. 191-275

(2) Schartl, M. and Barnekow, A. (1984) Devel. Biol. 105, 415-422.

(3) Schartl, M., Barnekow, A., Bauer, H., and Anders, F. (1982) Cancer Res. 42, 4222-4227.

(4) Vielkind, J., Haas-Andela, H., Vielkind, U., and Anders, F. (1982) Mol. Gen. Genet. 185, 379-389.

PROGRAMMED ALTERNATIVE SPLICING IN THE GENERATION OF Ly-5 ISOFORMS

Y. Saga, J.-S. Tung, F.-W. Shen, and E.A. Boyse

Memorial Sloan-Kettering Cancer Center, NY NY 10021

The mouse Ly-5 system is expressed by hematopoietic cells and comprises a series of glycoprotein isoforms that typify different hematopoietic cell lineages. The 200 kDa isoform of T cells (T200) and the 220 kDa isoform of B cells (B220) differ in peptide composition.

Ly-5 cDNA (Ly-5-68) for T200 was cloned (1) and sequenced (2). With this cDNA as a probe, RNA transfer blotting showed that B220 mRNA of Ly-5 is larger than T200 mRNA (1), and S1 protection mapping suggested that this size difference represents an extra B cell sequence at the 5' end of B220 mRNA following a leader sequence. This is supported by study of B cell cDNA clones generated by extension on B220 mRNA from an oligonucleotide primer synthesized on T cell cDNA near the putative extra B220 sequence.

(I) A 5' region of Ly-5 RNA distinguishing T cells from B cells and macrophages (Mϕ). In RNA transfer blotting with a pLy-5-68 probe, B220 Ly-5 RNA (~5 kb) is 0.3 kb longer than T200 Ly-5 RNA, which accords with the difference in protein size (190 kDa vs 160 kDa) (1).

To identify mRNA differences between T200 and B220, nine probes for S1 mapping were prepared from subclones of pLy-5-68, spanning pLy-5-68 from 5' to 3' (Fig. 1). Each was end-labeled and hybridized with total RNA of T cells (leukemia ISL-57; isoform T200), B cells (leukemia I.29; isoform B220) and macrophages (PU-5 cell line; isoform 205 kDa) at optimal temperature for each probe. After S1 digestion, DNA/RNA hybrids were electrophoresed on 1.5% agarose gels. Eight of the nine probes did not distinguish between the three sources of RNA. With the remaining probe, representing 705 nucleotides from the 5' end to the first downstream BamHI site of pLy-5-68, the protected fragment (PF) for RNA of I.29 and PU-5 cells was shorter than that of ISL-57.

To pinpoint the difference in sequence, a shorter probe (288 nucleotides, from the 5' end of pLy-5-68 to the first downstream XbaI site) was derived and PFs were analyzed on 6% polyacrylamide-7M urea DNA sequencing gels. At 52°C (Fig. 2B), ISL-57 (T200) RNA gave a complete PF of 288 nucleotides whereas I.29 (B220) and PU-5 RNAs gave a shorter PF of 230 nucleotides. I.29 and PU-5 cells also yielded a 288 PF, faint at 52°C but intense at 42°C (Fig. 2A).

These data signify that B220 Ly-5 mRNA differs from T200 mRNA by an extra sequence at about 58 nucleotides downstream from the 5' end of pLy-5-68, ie immediately following a putative leader sequence (Figs. 1 and 2).

(II) Analysis of B cell cDNA clones isolated from a primer-extended library. Since B220 mRNA appears to have an extra sequence, oligonucleotide primer was synthesized at 150 bp downstream from the breakpoint, and hybridized with B220 mRNA (I-29 B cell leukemia) to yield an extended cDNA library in Puc vector.

Screening with T200 cDNA probe, positive clones were isolated, restriction mapped, and three types of clones (pLy-5B-11, 12, 51) studied and sequenced. All three clones share the same (extra) B220 exon (Fig. 3). Immediately upstream, clone B-11 has the same leader sequence as T200,

B-12 has an intron (probably from incomplete splicing) and B-51 has an additional 147 bp sequence which we believe is another B220-specific exon. The reading frame for the B-11 B220 sequence (47 amino acids predicted) and for the more extensive B-51 sequence (47 and 49 amino acids) is in phase with T200 mRNA.

(III) Alternative splicing generates the isoform-related diversity of Ly-5 mRNA. Screening of a genomic DNA library with pLy-5-68 yielded ten overlapping clones. Restriction mapping of overlaps showed no discordant restriction site, consistent with a single Ly-5 gene spanning at least 60 kb. The Ly-5 genomic map thus constructed showed ten EcoRI sites. Cellular DNA was digested with EcoRI and Southern blotted with three cDNA fragments representing the entire sequence of pLy-5-68 as probes. The sizes and map positions of the hybridized fragments were as expected.

Conclusion: These data are consistent with the generation of Ly-5 protein isoforms by alternative splicing of the primary transcript in accord with the discrete programs of Ly-5$^+$ cells of different hematopoietic cell lineages.

Fig. 2. S1 mapping with a 5' end probe of the T cell cDNA clone pLy-5-68, showing different protected fragments (PFs) for T cells as compared with B cells and macrophages. T cells give the complete PF of 288 nucleotides. B cells (and macrophages) give PFs of 288 and 230 (relative proportions depending on hybridization temperature). The 230 PF is attributed to weak duplex bonding near the loop created by the putative B-cell typical extra B220 sequence (see below in the Figure), allowing S1 nuclease to cut through (more so at 52°C).

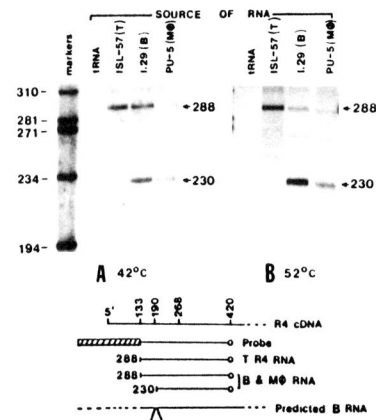

Fig. 3. Comparison between T200 cDNA (pLy-5-R4) and B220 cDNA clones pLy-5-B11, -B12 and -B51). All 3 B220 cDNA clones have a common B-cell-typical exon. Clone B-51 has a more extensive B-cell-typical exon sequence (hatched box).

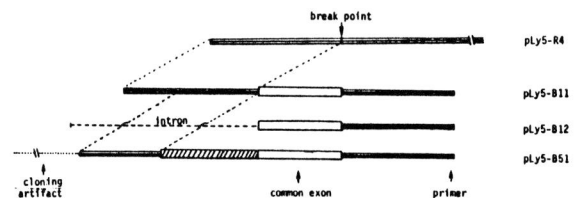

Fig. 1. Restriction maps of pLy-5-68 and pLy-5-R4. Clone pLy-5-R4, which contains the complete coding region of T200 Ly-5 protein, was isolated by rescreening the T cell cDNA library with a 5' probe of pLy-5-68.

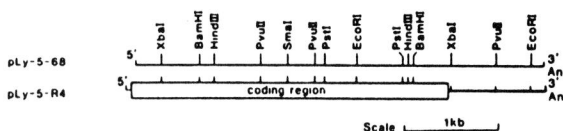

REFERENCES

(1) Shen, F.W., Saga. Y., Litman, G., Freeman, G., Tung, J.S., Cantor, H., and Boyse, E.A. (1985) Proc. Natl. Acad. Sci, USA 82, 7360-7363.

(2) Saga, Y., Tung, J.S., Shen, F.W., and Boyse, E.A. (1986), Proc. Natl. Acad. Sci., USA 83, 6940-6944.

DIFFERENTIAL INDUCTION OF HEPATIC ESTROGEN RECEPTOR AND VITELLOGENIN GENE TRANSCRIPTION IN XENOPUS

D. R. Schoenberg, A. T. Riegel, S. C. Aitken and M. B. Martin

Department of Pharmacology, Uniformed Services University of the Health Sciences, Bethesda, MD 20814-4799

INTRODUCTION. Estrogen receptor is present in embryonic Xenopus liver, however estradiol is unable to induce either vitellogenin (vit) synthesis or elevation in estrogen receptor content until larvae reach stage 62 in metamorphosis (1). In adult male Xenopus, estradiol administration results in the induction of hepatic estrogen receptor from 500 sites/cell to 2500 sites/cell concurrent with the de novo transcriptional induction of the vitellogenin genes. A single dose of estradiol is sufficient for a differentiative event leading to the prolonged acquisition of elevated estrogen receptor content and a vitellogenic response to a second dose of hormone that is indistinguishable from that seen in female animals (2). These observations have led to the concept that elevated levels of estrogen receptor facilitate the transcriptional induction of the vit genes. The present study used the potent estrogen receptor antagonist 4-hydroxytamoxifen (OHT) to examine this relationship in greater detail.

MATERIALS AND METHODS. Changes in vit gene transcription were measured by a transcription 'run-on' assay (3). To measure hepatic estrogen receptor (ER) an exchange assay was developed in which ER was stabilized by adsorption onto hydroxylapatite. Extracts for ER assay were prepared in buffers containing 0.5 M KCl to extract this protein and 1 mM dithiothreitol (DTT) to inactivate a middle-affinity cytoplasmic estrogen binding protein (EBP). Extracts for EBP activity were prepared in buffers lacking KCl and DTT.

RESULTS AND DISCUSSION. We have previously demonstrated (4) that OHT is the active form of the antiestrogen tamoxifen in Xenopus and micromolar concentrations of both drugs are maintained for prolonged periods of time in serum after a single dose. In that same report we demonstrated that OHT possesses weak agonist activity as measured by its ability to induce low levels of vit mRNA, however it is a potent antagonist of vit mRNA induction by estradiol. The present study extends these observations to the effects of OHT on the induction of ER and vit gene transcription.

Xenopus liver ER has the same affinity ($K_D = 2 \times 10^{-9}$M) for E and OHT. While unlabeled E is readily displaced by [^{3}H]E at all temperatures studied, OHT cannot be displaced from the receptor at temperatures below 15°C. Under these conditions receptor rapidly degrades in solution. To overcome this problem we developed a solid-phase ER exchange assay in which ER-containing liver extracts were adsorbed onto hydroxylapatite prior to displacement of bound ligand with [^{3}H]E. Under these conditions full ER binding activity is maintained for >24 hr at 22°C.

The data in Figure 1 demonstrate the effects of E and OHT on the induction of hepatic ER. A single dose of OHT results in a net loss of ER, from 800/cell to 250/cell. E alone causes the induction of ER from 800/cell to 2700/cell. Surprisingly, administration of E after OHT results in induction of ER from the suppressed level of

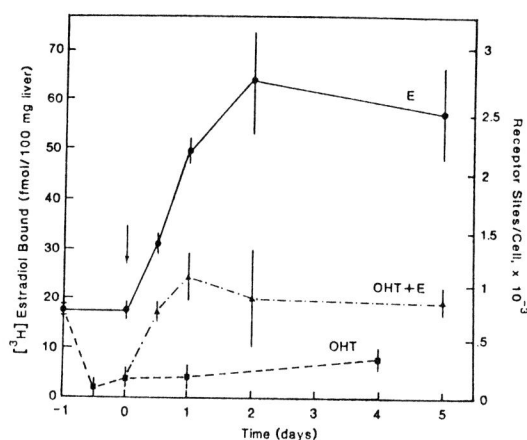

Figure 1. Effect of E and OHT on ER induction. Animals were injected with OHT on day 0 and E on day 1 (arrow)

250/cell back to that seen in control animals (1000/cell). To determine the functional nature of these receptor complexes vit gene transcription was examined. Although E alone maximally induced vit transcription, pretreatment with OHT completely blocked this response. By this criterion OHT rendered ER functionally inactive.

The ability to up-regulate ER under conditions in which ER is functionally inactive suggested that a moiety other than ER mediates ER induction. One possible candidate for this is EBP, a middle-affinity cytoplasmic estrogen-binding protein present only in estrogen-responsive tissues in Xenopus (5). Competitive binding experiments demonstrated that OHT has no effect on E binding to EBP. Therefore, EBP is a likely candidate for a protein that regulates ER induction by E.

The fact that OHT caused a net loss in hepatic ER content when administered to hormonally-naive animals suggests a complex interplay between factors that maintain ER levels and factors responsible for ER induction in response to hormone. Our data are consistent with a model in which the elevation of ER content is a necessary, but not sufficient, component of the massive induction of vitellogenin gene transcription caused by E.

REFERENCES

(1) May, F.E.B. and Knowland, J. (1981) Nature 292, 853.
(2) Brock, M.L. and Shapiro, D.J. (1983) J. Biol. Chem. 258,5449.
(3) Martin, M.B., Riegel, A.T. and Schoenberg, D.R. (1986) J. Biol. Chem. 261,2355.
(4) Riegel, A.T., Jordan, V.C., Bain, R.R. and Schoenberg, D.R. (1986) J. Steroid Biochem. 24,1141.
(5) Hayward, M.A. and Shapiro, D.J. (1981) Dev. Biol. 88,333.

ACKNOWLEDGMENTS

This work was supported by grants from the NIH, the Uniformed Services University of the Health Sciences, and the Damon Runyon-Walter Winchell Cancer Fund.

98

TRANSCRIPTION AND REGULATION OF EMBRYONIC MYOSIN LIGHT CHAIN GENE EXPRESSION

M.A.Q. Siddiqui, A. M. Zarraga, K. Danishefsky, M. Noe, and C. Mendola

Roche Institute of Molecular Biology, Roche Research Center, Nutley, N.J. 07110

Myosin polypeptides, the major constituents of muscle proteins, are expressed differentially in different cells and even in the same cell at different stages of development. Recent work on characterization of muscle specific cloned DNAs has revealed that myosin polypeptides are encoded by multigene families, the members of which are tissue specific and developmentally regulated (1). The multigene family of cardiac myosin is particularly attractive for elucidation of the mechanisms underlying control of embryonic muscle gene expression, since cardiac muscle formation is an early event during fetal cell development. Furthermore, the expression of cardiac myosin isoforms changes not only during normal development, but also in response to various physiological and pathophysiological stimuli that results in variations of muscle phenotypes. In most vertebrates, the cardiac myosin light chain (MLC) is made of two isoforms; one is classified as regulatory, phosphorylatable MLC_2, and the other non-phosphorylatable, alkali MLC_1. The physiological role of MLCs is not well understood, although their involvement in regulation of myosin ATPase activity and in myosin/actin interaction has been the subject of a few studies (2). It is, however, clear that distinct isoforms of MLC are synthesized in different muscle types and that their expression in normal and diseased muscle cells is controlled by a complex program of gene regulation.

In order to understand the molecular mechanisms involved in control of tissue and stage specific MLC gene expression, we isolated and characterized recombinant clones containing cDNA and genomic sequences corresponding to chicken and rat heart MLC_2 (3-5). Two recombinant phages, $\lambda LC5$ and $\lambda LC13$, encompassing the entire MLC_2 gene of chicken heart muscle, were characterized by partial nucleotide sequence analysis. Based on the primer extension analysis with a synthetic 20-mer corresponding to the 5'-leader sequence, the transcription initiation site on the gene was located. The gene promoter activity was demonstrated following transient expression of recombinant genomes containing the chicken upstream sequence fused to the bacterial chloramphenicol acetyltransferase (CAT) or to the rat preproinsulin II genes. Construction of one such recombinant plasmid, where rat preproinsulin gene is placed under the control of the putative chicken MLC_2 promoter, is shown in Figure 1. Transfection of Quail fibroblasts (QT35) cells with a CAT/MLC_2 recombinant and COS cells with a preproinsulin II/MLC_2 recombinant respectively resulted in an efficient expression of CAT and rat preproinsulin genes initiated from the chicken promoter. The appearance of preproinsulin mRNA was assayed by S1 nuclease protection method using a sequence specific DNA probe.

An examination of the 5'-flanking sequence of MLC_2 gene revealed the presence of two distinct inverted repeat sequences within the promoter region which can form remarkably stable hairpin like structures. Whether such sequences are biologically significant is being tested by introducing specific deletions into the upstream region for assaying the transcription activity of the gene. The upstream sequence of MLC_2 gene, unlike the coding sequence, shares little homology with the sequence at the same region of chicken skeletal muscle MLC_2 gene, suggesting that heart muscle gene promoter has diverged considerably from its counterpart in skeletal muscle.

Although many details remain to be established, the machinery responsible for gene transcription provides the cell with the means to modulate the pattern of RNA synthesis in response to specific signals. The binding of protein (and/or RNA) to DNA with the ability to discriminate between various DNA sequence elements in the promoter region and in other functionally important gene domains is an important step for modulation of gene expression. Identification of such factors and of the DNA sequence elements and their characterization and the knowledge on how muscle gene expression is controlled this way will help in understanding the performance of myofibrils in different physiological and pathophysiological states.

Fig. 1. Construction of expression vector containing rat preproinsulin II gene placed under the control of chicken MLC_2 gene promoter. For details see reference 5.

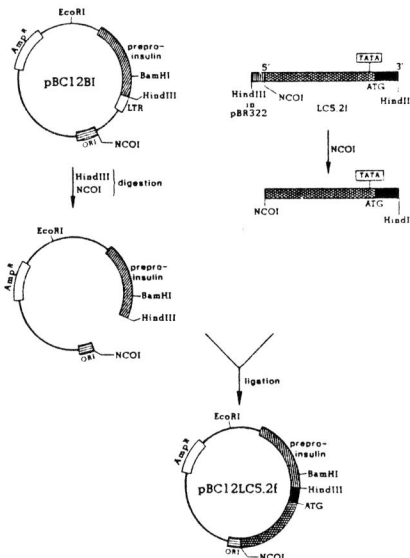

REFERENCES

(1) Pearson, M.L. and Epstein, H.F. (1982) in Molecular and Cellular Control of Cell Development, Cold Spring Harbor Laboratory, Cold spring Harbor, NY.

(2) Wagnor, R.D. and Weeds, A.G. (1977) J. Mol. Biol. 109, 455-473.

(3) Arnold, H.H., Kranskopf, M. and Siddiqui, M.A.Q. (1983) Nucleic Acids Res. 111, 1123-1131.

(4) Kumar, C.C., Cribbs, L., Delaney, P., Chien, K. and Siddiqui, M.A.Q. (1986) J. Biol. Chem. 261, 2866-2872.

(5) Zarraga, A.M., Danishefsky, K., Deshpande, A.K.,Nicholson, D., Mendola, C.and Siddiqui, M.A.Q. (1986) J. Biol. Chem. (in press).

MIDGESTATION HEMATOPOIETIC DEVELOPMENT: COMPARISON OF NORMAL PRECURSORS AND ABELSON MURINE LEUKEMIA VIRUS-TRANSFORMED CELLS

Michael Siegel, Charles Brown and Edward Siden

Department of Immunology and Medical Microbiology, University of Florida, College of Medicine

An important parameter in the analysis of hematopoietic cell development is the definition of specific growth factor requirements during the maturation of progressively restricted progenitor cells. We have identified 2 waves of hematopoietic precursors in the midgestation murine embryo between 8 and 15 days of gestation that differ in both growth factor requirements and developmental potential. We began these studies to probe for the early embryonic source of B lymphocyte progenitors. Melchers (1) originally identified cells of the murine placenta during midgestation that could develop in vitro into antibody secreting cells.

To begin to characterize these very early lymphoid cells, we infected placenta and embryo with Abelson murine leukemia virus (Ab-MuLV), a retrovirus with a known tropism for cells of the B lineage. The virus sensitive target cells (cfu v-abl) were quantitated by clonal growth in agarose (2) and were most frequent in the embryonic tissues at 10 days of gestation. We established continuous cell lines from these colonies and have found that although some of the lines express certain differentiative properties of fetal lymphoid cells, all of the lines express multiple characteristics of immature mast cells. To identify analogous normal cells in the murine embryo, we probed for mast cell progenitors that could proliferate in response to the multi-lineage hematopoietic growth factor interleukin 3 (IL-3). Surprisingly, the frequency of IL-3 responsive mast cell progenitors (cfu mast) did not coincide with the appearance of virus-sensitive cells in the embryo (Figure 1). The sequential appearance of the two populations (Ab-MuLV and IL-3 sensitive) indicate either an ordered acquisition of growth factor sensitivity in a single developing lineage or the presence of two discrete populations with already restricted developmental potential.

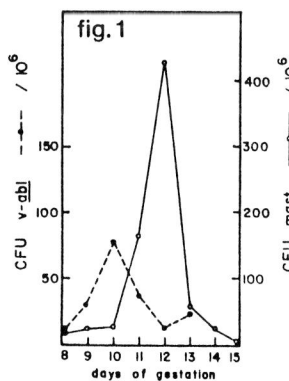

fig. 1

The capacity of the two populations of cells to develop into antibody secreting cell was analyzed by detecting embryonic immunolgobulin production after liquid culture in the presence of the B cell mitogen, lipopolysaccharide. The cells were isolated from F1 embryos at 10 or 12 days of gestation and cultured until the equivalent of 18 days of gestation. The cell culture supernatants were then analyzed using an ELISA specific for the paternal immunoglobulin genotype (IgM of the IgH-C^a allotype). Figure 2 shows the results of this competitive ELISA. IgM in the supernatant of cultured spleen cells of the paternal strain, CBA/J was able to effectively compete against solid phase paternal genotype Ig for binding of a rat monoclonal antibody (Bet 1) specific for that allotype (-O-). Spleen cell supernatant from the maternal strain C.AL 20 does not compete in this assay (-●-) nor does anything in the supernatant from cells isolated at 12 days of gestation from the placenta of (CBA/J x C.AL 20) F1 embryos (++●++). However, in confirmation of Melchers original studies, IgM in the

supernatant of cells isolated at 10 days of gestation was a product of the paternal genes of the embryo (--●--).

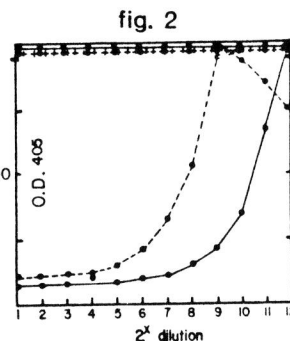

fig. 2

To further characterize the B cell progenitor of the first wave of the midgestation hematopoietic cells, we continued our analysis of the Ab-MuLV transformed lines from this period of gestation. We probed poly-A$^+$ RNA from a number of the lines with a ^{32}P-UTP-labeled probe that can detect transcription of unrearranged variable region genes. This product is thought to be characteristic of the earliest heavy chain gene-rearranging cells in the murine embryo (3). Two of the mast cell lines, 10P12-2 and 11P62-4 make this product, although at significantly lower levels than the pre-B cell line FLE1-4 (Figure 3).

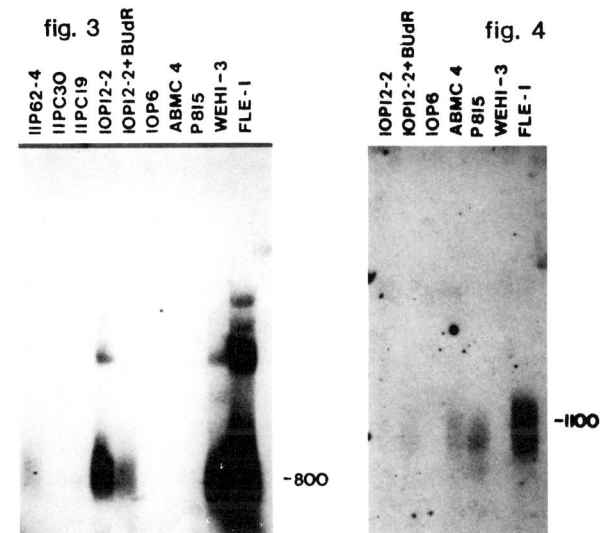

fig. 3

fig. 4

We also tested a number of known differentiation promoting agents for their ability to induce further B cell maturation in the mast cell lines. The halogenated pyrimidine, BUdR, can significantly increase the number of cells in the mast cell line 10P12-2.1 that express the B cell differentiation antigen B220. When we probed these cells with the monoclonal antibody RA2-3A1, we saw an increased number of B220$^+$ cells. Coincident with this induction was the detection of an RNA product of the gamma T cell antigen receptor gene (Figure 4). This may suggest an order of early B cell progenitor maturation of germline V$_H$ transcription, followed by C transcription.

In summary, we have defined two hematopoietic progenitor populations in the midgestation embryo. The virus-sensitive population may be early lymphoid cells and the response to the v-abl oncogene product or c-abl protooncogene product may be both ontogenetically and phylogenetically more primitive than the response to IL-3.

References:
1. Melchers, F. (1979) Nature 277:219-221.
2. Siegel, M., R. Horwitz, T. Morris, R. Divenere, and E.J. Siden. (1985) European J. of Immunol. 15:1136-1141.
3. Yancopoulos, G.D., and F.W. Alt. (1985) Cell 40:271-281.

100

TRANSFORMATION OF QUAIL HEMATOPOIETIC CELLS BY AVIAN ERYTHROBLASTOSIS VIRUS (ES4) RESULTS IN A PHENOTYPE DISTINCT FROM SIMILARLY INFECTED CHICKEN CELLS

M.L. Siegel, D. Franzini, and M.G. Moscovici
Dept. of Pathology, University of Florida, Gainesville, FL 32610 and Veterans Administration Medical Center, Gainesville, FL 32620

Avian erythroblastosis virus (AEV) is a replication-defective retrovirus which is capable of rapid induction of erythroblastosis and fibrosarcomas in susceptible chickens (1, 2) and of transformation of chicken erythroid cells and fibroblasts in vitro. Moscovici et al. (3) performed a comparative study of the oncogenic potential of AEV which indicated that the oncogenic response of Japanese quail (Coturnix coturnix japonica) was different from that of the chicken. In the quail, myeloblastic leukemia and erythroblastosis were induced by AEV; these observations were later confirmed by Morikawa et al. (4). In this report, we describe further characteristics of both in vivo- and in vitro-derived AEV-quail transformants and present evidence that the non-erythroid cells are (1) morphologically, biochemically, and histologically identical to mast cells and (2) synthesize large quantities of the erb A gene product but not the erb B gene product.

DERIVATION AND CHARACTERISTICS OF QUAIL CELL LINES

AEV-transformed quail cell lines were derived by both in vivo and in vitro infection. Five cell lines were obtained and all were found to be non-producers. Blood smears from selected infected quail exhibited proliferation of a population of cells initially identified as leukemic myelocytes as well as erythroblasts; in contrast to the quail, similarly infected chickens develop only erythroblastosis. Similar types of cells were observed in in vitro-infected cell cultures. Although the cultures initially contained erythroid and nonerythroid elements, prolonged propagation resulted in selection of only the latter as erythroid cells continued to differentiate. Upon more extensive scrutiny, however, we observed that some of the "myeloid" cells contained basophilic granules and morphologically resembled mast cells. Electron microscopy of the cells indicated scrolled granule structure and other details characteristic of mast cells described in other species (Figure 1). Furthermore, the granules are stained with acidic toluidine blue (Figure 2) as well as alcian blue, conventional histochemical indicators of mast cells. We also detected histamine in some of the cells by an isotopic-enzymatic microassay. The quantities of histamine in the cells was similar to that reported for mouse culture-derived and Abelson murine leukemia virus-infected mast cells (5) biochemically confirming the mast cell phenotype.

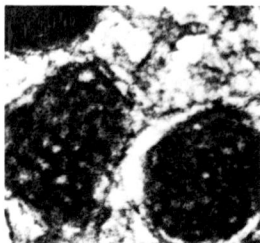

Fig 1. Electron micrograph of AEV-transformed quail cell.

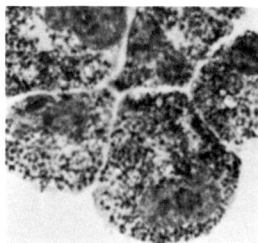

Fig 2. Light micrograph of AEV-transformed quail cell (toluidine blue, pH 2.5).

EXPRESSION OF V-ERB-A AND V-ERB-B GENE PRODUCTS

The genome of AEV contains two domains that are involved in oncogenesis. The product of the v-erb A locus, which is encoded by a genomic transcript, is a fusion protein ($p75^{gag-erb\ A}$) whose function is unclear: absence of the v-erb A gene product in mutant and engineered viruses is still permissive to transformation of chicken fibroblasts and erythroid cells, but such cells tend to escape the transformed phenotype of wild type AEV-transformed cells (6). The v-erb B gene product, on the other hand, appears to be necessary and sufficient for AEV-induced transformation of chicken fibroblasts and erythroid cells. The product of the v-erb B locus is initially translated as a protein of 61,000 daltons ($p61^{v-erb\ B}$) which is subsequently glycosylated into higher molecular weight forms ($gp65^{v-erb\ B}$, $gp68^{v-erb\ B}$, $gp74^{v-erb\ B}$) (7, 8, 9). Proteins of quail AEV-transformed and of chicken control cell lines were biosynthetically labeled with ^{35}S-L-methionine, lysed in detergent buffers, and immunoprecipitated with polyclonal anti-erb-A, anti-erb-B, and normal rabbit serum after the method of Privalsky et al. (9). Treatment of lysates of the AEV-transformed chicken cell line 6C2 confirmed previous reports of erb A and erb B gene product expression (Figure 3). In contrast similar treatment of quail yolk sac-derived and bone marrow-derived, AEV-transformed mast cells resulted in the observation of the erb A but not the erb B gene product. We are presently investigating the nature of the latter phenomenon at the proviral, transcriptional, and translational levels.

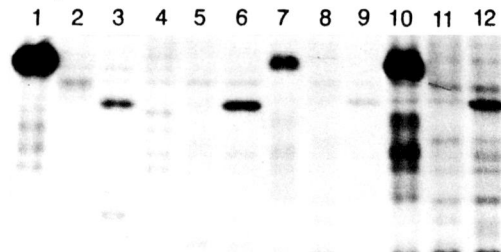

Fig 3. Autoradiograph of immunoprecipitated viral proteins. Lysates from 6C2 (Lanes 1, 2, 3), BM2 (4, 5, 6,), AEV-transformed quail yolk sac (7, 8, 9) and bone marrow (10, 11, 12) were adsorbed with anti-erb A (1, 4, 7, 10), anti-erb B (2, 5, 8, 11), and normal rabbit serum (3, 6, 9, 12).

REFERENCES
(1) Meyer, R.A., and Engelbreth-Holm, J. (1933). Acta Pathol. Microbiol. Scand. 10, 380-428.
(2) Graf, T., Royer-Pokora, B., Schubert, G.E., and Beug, H. (1976). Virology 71, 423-433.
(3) Moscovici, C., Samarut, J., Gazzolo, L., and Moscovici, M.G. (1981). Virology 113, 765-768.
(4) Morikawa, S., Yoshikawa, Y., Mizutani, M., Taniguchi, H., Morita, M., and Yamanouchi, K. (1984). Japan J. Med. Sci. Biol. 37, 105-116.
(5) Siegel, M.L., Horwitz, R.J., Morris, T.D., DiVenere, R.M., and Siden, E.J. (1985). Eur. J. Immunol. 15, 1136-1141.
(6) Frykberg, L., Palmieri, S., Beug, H., Graf, T., Hayman, M.J., and Vennstrom, B. (1983). Cell 32, 227-238.
(7) Privalsky, M.L., and Bishop, J.M. (1982). Proc. Natl. Acad. Sci. USA 79, 3958-3962.
(8) Hayman, M.J., Ramsey, G., Savin, K., Kitchener, G., Graf, T., and Beug, H. (1983). Cell 32, 579-588.
(9) Privalsky, M.L., Sealy, L., Bishop, J.M., McGrath, J.P., and Levinson, A.D. (1983). Cell 32, 1257-1267.

PROTO-ONCOGENES C-MYC, N-MYC AND C-SRC ARE DIFFERENTIALLY EXPRESSED IN F9 AND PCC7 EMBRYONAL CARCINOMA CELLS

J.Sumegi,# T.Sejersen,* N.R.Ringertz* and G.Klein#

#Department of Tumor Biology, *Department of Medical Cell Genetics, Karolinska Institute, S-104 01 Stockholm, Sweden

Recent studies have implicated the products of several proto-oncogenes in cellular differentiation processes. The c-src, the cellular homologue of the Rous sarcoma virus oncogene is expressed at high levels in differentiated neural tissues (1). The expression of proto-oncogenes c-myc and c-myb is diminished and the expression of c-fos is induced during differentiation of human promyelocytic leukemia cell line HL60 (1). Because of their potential to differentiate into a wide variety of cells, the embryonal carcinoma cells represent an interesting system to examine the regulation of proto-oncogene expression during terminal differentiation.

C-MYC AND N-MYC ARE EXPRESSED IN EMBRYONAL CARCINOMA (EC) CELLS. Both c-myc and N-myc are expressed at high levels in three undifferentiated EC cell lines (PCC3, PCC4 and F9), neither of which showed amplification of the two proto-oncogenes (2). The detection of N-myc transcripts in the F9 EC cells committed to differentiate into extra-embryonal endoderm and in two multipotent stem cell lines (PCC3 and PCC4) indicates that the N-myc is not simply a neuroectoderm specific gene. In PCC7, the fourth EC line included in this study, the c-myc gene was found to be suppressed as indicated by Northern-blot hybridization and nuclear run off experiments. It has been reported that mitotically and meiotically dividing germ cells have very few myc transcripts and appear to proliferate at least for a few divisions in the absence of c-myc transcription (3). It seems that the proliferation of certain cell types does not require the c-myc expression as a mandatory requirement. The N-myc was, however in PCC7 expressed at a level comparable to that of three other EC lines. The c-src was not expressed at a detectable level in any of the four undifferentiated EC lines studied. We determined the intracellular turnover of c-myc and N-myc mRNAs in proliferating F9 and PCC7 cells by following the decline in the steady state levels of both transcripts after addition of Actinomycin D. A quantitative analysis of the turnover kinetics of the two transcripts revealed half-life approximately 30 min for c-myc and 120 min for N-myc.

C-MYC AND N-MYC EXPRESSION IN PROLIFERATING EC CELLS. The effect of serum deprivation on c-myc and N-myc expression was studied by exposing subconfluent cultures of F9 cells to 1% fetal calf serum (FCS) for 48 hours. This treatment caused 84% reduction in the amount of 3H-thymidine incorporated into DNA during one hour pulse. The relative abundance of c-myc and N-myc also decreased. Actin and alpha-fetoprotein expressions were not affected. When the serum deprived F9 cells were stimulated by addition of 15% FCS, the c-myc expression sharply increased and reached peak level four hours after serum addition. The N-myc expression was moderately enhanced. Serum contains many growth factors, hormones and other growth supporting components. Therefore, stimulation of F9 cells by serum may conceal differences in regulation of c-myc and N-myc. Insulin and transferrin have been reported to induce proliferation of F9 cells (4). When serum free medium supplemented with insulin (1ug/ml) and transferrin (5ug/ml) were added to F9 cells a dramatic increase in the expression of c-myc was observed. The expression of N-myc gene was not stimulated. Nuclear run off transcription demonstrated that the level of c-myc transcription was higher in serum stimulated than in serum deprived F9 cells.

C-MYC AND N-MYC EXPRESSION IN DIFFERENTIATING EC CELLS. F9 cells can be induced to differentiate to either parietal or visceral endoderm (5). Parietal endoderm was formed from F9 cells cultured in the presence of retinoic acid (RA) and cAMP. The formation of parietal endoderm was accompanied by a reduction in the expression of both c-myc and N-myc genes. Northern-blot hybridization showed that the differentiation associated down-regulation of c-myc and N-myc followed different kinetics. The decline in the c-myc expression in F9 cells was the earliest change, proceeded both the cessation of DNA synthesis and the appearance of the differentiation specific alpha-fetoprotein gene expression. The decrease in the N-myc expression was not detectable before the first 72 hours after addition of RA and cAMP, suggesting that its decline possibly was a consequence of cell cycle changes.
Cells of the PCC7 EC line differentiate into cholinergic neurons after addition of RA and cAMP. The differentiation resulted in gross changes in cell morphology (2). EC cells growing as dispersed cells changed into aggregates of cells with massive neurite outgrowth. Differentiation into nerve cells resulted in a decrease in DNA synthesis. Densitometry of Northern-blots showed that the abundance of N-myc transcripts in total RNA was 85% lower in differentiated PCC7 cells than in proliferating undifferentiated PCC7 cells. The decline of the N-myc expression in differentiating PCC7 cells occurred before the phenotypic changes and much before the appearance of neurofilaments.

INDUCTION OF C-SRC EXPRESSION IN DIFFERENTIATING PCC7 CELLS. The formation of neurofilament positive cells of PCC7 resulted in an enhanced expression of the c-src gene. This induction was due to a transcriptional activation of the gene as indicated by data obtained by nuclear run off transcription assay.

(1) Müller,R.(1986) TIBS 11, 129-132
(2) Sejersen,T. Björklund,H. Sumegi,J. and Ringertz,N.R.(1986) J.Cell Phys. 127, 274-280
(3) Stewart,T.A. Bellve,A.R. and Leder,P. (1984) Science 226, 707-710
(4) Rizzino,A. and Crowley,C. (1980) Proc. Natl.Acad.Sci.USA 77, 457-461
(5) Hogan,B.L.M. Taylor,A. and Adamson,E. (1981) Nature 291, 235-237

BRAIN GLUTAMATE DEHYDROGENASE mRNA LEVELS ARE DECREASED IN AN ANIMAL MODEL OF HEPATIC ENCEPHALOPATHY

J. W. Thomas*, J. Whitman*, K. Mullen**, and C. Banner*
* Laboratory of Molecular Biology, NINCDS, NIH, Bldg. 36, Rm. 3D02, Bethesda, MD 20892
** Digestive Diseases Branch, NIADDKD, NIH, Bethesda, MD 20892

INTRODUCTION - Hepatic encephalopathy is a neurological disorder caused by liver dysfunction and generally associated with brain hyperammonemia. Ammonia is neurotoxic, but this fact alone does not seem adequate to explain the clinical encephalopathy. Proposed pathogenic hypotheses include the synergistic effects of ammonia with other neurotoxins such as mercaptans and fatty acids or alterations in brain neurotransmitters such as serotonin, dopamine, norepinephrine, or GABA (gamma-aminobutyric acid) (1).

The brain is not able to detoxify ammonia through the production of urea, but instead must fix ammonia via glutamine synthetase and glutamate dehydrogenase (GDH). These enzymes are also significant in the metabolism of glutamate and GABA, two of the most important amino acid neurotransmitters. It therefore seems plausible that hyperammonemia might interfere with brain function indirectly as well as via direct neurotoxicity.

To examine this possibility we have been studying the control of enzymes in the glutamate-glutamine-ammonia cycle at the level of gene expression. This brain pathway is thought to be compartmentalized (2). Neuronal cells synthesis the amino acid transmitters while glial elements may be responsible for transmitter inactivation via uptake followed by conversion to glutamine. Glutamine is transported back to the neuronal cell for re-synthesis of glutamate or GABA to complete the cycle. A cDNA for glutamate decarboxylase (GAD) has been previously isolated from cat brain (3). This enzyme which catalyzes the conversion of glutamic acid to GABA is believed to be a reliable marker for GABAergic neurons. Recently we isolated a cDNA from human brain for GDH (4). This mitochondrial enzyme which catalyzes the reversible reaction, alpha-ketoglutarate + NH_4^+ + NAD(P)H <---> glutamate + H_2O + $NAD(P)^+$, is thought to be localized predominantly in astroglial cells in brain. Using these cDNA's as a probes, we examined GAD and GDH mRNA levels in an animal model of hepatic encephalopathy in which ammonia levels are highly elevated. We now report that GDH mRNA levels are specifically decreased in the three brain regions examined in the HE animals, while GAD mRNA levels are not altered in these same areas.

MATERIALS and METHODS - The encephalopathy was experimentally induced in rats using a potent liver toxin, thioacetamide (5). Animals were injected with thioacetamide (300 mg/kg, i.p.) or normal saline on day one and day two, and sacrificed on day three, 24 h after the second injection. The brain tissue was removed, dissected, frozen immediately, and stored at -70° C until assayed.

RNA was isolated from cortex, hippocampus, and striatum using the guanidine isothiocyanate procedure of Chirgwin et al. (6). The RNA was fractionated by electrophoresis through formaldehyde-agarose gels (7), blotted to nitrocellulose, and probed with either GAD or GDH cRNA. The GDH RNA probe was prepared by in vitro transcription from a Sma I/EcoR V fragment of the modified cDNA subcloned into pBLSCRB (Vector Cloning Labs), a plasmid bearing promotors for T3 and T7 RNA polymerases (8). The GAD probe was prepared by in vitro transcription from cDNA subcloned into pSP64 (Promega Biotec) as provided by D. Kaufman et al. (3). The relative levels of mRNA were determined from the resulting autoradiographs by scanning laser densitometry, and the statistical comparisons were made using Student's t-test. Comparisons of relative GAD and GDH mRNA levels were done using either identical Northern blots run in parallel on the same day or using the same Northern blots which had been stripped of the first probe before they were rehybridized.

RESULTS and DISCUSSION - The level of GDH mRNA was significantly decreased compared to control in each brain region: hippocampus...50 $\pm$ 4%; striatum...58 $\pm$ 10%; cerebellum...70 $\pm$ 17% (% of control, n=6). The level of GAD mRNA was not altered compared to control in these three brain regions.

We conclude that some effect of liver damage, possibly brain hyperammonemia, has specifically affected the expression of a gene in the brain associated with ammonia metabolism. Since the alteration affects the RNA level for a predominantly glial enzyme with no effect on the RNA level for a neuronal enzyme involved in the same cycle, it will be of interest to determine if the expression of other genes for enzymes in the glutamate-GABA-ammonia cycle is similarily altered. Such data may give insights into the role of coordinate neuronal-glial gene expression in the control of amino acid neurotransmitter levels in brain and as well as providing insight into the mechanism of hepatic encephalopathy and hepatic coma.

REFERENCES

(1) Fraser, C.L., and Arieff, A.I. (1985) New England J. Med. 313, 865.
(2) Hertz, L., Kvamme, E., McGeer, E.G., and Schousboe, A., Eds. (1983) Glutamine, glutamate, and GABA in the central nervous system, A.R. Liss, Inc., NY.
(3) Kaufman, D.L., MFcGinnis, J.F., Krieger, N.R., and Tobin, A.J. (1986) Science 232, 1138.
(4) Banner, C., Silverman, S., Vitkovic, L., Thomas, J.W., Lampel, K., and Wenthold, R.J. (1986) Soc. Neurosci. Abst. 12, 980.
(5) Hilgier, W., Albrecht, J., and Krasnicka, Z. (1983) Neuropat. Pol. 21, 487.
(6) Chirgwin, J.M., Przybyla, A.E., MacDonald, R.J., and Rutter, W.J. (1979) Biochem. 18, 5294.
(7) Rave, N., Crkvenjakov, R., Boedtker, H. (1979) Nucleic Acids Res. 6, 3559.
(8) Melton, D.A., Krieg, P.A., Rebagliati, M.R., Maniatis, T., Zinn, K., and Green, M.G. (1984) Nucleic Acids Res. 12, 7035.

DEVELOPMENTAL EXPRESSION OF OPSIN, A PHOTORECEPTOR-SPECIFIC GENE

J.E. Treisman, M.A. Morabito and C.J. Barnstable
Laboratory of Neurobiology, The Rockefeller University, New York, NY 10021.

INTRODUCTION: Rhodopsin, the visual pigment protein, initiates the conversion of light energy into nerve signals. The appearance of rhodopsin detectable immunologically or by lectin binding in the rat retina correlates approximately with the final mitosis of photoreceptor precursor cells at postnatal days 1-4 (1,2). In the mouse absorbance at 500nm attributable to rhodopsin appears slightly later, at P6-7 (3). The present study examines accumulated transcripts of the rat opsin gene using both filter and in situ hybridization.

RESULTS: The probe used is diagrammed in Fig.1. It is a SmaI - PstI fragment of an opsin cDNA clone spanning the opsin coding sequence (4), subcloned into the transcription vector pT3/T7-18.

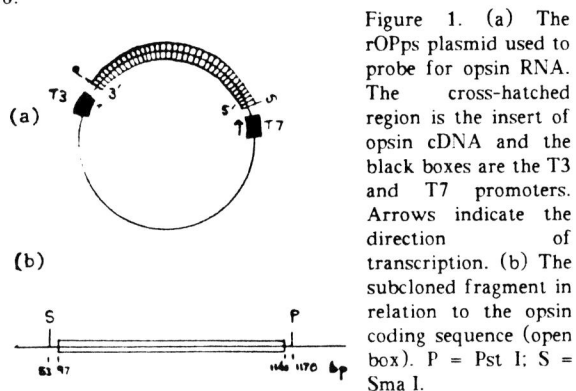

Figure 1. (a) The rOPps plasmid used to probe for opsin RNA. The cross-hatched region is the insert of opsin cDNA and the black boxes are the T3 and T7 promoters. Arrows indicate the direction of transcription. (b) The subcloned fragment in relation to the opsin coding sequence (open box). P = Pst I; S = Sma I.

Dot blots of total RNA from rat retinas of various ages were probed with the nick-translated rOPps plasmid (Fig. 1). RNA was prepared by guanidine thiocyanate extraction and centrifugation through cesium chloride. Dot blotting and filter hybridization were done by standard methods and quantified by autoradiography and scintillation counting. Each point represents the average of duplicate spots at several RNA concentrations, with the signal from a sample treated with RNase subtracted. Different experiments were normalized by measuring the radioactivity bound to homologous plasmid DNA on the filter; this also served as a concentration standard. Newborn and P2 RNA samples give only background hybridization; the percentage of complementary RNA first becomes detectable above the background level at postnatal day 3 (5.4×10^{-6}%) and rises steadily to the adult level of 8.3×10^{-3}%.

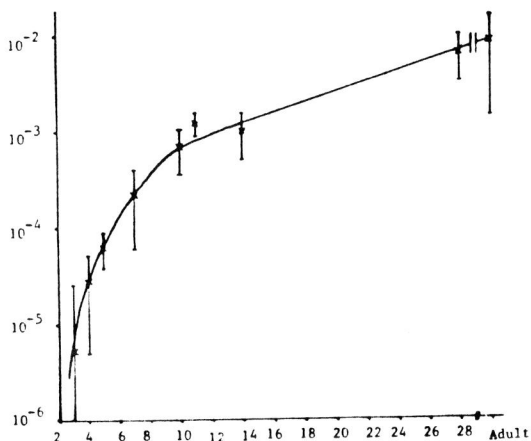

Figure 2. Opsin RNA present at various ages (in postnatal days) expressed as a percentage of total RNA (log scale).

In situ hybridization studies were done using ^{35}S-UTP labelled transcripts of the plasmid in Fig.1; T3 RNA polymerase was used to synthesize antisense and T7 polymerase to synthesize sense RNA. In both cases the plasmid was linearized at the other end of the insert to avoid transcription of vector sequences, and probes were hydrolysed 40 min in 40mM $NaHCO_3$/ 60mM Na_2CO_3, pH 10.2, to approximately 100 bases. Frozen retinal sections were cut onto poly-lysine coated slides and prefixed to the slide with 4% paraformaldehyde; tissue pretreatment, hybridization and washes were then performed as described in (5) except that 3 1-hr washes at 37°C in 50% formamide/ 0.6M NaCl/ 20mM Tris.Cl pH 7.4/ 1mM EDTA were substituted for their washes in SSC. Slides were exposed to emulsion for 6-10 days at 4°C, developed and stained with toluidine blue. Fig.3 shows an adult retinal section. An area of intense labelling can be seen over the layer corresponding to the photoreceptor inner segments. Some silver grains are also visible over the photoreceptor cell bodies, reflecting heterogeneous nuclear opsin transcripts; the background over the rest of the retina is low. Hybridization to sense RNA gave no specific signal; in P4 sections both antisense and sense probes gave background levels of hybridization (data not shown).

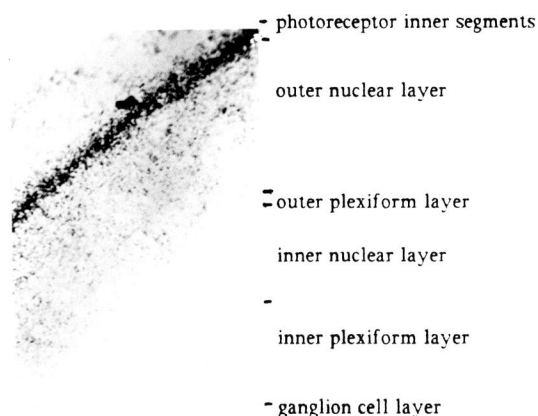

- photoreceptor inner segments

outer nuclear layer

- outer plexiform layer

inner nuclear layer

- inner plexiform layer

- ganglion cell layer

Figure 3. An adult retinal section hybridized with antisense RNA transcribed from rOPps and exposed to emulsion.

DISCUSSION: This study shows that opsin transcripts first become detectable at P3 at a level 1000-fold lower than that of the adult. This suggests that opsin expression is determined by transcriptional initiation or mRNA stabilization, rather than by translational initiation or post-translational modification. In the adult retina opsin transcripts are localized to the rod inner segment, the cellular region containing most of the translation apparatus (6). However, in the 4-day rat this method is not sensitive enough to observe the distribution of opsin RNA and compare it to the discontinuous staining seen with the antibody RET-P1 (1). The cellular specificity of labelling suggests that restriction to the photoreceptor is also achieved at a pre-translational level.

REFERENCES:
(1) Hicks, D. and Barnstable, C.J. (1986) J. Neurocytology 15,219-230.
(2) Barnstable, C.J. (1985) in "Hybridoma Technology in the Biosciences and Medicine" (Springer, T., ed.) pp. 269-289. Plenum Press, New York.
(3) Carter-Dawson, L., Alvarez, R.A., Fong, S.-L., Liou, G.I., Sperling, H.G. and Bridges, C.D.B. (1986) Dev. Biol. 116, 431-438.
(4) Nathans, J. and Hogness, D.S. (1983) Cell 34, 807-814.
(5) Cox, K.H., DeLeon, D.V., Angerer, L.M. and Angerer, R.C. (1984) Dev. Biol. 101, 485-502.
(6) Young, R.W. (1967) J. Cell Biol. 33, 61-72.

ACKNOWLEDGEMENTS: This work was supported in part by the Lucille P. Markey charitable foundation and by NIH grants EY05206, NS20483 and NS22789. We thank M. Applebury, Purdue University, for the original opsin cDNA clone.

REGIONALLY LOCALIZED EXPRESSION OF MOUSE HOMEO BOX GENES WITHIN THE DEVELOPING CNS

M. F. Utset[1], A. Awgulewitsch[2], F. H. Ruddle[1,2], and W. McGinnis[3]

Departments of [1]Human Genetics, [2]Biology, and [3] Molecular Biophysics and Biochemistry, Yale University, New Haven, CT 06510

Our laboratories are currently studying mouse homeo box genes to determine if they perform morphogenetic functions similar to those of the Drosophila homeotic genes. As a preliminary test of this hypothesis, we are examining the spatial distribution of transcripts from mouse homeo box genes. Drosophila homeotic genes display regionally localized patterns of expression[1], and we anticipate similar patterns from mouse homeo box genes with similar functions. We have recently reported the regionally localized expression of a mouse homeo box gene, Hox-3.1, within the CNS of the newborn mouse[2]. Here we discuss evidence for the regionally localized expression of this and another mouse homeo box gene, Hox-2.1, throughout the final third of prenatal development.

(I) NEWBORN CNS. We have studied the regional distribution of Hox-3.1 transcripts within the newborn mouse CNS by Northern blot analysis and *in situ* hybridization, as described[2]. These experiments demonstrate that at this stage Hox-3.1 transcripts within the CNS are restricted to cells caudal to the third cervical vertebra[2]. In contrast, similar experiments using a Hox-2.1 probe indicate that within the newborn CNS Hox-2.1 transcripts accumulate within and caudal to the medulla, the most caudal region of the hindbrain (in preparation).

(II) 13.5 pc CNS. We have also examined the distribution of Hox-3.1 and Hox-2.1 transcripts within the CNS at earlier stages of development. The figure shows the results of hybridizing a Hox-3.1 probe to sections of mouse embryos 13.5 days post coitum (pc). The upper and lower panels depict a sagittal section hybridized with a Hox-3.1 probe and the corresponding autoradiograph, respectively. At this stage Hox-3.1 expression within the CNS again appears restricted to cells caudal to the third cervical vertebra (arrows). Similar experiments indicate that within the CNS of the 13.5 pc embryo Hox-2.1 transcripts are again detected only within and caudal to the medulla (in preparation).

The observed spatial distribution of transcripts from these two genes is highly reminiscent of the patterns of expression of Drosophila homeotic genes. Therefore, we consider our data consistent with, although not a proof of, the hypothesis that Hox-3.1 and Hox-2.1 perform region specific determinative functions along the rostro-caudal axis of the developing murine CNS.

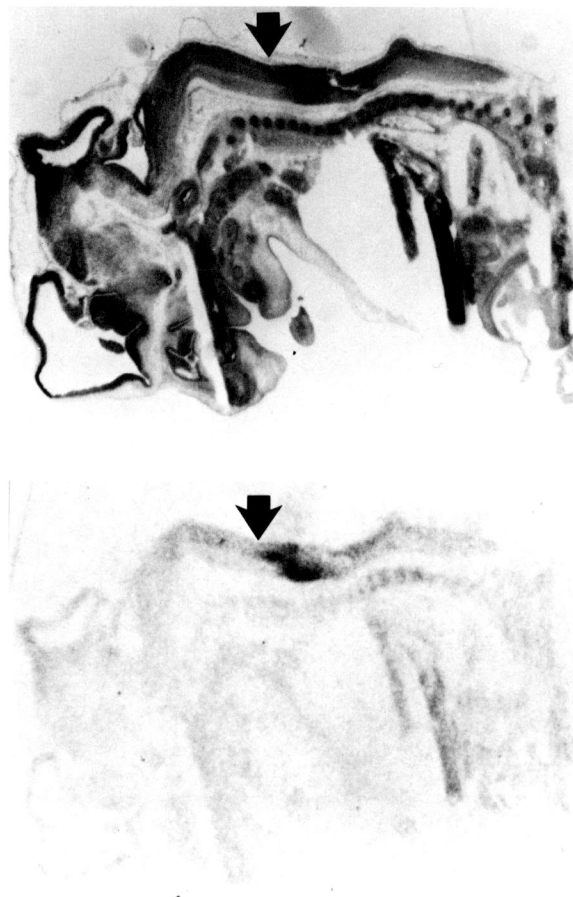

REFERENCES.

(1) Harding, K., Wedeen, C., McGinnis, W., and Levine, M. (1985) Science *229*, 1236-1242.
(2) Awgulewitsch, A., Utset, M. F., Hart, C. P., McGinnis, W., and Ruddle, F. H. (1985). Nature *320*, 328-335.

(3) ACKNOWLEDGEMENTS: M.F.U. receives support from the Life and Health Insurance Medical Research Fund. This research is supported in part by grants from the N.I.H. to F.H.R. and the N.S.F. to W.M.

GLUCOCORTICOID-MEDIATED INDUCTION OF HO-MEO BOX AND C-FOS mRNAs IN THE DEVELOPING RAT LIVER, SMALL INTESTINE AND COLON

M. J. Walsh[1], N. S. LeLeiko[1] and K. M. Sterling, Jr.[1,2].

Departments of Pediatrics[1] and Biochemistry[2], The Mount Sinai School of Medicine, One Gustave L. Levy Place, New York, N. Y. 10029.

INTRODUCTION

In the rat, hepatocytes, small intestinal epithelial cells and colonic epithelial cells differentiate at the time of weaning (1,2,3). Differentiation of these cells can be induced prematurely by administration of glucocorticoid to rats during the suckling period (2,3,4). The mechanism by which glucocorticoid can induce epithelial cell differentiation in these tissues is not known.

The homeotic genes of Drosophila appear to be essential for direction of the developmental pathways of the body segments of this organism, with the result that each segment has its own differentiated phenotype (5). These homeotic genes of the fruit fly all contain a conserved DNA sequence known as the homeo box (5).

The Hox 3 Locus homeo box gene of the mouse is expressed in the embryonic, newborn and adult mouse in a manner similar to the homeo box genes of Drosophila (6).

The cellular oncogene c-fos which is believed to play a role in development and differentiation is also expressed in a stage and tissue specific manner in the developing mouse (7).

We report here the glucocorticoid-mediated regulation of a putative rat homeo box gene and c-fos gene in neonatal rat liver, small intestine and colon.

MATERIALS AND METHODS:

Pregnant rats and their offspring were cared for as previously described (2).

Total RNA from rat intestine was prepared by the guanidine isothiocyanate procedure described by Chirgwin et al. (8).

Hybridization of ^{32}P-labeled DNA probes, pc-fos-1, mouse c-fos DNA and pMoEA, mouse homeo box DNA, (gifts respectively from Dr. T. Curran and Dr. W. McGinnis) to nitrocellulose bound RNA was done by the procedure of Thomas (9).

RESULTS:

Administration of dexamethasone (0.8 mg/kg) to 9 day old rats for three consecutive days resulted in significantly increased levels of homeo box gene transcripts and c-fos gene transcripts in the liver, small intestine and colon (Table 1 & 2). Adult rats treated with dexamethasone exactly as for neonates, had significantly less homeo box and c-fos gene transcripts than the untreated adult control (Table 1 & 2). Also, control adult rat liver contained less homeo box and c-fos gene transcripts than control neonatal liver (Table 1 & 2).

TABLE I
Relative Homeo Box mRNAs
(expressed as mean $\pm$ S.E. of ^{32}P-labeled pMoEA DNA hybridized to 10 µg each of total RNA from 3 separate animals)

Adult Liver	
Control	136 ± 20
Dexamethasone	57 ± 67
12 Day Liver	
Control	175 ± 25
Dexamethasone	817 ± 153
12 Day Small Intestine	
Control	125 ± 15
Dexamethasone	257 ± 47
12 Day Colon	
Control	447 ± 55
Dexamethasone	820 ± 36

TABLE II
Relative C-fos mRNAs
(expressed as mean $\pm$ S.E. of CPMS of ^{32}P-labeled pc-fos-1 DNA hybridized to 10 µg each of total RNA from 3 separate animals)

Adult Liver	
Control	246 ± 68
Dexamethasone	70 ± 20
12 Day Liver	
Control	270 ± 55
Dexamethasone	520 ± 130
12 Day Small Intestine	
Control	167 ± 65
Dexamethasone	697 ± 200
12 Day Colon	
Control	384 ± 191
Dexamethasone	780 ± 141

DISCUSSION:

The detection of putative rat homeo box mRNAs and c-fos mRNAs in suckling rat liver and intestine which were increased by glucocorticoid administration implies a possible role for the expression of these genes, mediated by glucocorticoids, in the glucocorticoid-regulated induction of liver and intestinal development and differentiation.

There are many aspects of the results presented here which must be addressed by further investigation in order to further identify the rat homeo box gene transcripts and the possible role of these and c-fos genes in development of these tissues.

REFERENCES:

(1) Grisham, J.W. (1969) In Normal and Malignant Cell Growth, ed. R.J.M. Fry, et al. pp. 28-43. Recent Results in Cancer Resarch. Vol. 17.

(2) Martin, G.R. and Henning, S.J. (1984) Am. J. Physiol. 246, G695-G699.

(3) Buts, J.-P., et al. (1983) Am. J. Physiol. 244, G469-G474.

(4) Wheatley, D.N. (1972) Exp. Cell Res. 74, 455-465.

(5) McGinnis, W., et al. (1984) Nature 308, 428-433.

(6) Awgulewitsch, A., et al. (1986) Nature 320: 328-335.

(7) Muller, R., et al. (1982) Nature 299, 640-644.

(8) Chirgwin, J.M., et al. (1979) Biochemistry 18: 5294-5299.

(9) Thomas, P. (1980) Proc. Natl. Acad. Sci. USA 77, 5201-5205.

CHANGES IN CONSTITUTIVE AND HEAT SHOCK-INDUCED GENE EXPRESSION DURING ERYTHROID DIFFERENTIATION IN XENOPUS LAEVIS

Robert S. Winning*, and Leon W. Browder.

Department of Biology, University of Calgary, Calgary, Alberta, Canada, T2N 1N4 (*Present address: Department of Biology, University of Waterloo, Waterloo, Ontario, Canada N2L 3G1).

Vertebrate erythrocytes are examples of extreme specialization to a singular role in organismic function. A consequence of this specialization is the repression of macromolecular synthetic capacity. The Xenopus erythrocyte has been described as one of the most synthetically repressed cells thus far studied; these nucleated cells are reportedly inactive in DNA and RNA synthesis, and protein synthesis occurs at low levels in a minority of cells (1).

One of the most universal cell functions is the heat shock response (2). The synthesis of heat shock proteins (hsps) in response to a sudden increase in temperature normally requires transcription. Therefore, we inquired whether erythrocytes can synthesize heat shock proteins in response to heat shock and whether heat shock gene transcription can be elicited in these apparently inert cells. If so, it would indicate that Xenopus erythrocytes have latent transcriptional potential. We were also able to study changes in the heat shock response during erythroid differentiation by making frogs anemic with phenylhydrazine injections and heat shocking the cells at progressive stages of differentiation.

(I) HEAT SHOCK PROTEIN SYNTHESIS. To determine the effects of elevated temperatures on protein synthesis, newly synthesized proteins were labeled with ^{35}S-methionine at ambient temperature or at one of various heat shock temperatures and analyzed by electrophoresis and fluorography. Xenopus erythrocytes are engaged in protein synthesis at ambient temperature. Globin is the major protein synthesized, but many other proteins are synthesized as well. The synthesis of a 70 kDa protein (hsp 70) is strongly induced at 33°C, 10° above ambient. Synthesis of this protein is induced over a very narrow range of temperatures; it is synthesized slightly at 30°C and not at all at 35°C or 37°C. All further experiments with heat shock were done at 33°C. During erythropoiesis, hsp 70 synthesis was induced at all stages (the progression of erythropoiesis proceeds from [1] basophilic, [2] polychromatophilic, [3] orthochromatic erythroblast to [4] erythrocyte). Interestingly, orthochromatic erythroblasts also synthesize a 70 kDa protein at ambient temperature; heat shock-induced synthesis of hsp 70 in these cells is slightly higher than that of the corresponding constitutive protein.

Actinomycin D (10 µg/ml) was used to test for the requirement of transcription for heat shock-induced protein synthesis. Induction of hsp 70 synthesis in erythrocytes was inhibited by actinomycin, suggesting that transcription is induced by the heat shock response. Interestingly, actinomycin reduced synthesis of hsp 70 in orthochromatic erythroblasts down to the constitutive levels. This result implies that transcription is required for the heat-induced elevation of hsp 70 synthesis, whereas constitutive synthesis utilizes pre-existing messenger RNA. These data do not, however, reveal whether heat shock activates transcription of the hsp 70 gene.

(II) HSP 70 GENE TRANSCRIPTION. To determine if hsp 70 gene transcription is induced by heat shock, we measured transcription of this gene by nuclear runoff transcription. Isolated nuclei were incubated with nucleoside triphosphates (including ^{32}P-UTP) under conditions in which previously-initiated transcripts are elongated. Labeled RNA was then hybridized to a cloned Xenopus hsp 70 gene. For comparison, RNA was also hybridized to cloned Xenopus ribosomal and globin genes. To establish a base line for legitimate transcription, RNA was also hybridized to the Xenopus vitellogenin gene, which is transcribed only in hepatocytes.

Erythrocytes (Table I) transcribe β-globin and ribosomal genes at ambient temperature; hsp 70 gene transcription, which is absent at ambient temperature, is strongly induced at 33°C. α-Amanitin (5 µg/ml) prevents the heat-induced elevation of hsp 70 synthesis but does not inhibit ribosomal transcription. Thus, the hsp 70 transcripts were apparently transcribed by RNA polymerase II.

Table I. Relative transcription rates (in parts per million; ppm) in erythrocyte nuclei.

β-globin	130	hsp 70	
ribosomal	672	ambient	7
		33°	68
		37°	7
vitellogenin	20		

During erythropoiesis (Table II), the trend is toward reduced transcription of these genes as differentiation proceeds, although there are stage-specific exceptions to that rule. Ambient temperature transcription of the hsp 70 gene is high during early erythropoiesis and is eventually terminated. These transcripts may be stored to be used in constitutive hsp 70 synthesis in the orthochromatic erythroblasts. Heat shock induces transcription of the hsp 70 gene at all stages. However, inducibility of the gene is reduced during differentiation.

Table II. Relative transcription rates (in ppm) during erythropoiesis.

Stage	Gene			
	β-globin	ribosomal	hsp 70 ambient	33°
Basophilic erythroblasts	36,268	2,242	192	1,137
Polychromatophilic erythroblasts	4,576	754	40	160
Orthochromatic erythroblasts	10,518	127	0	72
Erythrocytes	130	672	7	68

(III) SUMMARY. Although they become synthetically repressed during differentiation, erythrocytes retain constitutive and inducible transcription and protein synthesis.

REFERENCES

(1) Thomas, N. and Maclean, N. (1975) J. Cell Sci. 19, 509-520.
(2) Schlesinger, M.J., Ashburner, M., and Tissieres, A., eds. (1982) Heat Shock: From Bacteria to Man. Cold Spring Harbor Lab., Cold Spring Harbor, N.Y.

EXPRESSION OF U1 GENES IN SEA URCHIN NUCLEI AND IN NUCLEAR EXTRACTS

Jin-Chen Yu, Carlous Santiago and William Marzluff.
Dept. of Chemistry, Florida State University, Tallahassee, Fla. 32306.

There are at least two separate tandemly repeated units encoding U1 genes in the sea urchin L. variegatus. We have isolated and completely sequenced two different repeat units, pU1.1 and pU1.2 (1). Each 1400 base repeat unit encodes a single U1 gene. One of the repeats (U1.2) is present in four times the copy number of the other. The sequences of the two repeats are nearly identical (95%) for 500 bases 5 to the gene. The sequences 3 of the gene are highly divergent starting immediately after the CAAGAAGAA sequence required for formation of the 3 end of U1 RNA. The boundary of the conserved and divergent regions is a long polypyrimidine sequence. It is likely that gene conversion is maintaining the conservation of the 5 flanking region although it is possible that there is extensive selective pressure on this region.

Since the transcription unit of the U1 genes in isolated nuclei extends 3 to the genes, it is possible to determine whether a particular gene is expressed by analyzing the expression of a sequence 3 to the U1 gene. Subclones were constructed from unique regions 3 to the cloned U1 repeat units. Using these as probes we have shown that in blastula nuclei there are 4-5 times more pU1.2 than pU1.1 transcripts(Fig. 1). This correlates exactly with the gene copy number and indicates that the two repeats are transcriptionally equivalent. The expression of pU1.1 and pU1.2 genes and total U1 RNA all peak at 2 hours after hatching, but the transcripts of pU1.1 and pU1.2 at this stage are much less than the total U1 RNA. This implies that there may be other U1 gene repeats which encode most of the U1 RNA at this stage. In gastrula nuclei both of the U1 repeats are transcriptionally inactive. At this stage transcription from the nonrepeated genes first becomes apparent. The data indicate that different U1 gene sets are expressed differentially during sea urchin development.

We have previously described the synthesis of U1 RNA in a DNA-dependent extract from sea urchin embryos (2). In this extract the U1 RNA is initiated accurately, and at least a portion of the transcripts have the correct 3' end. A template containing only 203 bases 5' of the gene gave as efficient transcription as did the whole U1 gene. A similar type of extract was prepared from blastula nuclei of L. variegatus sea urchin embryos. Using this extract in conjunction with a series of BAl 31 deletion templates we have started to delineate the sequence requirements for efficient transcription of the U1 gene. The results of this analysis suggest that the U1 RNA gene also contains sequences involved in repressing transcription. The transcription of DNA templates containing these sequences is inversely related to the protein concentration. However, templates that do not contain these sequences are relatively insensitive to the protein concentration. These sequences may be involved in the transcriptional inactivation of the repeated genes which occurs after blastula stage. In addition, this extract efficiently processes most of the U1 RNA transcripts made in vitro as indicated by the absence of "runoff" products. However, exogenously added U1 RNA (riboprobe constructs) was not processed under any of several conditions tested. Isolated nuclei which normally make mature size U1 RNA are also incapable of processing exogeneously added U1 RNA. These observations suggest that processing and transcription of U1 RNA are coupled.

REFERENCES

(1) Jin-Chen Yu, Michael Nash, Carlous Santiego, and William F. Marzluff. (1986) Nuc. Acid Res. , Submitted.
(2) Morris, G.M., Price, D., and William F. Marzluff (1986), PNAS 83 : 3674-3678.

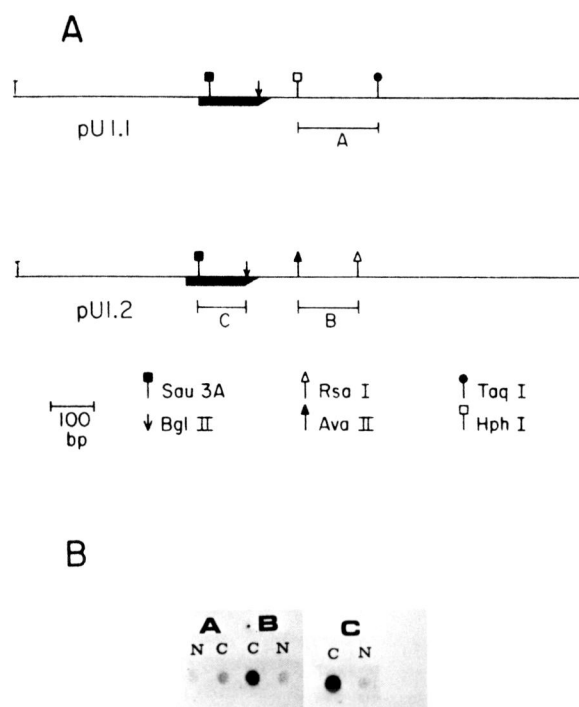

Fig. 1. Expression of the U1.1 and U1.2 repeats. A. Fragments A and B specific for the U1.1 and U1.2 repeats respectively were subcloned into pUC118 and pUC119 to generate single-strand probes. Subclone C is the coding region. B. RNA synthesized in isolated nuclei in the presence of $-^{32}PO_4$ CTP was hybridized to the immobilized DNA of the subclones in A. The filters were washed and the hybridized RNA detected by autoradiography.

ISOLATION AND PARTIAL CHARACTERIZATION OF A GENOMIC CLONE CORRESPONDING TO THE GENE FOR RAT BONE ALKALINE PHOSPHATASE.

J. Zernik, M. Noda*, M.A. Thiede*, K. Yoon*, G.A. Rodan*, A.C. Lichtler+ and W.B. Upholt

University of Connecticut Health Center, Dept. of Oral Biology and Dept. of Pediatrics+, Farmington CT 06032; and Merck Sharp & Dohme Research Laboratories, Dept. of Osteoporosis and Bone Research *, West Point PA 19486.

INTRODUCTION

Alkaline phosphatase is a membrane glycoprotein and a specific phenotypic marker of osteoblasts and hypertrophic chondrocytes. Its function in these cells is not clear; however, it is assumed to have a role in mineralization (1), and thus in skeletal development. Deficiency in alkaline phosphatase activity in the human inherited disease infantile hypo- phosphatasia is correlated with severe lack of calcification of skeletal tissues (reviewed in ref. 2).

Cloning the gene for rat bone alkaline phosphatase will facilitate studies of the differentiation of bone and cartilage cells and the development of skeletal tissues.

RESULTS & DISCUSSION

A rat bone alkaline phosphatase cDNA clone containing the 3' noncoding region was used to screen 2 Lambda Charon 4A rat genomic libraries (3).

Fig. 1. **Southern Blot Analysis of the Genomic Clone in Charon 4A.**

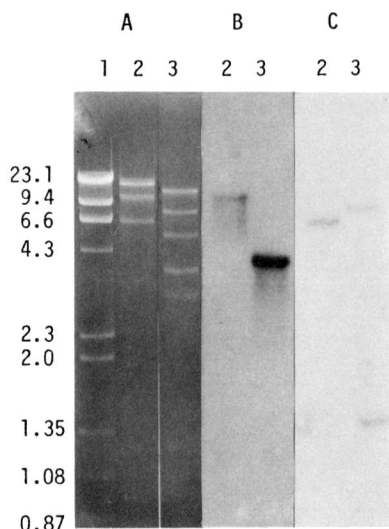

A. Ethidium bromide stained 1% agarose electrophoresis gel. B. Filter hybridized to a 3' noncoding region cDNA clone. C. Filter hybridized to a 1050 bp fragment from the 5' end of a 2.4 kb cDNA clone.
1. DNA molecular weight markers, size indicated in kilobases.
2. Eco RI digestion of the phage clone.
3. Bam HI digestion of the phage clone.

One positive plaque was identified in the library constructed by inserting the fragments resulting from Hae III partial digestion into the Eco RI site of the vector. This plaque was purified, yielding a phage with a 17 kb insert, consisting of two Eco RI fragments (10.6 kb and 6.6 kb).

The 10.6 kb fragment hybridizes strongly (Tm>85°c in 1 M sodium salt solution) with the rat cDNA fragment. Subsequently, a 2.4 kb cDNA clone was isolated. A 1050 bp fragment, derived from the 5' end of this clone, hybridizes to the 6.6 kb Eco RI fragment (Fig. 1).

Both fragments were subcloned for further analysis and additional restriction sites were mapped.

An oligonucleotide was synthesized based on the cDNA sequence. This was used as a primer for dideoxy sequencing, using the double-stranded subclone as a template. The resulting sequence matched the cDNA sequence.

These results suggest that the 6.6 kb fragment is 5' to the 10.6 kb fragment with respect to the gene.

We will use the currently available clones to isolate the entire gene for alkaline phosphatase and its regulatory sequences. These clones will be used in studying the regulation and tissue specificity of gene expression during differentiation and development of bone and cartilage tissues.

REFERENCES

(1) Robison, R.A. (1923) Biochem. J.17:286-293.
(2) Rassmussen, H. (1983) in The Metabolic Basis of Inherited Disease (Stanbury, J.B. et al. eds.) pp.1497-1507, McGraw Hill, N.Y.
(3) Sargent, T.D., Wu, J.-R., Sala-Trepat, J.M., Wallas, R.B., Reyes, A.A., and Bonner, J. (1979) Proc. Natl. Acad. Sci. USA 76:3256-3260.

ACKNOWLEGEMENTS

We thank M.L. Stover for technical assistance, Dr. G. Carmichael for synthesis of oligonucleotides, and Dr. D.W. Rowe for helpful discussions.

CELL SURFACE AND TISSUE INTERACTIONS

ENVIRONMENTAL REGULATION OF NEURAL CREST DEVELOPMENT

J. A. Weston and K. S. Vogel

Dept. of Biology, University of Oregon, Eugene, OR

Analysis of the cellular and genetic mechanisms, which regulate cell differentiation and tissue interactions during embryonic development, addresses two major issues: How phenotypic differences arise and become stabilized within cell lineages, and what role is played by environmental cues that are normally encountered by cells expressing these differences.

To deal productively with these issues, systems must be identified where we can study the developmental abilities of individual cells and, at the same time, determine with certainty the extent of stable, propagable restrictions in developmental ability. We must then be able to establish when such restrictions are imposed on a cell or its progeny, so that the nature of the environmental cues impinging on them can be accurately defined.

The vertebrate neural crest gives rise to a diversity of cellular phenotypes.

The neural crest, a transient stem cell population, can be recognized early in vertebrate embryogenesis. Its cells disperse in embryonic interstitial spaces, proliferate rapidly, and give rise to a variety of cell types. These include the pigment cells of the integument and iris, neurons and glial cells of the peripheral (sensory, autonomic and enteric) nervous systems, and neurosecretory tissues in the adrenal, heart, lungs, and pharyngeal structures (e.g. carotid body and the C-cells of the thyroid). In addition, so-called ectomesenchyme from the cranial neural folds ultimately differentiates as skeletal and connective tissue of the head and face (1-4).

Heterotopic transplantation of the neural crest has revealed that crest cell populations from every axial level maintain a wide range of developmental potentialities and differentiate in response to local environmental cues (1,3). The results of such heterotopic grafting experiments can be interpreted in two distinct ways: First, environmental cues might act on developmentally labile cells within the crest population to change the expression of genetic determinants by the responding cells. Alternatively, local environmental factors might promote changes in survival, proliferation and maturation of developmentally distinct crest cell subpopulations in which developmental decisions about the use of genetic information have already been taken. In contrast to the conventional hypothesis invoking crest cell pluripotentiality (see 5), the latter interpretation predicts that neural crest-derived subpopulations are already present amongst migrating crest cells. Before we can understand the role of environmental cues in the differentiation of the various crest-derived cellular phenotypes, therefore, it will be essential to determine when developmental restrictions occur in identified crest-derived cells (6,7).

Cells with partial developmental restrictions are present in migrating crest cell populations.

Since the distinction between the two alternatives mentioned above rests on how early in development the responding cells undergo restrictions of their developmental abilities, evidence for the presence of developmentally distinct neural crest subpopulations at early migratory stages would have an important bearing on interpreting the role of the environment in the appearance of crest derivatives.

The presence of phenotypically distinct neural crest subpopulations at early migratory stages has now been directly demonstrated by the use of cell type-specific monoclonal antibodies (8-14). These distinct cell types probably arise at very early stages of crest cell dispersal, possibly even before they are detected by sensitive immunological criteria. Thus, although local environmental cues encountered at the earliest stages of crest cell dispersal may influence developmental fates, the possibility remains that intrinsic mechanisms generate subpopulations within the premigratory crest upon which environmental factors act (14).

It should be emphasized that although early phenotypic diversity has been demonstrated, the developmental abilities of most of these subpopulations still remain to be elucidated. There is some convincing evidence, however, that heterogeneity of developmental potential also exists in early neural crest cell populations. For example, although cloning experiments demonstrated that some progenitor crest cells were at least bipotent, other clonal progenitors produced colonies that seemed to express a single cell phenotype (15). Likewise, heterochronic grafts of crest-derived structures into the crest migratory spaces of younger host embryos indicate that although some cells are able to migrate, localize and differentiate appropriately, other crest-derived cells in the grafted tissues are unable to do so (16-18).

A limited number of crest-derived precursors express neuronal traits at very early migratory stages.

Recently, we have confirmed the inference (see 16, 19) that some crest-derived neurons arise very early in embryogenesis. Using the neuron-specific antibody, A2B5 (20) on cultures of neural crest cells, about 1% of the crest cells with neuron-specific A2B5-immunoreactivity can be detected during the first day of culture (14). When such cultures are maintained under appropriate conditions, additional A2B5$^+$ cells appear (Fig. 1).

The observed increase in the number of A2B5$^+$ cells could occur in at least two ways. First, cells with A2B5 (neuronal) immunoreactivity could be recruited from a homogeneous, dividing population of crest cells, all of which have the ability to express neuronal traits if subjected to suitable environmental cues. Alternatively, A2B5$^+$ cells might arise from a specific, developmentally restricted, A2B5$^-$ lineage antecedent within the crest population. Thymidine-incorporation studies indicate that the initial A2B5$^+$ cells in crest cultures are postmitotic, and that most, if not all of the A2B5$^+$ cells, which appear subsequently, arise from cells that incorporated label early in the culture period (14).

If, as suggested by our thymidine-incorporation studies, the number and proliferative potential of specific, developmentally restricted lineage antecedents were limited, the crest cell cultures should progressively lose the ability to generate new A2B5$^+$ cells. To test these predictions, A2B5-immunoreactive cells were selectively ablated by adding complement plus the A2B5-antibody to progressively older secondary cultures of crest cells (21). When this is done during the first few days of secondary culture (e.g. days 2-4), A2B5$^+$ cells reappear after 1-2 days of additional culture. In contrast, when ablation of A2B5$^+$ cells is performed on older subcultures, when the proportion of A2B5$^+$ cells approaches its maximum (e.g. days 6-7), the ability to replace the ablated A2B5$^+$ cells is dramatically and consistently reduced (Fig. 1).

Fig. 1. Changes in the proportion of A2B5[+] cells in secondary crest cell cultures (○), and progressive loss of ability of cultures to produce A2B5[+] cells after complement-mediated ablation, expressed as percent of untreated controls (●).

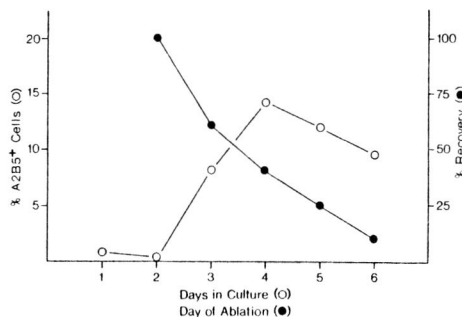

Taken together, the labeling and ablation studies indicate that a specific precursor of A2B5-immunoreactive cells exists as a subpopulation in crest cultures, and that the ability to regenerate A2B5[+] cells diminishes as proliferating precursors mature into postmitotic neurons in the culture population.

Culture conditions can alter the proportions of cells that express alternative crest-derived phenotypes.

The increase in number of A2B5-immunoreactive cells in crest cell cultures (see above, Fig. 1) is markedly dependent on early culture conditions. Thus, when crest cells are cultured as coherent clusters for 32 hr or longer, only the initial population of A2B5[+] cells, distinguished by their uni- or bipolar morphology, could be detected, and new A2B5[+] cells fail to appear in cultures from which A2B5[+] cells have been ablated. In contrast, if crest cells remain associated in culture for shorter periods (24 hr or less), large numbers of A2B5[+] cells appear (14, 21, and in preparation).

It is of considerable interest that differentiation of crest-derived pigment cells in these cultures exhibits a reciprocal pattern. Thus, the proportion of melanocytes appearing in cultures of crest cells that contain A2B5-immunoreactive cells is much lower (10-20%) than in cultures that lack cells expressing A2B5-immunoreactivity (usually > 90%; 21, 22). This reciprocal expression of neuronal and pigment cells might be reconciled by what we have observed in cultures of nascent, crest-derived sensory ganglia, where the choice between melanogenesis and gliagenesis seems to depend on the interaction between neurons and glial cell precursors (23, 24). Thus, in the case of cultures of early crest-derived cells, melanogenesis seems to be favored when environmental conditions in crest cell cultures prevent or do not permit survival or differentiation of neurons, and, in consequence, the interaction between nascent neurons and the intermediate glial/melanocyte progenitors cannot occur.

The role of autonomous cellular processes in the generation of diverse neural crest-derived subpopulations.

The general inference from recent studies of neural crest cell development is that some differentiative events occur very early and may be regulated autonomously (6, 7). We have argued, moreover, that neural crest cells probably undergo phenotypic diversification by a sequence of developmental restrictions in the crest lineage (4, 7). Such restrictions would be expected to produce

subpopulations of cells with different developmental abilities. Within each partially restricted, developmentally intermediate subpopulation, local environmental stimuli, including interactions with other crest-derived subpopulations, could presumably elicit expression of one or more of the phenotypes that remain in its repertoire. Conversely, in the absence of appropriate local cues, some intermediate cell types might fail to thrive. If such cells survive, however, parts of their repertoire might remain latent, or be further segregated by subsequent developmental events into more highly restricted populations (6, 7).

References

1. Le Douarin, N. (1982). The Neural Crest. Cambridge Univ. Press.
2. Noden, D. (1983). Devel. Biol. 96: 144-165.
3. Noden, D. (1984). Anat. Rec. 206: 1-13.
4. Weston, J. (1982). In: Cell Behaviour (ed. R. Bellairs, A. Curtis, G. Dunn), Cambridge Univ. Press, pp. 429-470.
5. Weston, J. (1970). Adv. Morphogen. 8: 41-114.
6. Weston, J., Ciment, G., Girdlestone, J. (1984). In: The Role of Extracellular Matrix in Development (ed. R. Trelstad), A.R. Liss, NY, pp. 433-460.
7. Weston, J., Vogel, K., Marusich, M. (1987). In: From Message to Mind (ed. B. Carlson, K.Barald, S. Easter), Sinauer Press, NY. (in press).
8. Barald, K. (1982). In: Neuronal Development (ed. N. Spitzer), Plenum Press, NY, pp. 101-119.
9. Ciment G., Weston, J. (1982). Devel. Biol. 93: 355-367.
10. Ciment G., Weston, J. (1985). Devel. Biol. 111: 73-83.
11. Payette R., Bennett, G., Gershon, M. (1984). Devel. Biol. 105: 273-287.
12. Vincent M., Thiery, J-P. (1984). Devel. Biol. 103: 468-481.
13. Barbu, M., Ziller, C., Rong, P., Le Douarin, N. (1986). J. Neurosci. 6: 2215-2225.
14. Girdlestone J., Weston, J. (1985). Devel. Biol. 109: 274-287.
15. Sieber-Blum M., Cohen, A. (1980). Devel. Biol. 79: 170-180.
16. Erickson C., Tosney, K., Weston, J. (1980). Devel. Biol. 77: 142-156.
17. Le Lievre, C. Schweitzer, G., Ziller, C., Le Douarin, N. (1980). Devel. Biol. 77: 362-378.
18. Ciment, G., Weston, J. (1983). Nature 305: 424-427.
19. Ziller, C., Dupin, E., Brazeau, P., Paulin, D., Le Douarin, N. (1983). Cell 32: 627-638.
20. Eisenbarth, G., Walsh, F., Nirenberg, M. (1979). Proc. Nat. Acad. Sci. USA 76: 4913-4917.
21. Vogel K., Weston, J. (1986). J. Cell Biol. (suppl.) (in press).
22. Glimelius, B., Weston, J. (1981). Cell Diff. 10: 57-67.
23. Nichols, D., Kaplan, R., Weston, J. (1977). Devel. Biol. 60: 226-237.
24. Holton, B., Weston, J. (1982). Devel. Biol. 89: 64-81.

Acknowledgements

Research in the authors' lab has been supported by PHS Grant DE-04315, NSF Grant PCM-8218899, and grants from the National Neurofibromatosis Foundation and the Dysautonomia Foundation, Inc.

ROLE OF TYROSINE KINASES IN EMBRYONIC DEVELOPMENT

P.A. Veno, N. R. Barton, W. Jiang, and
W. H. Kinsey

Dept. of Cell Biol. & Anatomy, Univ. of Miami
School of Medicine, Miami, FL.   33101

INTRODUCTION Fertilization results in transformation of the quiescent egg into a rapidly dividing embryo capable of undergoing differentiation and morphogenesis. Studies in invertebrate eggs have demonstrated that egg activation involves the stimulation of one or more protein tyrosine kinases, a process which appears to occur in two phases. The initial phase begins shortly after fertilization, is apparently an indirect response to the calcium transient, and occurs independently of protein synthesis (1,2). This response can be triggered parthenogenetically by treating eggs with the calcium ionophore A23187, which duplicates many aspects of egg activation but which is inadequate to stimulate cell division. A later, more dramatic response begins during the cleavage stages and continues through gastrulation and the development of differentiated structures (3). The analogy between fertilization and growth factor or viral transformation induced growth suggests that the tyrosine kinases acting during egg activation and early development may play important regulatory roles in embryogenesis. Since many protein tyrosine kinases could be active during these stages, it is necessary to identify the kinases that are present in the early embryo and determine how their activity changes with development.

In order to identify the protein tyrosine kinases which are present in the egg and embryo we have begun to make use of antisera to identified enzymes. Site directed antisera against the trpE-abl fusion proteins pEX-2 and pEX-5 (4) have been used to immunoprecipitate the c-abl protein kinase and measure its activity in an immune complex assay (5). We have adapted this assay to measure the low levels of c-abl kinase activity present in microgram quantities of mouse embryonic tissue. We have been able to detect c-abl kinase activity as early as the eight cell stage, suggesting that this enzyme may play an active role in early embryonic development.

IDENTIFICATION OF C-ABL IN MOUSE EMBRYOS: Site directed antisera produced against segments of the murine v-abl gene product have been shown to immunoprecipitate P150, the protein tyrosine kinase encoded by the c-abl gene (5,8). As is true for many protein tyrosine kinases, P150 is capable of phosphorylating itself when purified in the form of an immunoprecipitate and then incubated with γ-[32P]-ATP. We have taken advantage of this property to demonstrate the specificity of these antisera in the mouse embryo system. We used anti-pEX-5 serum (4) to immunoprecipitate P-150 from a detergent extract of day eight mouse embryos. Embryos were recovered from the uterus on the eighth day after mating and dissected free of decidua and extraembryonic membranes. They were then homogenized in a detergent containing buffer, centrifuged, and aliquots representing 100ug of embryo protein were incubated with either normal rabbit serum or anti-pEX-5 serum for six hours. The immune complexes were then precipitated with protein A-Sepharose, washed by centrifugation, and incubated with γ-[32P]-ATP in an autophosphorylation reaction (5). As seen in Fig. 1, a 150 kDa phosphoprotein was detected in immunoprecipitates

from anti-pEX-5 but not from control serum. This indicates that the site directed antisera is binding to the c-abl protein with a high degree of specificity.

Fig. 1. Immunoprecipitation and autophosphorylation of P150 c-abl from day eight embryos. 100 ug of embryo protein was immunoprecipitated with control (A) or anti-pEX-5 (B) serum, then autophosphorylated with γ-[32P]-ATP (1Ci/umol) in 20mM PIPES, 20mM MnCl2, pH 7.0 at 25 C for 90 min (5). Samples were analyzed on an 8% SDS-polyacrylamide gel and detected by radioautography.

QUANTITATION OF C-ABL KINASE ACTIVITY DURING DEVELOPMENT.

Protein tyrosine kinase activity immunoprecipitated by anti-pEX-5 serum was quantitated by measuring the phosphorylation of neurotensin, a small peptide that contains tyrosine residues as the only potential phosphorylation sites. The low level of activity evident in immunoprecipitates from control serum was subtracted to obtain abl-specific activity. Under our assay conditions, the phosphorylation of neurotensin was linearly dependent on the amount of protein in the detergent extract (Fig. 2).

Fig. 2. Enzyme dependence of the c-abl immune complex assay. Neurotensin was phosphorylated by immune complexes prepared from different amounts of a detergent extract of adult testis, a convenient source of this enzyme.

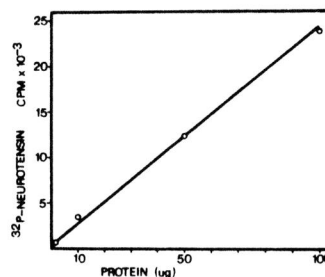

In order to determine whether the activity of c-abl changed during early development, we have immunoprecipitated extracts prepared from unfertilized eggs, eight cell (day 3), day 5, and day 8 embryos. The c-abl specific kinase activity was then quantitated as in Fig. 2. While we were unable to detect kinase activity in samples prepared from 15ug of unfertilized egg protein, significant activity was present in equal amounts of protein from eight cell embryos as well as in the more advanced embryos (Table I). Since the protein content of the eight cell embryo is slightly less than that of the unfertilized egg (6), it is clear that an increase in both the total and the specific activity of the c-abl kinase occurs during the first three cleavages.

TABLE I. C-ABL KINASE ACTIVITY IN THE MOUSE EMBRYO

Stage	Activity/embryo (fmol/100embryos/hr)	Specific Activity (fmol/mg prot./hr)
Egg	< 0.10	<43.0
8 cell	0.44	193.1
Day 5	1.30	na
Day 8	830.0	813.7

values represent the average from two experiments

The minimum level of detection is dictated largely by the level of background activity in the assay representing contamination picked up from both the control and pEX-5 serum. This background was less that 30% of the activity measured in eight cell embryos and was insignificant relative to the activity in the more advanced stages. Unfertilized eggs never exhibited any activity above background. The specific activity value which we obtained for the day eight embryos is comparable to that found in adult tissues rich in c-abl such as the spleen, and is three to four fold higher than that of adult liver and striated muscle.

DISCUSSION

We have identified c-abl as one of the protein tyrosine kinases which are active during the early cleavage stages of mammalian development. This enzyme is either synthesized or activated between fertilization and the third cleavage. The c-abl kinase activity continued to increase during subsequent stages of development involving gastrulation, neural tube and somite formation. While the specific activity of c-abl expressed per mg protein increased throughout development, the activity per cell remained fairly constant after the eight cell stage. Consistent with our findings, a previous study (7) demonstrated that the relative expression of c-abl specific mRNA increased several fold during days six through eleven of development. The complex nature of the more advanced embryos opens the possibility that c-abl is differentially expressed by different cell populations, a feature that would not be detected by our experiments.

Other protein tyrosine kinases known to be expressed during embryonic development include pp60c-src and the epidermal growth factor (EGF) recepeptor. The src gene product has been studied in the eggs and embryos of lower vertebrates, invertebrates, and insects (9,10,11) where it has been found to be expressed at low levels until the embryos begin organogenesis. The EGF receptor has been detected in day eleven mouse embryos where it displays considerable tissue specificity (12). It is not unexpected that the EGF and other growth factor receptors would be present in differentiated cells where they could promote the growth of that tissue in response to EGF. The role of other tyrosine kinases which have not been identified as hormone or growth factor receptors is far from being understood. The embryonic stages in which c-abl has been identified include the the eight cell stage during which the developmental potential of the blastomeres is just beginning to be restricted, as well as the more advanced stages which are characterized by the rapid proliferation of both pluripotent and committed stem cell populations. Further experiments involving measurement of enzyme activity in earlier, undifferentiated embryos and histochemical localization of c-abl at later stages of development would help to identify the developmental processes in which this enzyme is active.

REFERENCES

(1) Dasgupta, J.D. and Garbers, D.L. (1983) J. Biol. Chem. 258, 6174-6178.
(2) Kinsey, W.H. (1984) Devel. Biol. 105, 137-143.
(3) Satoh, N., and Garbers, D.L. (1985) Devel. Biol. 111, 515-519.
(4) Konopka, J.B., Davis, R.L., Watanabe, S.M., Ponticelli, A.S., Schiff-Maker, L., Rosenberg, N., and Witte, O.N.(1984) J.Virol. 51, 223-232.
(5) Konopka, J.B., and Witte, O.N. (1985) Mol. Cell Biol. 5, 3116-3123.
(6) Brinster, R.L. (1967) J. Reprod. Fert. 13, 413-420.
(7) Muller, R., Slamon, D.J., Tremblay, J.M., Cline, M.J., Verma, I.M. (1982) Nature (Lond.) 299, 640-644.
(8) Schiff-Maker, L., Burns, M., Konopka, J., Clark, S., Witte, O.N., Rosenberg, N. (1986) J. Virol. 57, 1182-1186.
(9) Schartl, M., and A. Barnekow (1984) Devel. Biol. 105, 415-422.
(10) Simon, M.A., Drees, B., Kornberg, T., and Bishop, J.M. (1985) Cell 42, 831840.
(11) Kamel, C., P.A. Veno, and W.H. Kinsey (1986) Biochem. Biophys. Res. Comm. 138, 349355.
(12) Adamson, E.A., M.J. Deller, and J.B. Warshaw (1981) Nature (Lond.) 291, 656659.

ACKNOWLEDGEMENTS We wish to acknowledge Dr. O. Witte for making available to us the site directed antisera used in this study. This work was supported by HD-14846 and RCDA HD-00620.

CELL SURFACE PROTEINS IN GASTRULATION AND SKELETON FORMATION

W. J. Lennarz, G. L. Decker, M. C. Farach,
S. Grant and H. Woodward

Department of Biochemistry and Molecular Biology
The University of Texas System Cancer Center
M. D. Anderson Hospital and Tumor Institute
6723 Bertner Avenue
Houston, Texas 77030

Introduction

As an outgrowth of our interest in the mechanism of synthesis of N-linked glycoproteins, we have undertaken to study their possible role in cellular recognition in differentiation and development. Much of our effort has concentrated on the sea urchin embryo because one can synchronously cultivate large numbers of embryos in sea water and there is much information on the fate of cell types and the morphogenetic changes that occur over the course of early development of this organism.

Initially, our efforts focused on the process of gastrulation. This process, in which the germ layers are produced, begins shortly after the blastula stage embryo hatches. At this time, descendants of the 16 cell-stage micromeres enter the blastocoel, where they become the primary mesenchyme cells. After a period of amoeboid exploratory movements in the blastocoel, these cells become relatively stationary. Shortly after the ingression of the primary mesenchyme cells into the blastocoel, the blastoderm invaginates, beginning the formation of the archenteron. Secondary mesenchyme cells at the tip of the archenteron extend pseudopodia to the inner surface of the blastocoel. Where the tip of the archenteron makes contact with the blastocoel wall, the mouth forms and a continous tube, the gut, is produced. During the period of invagination of the archenteron, the filopodia of the primary mesenchyme cells fuse to form a syncytial cable-like structure. This structure is the site for deposition of $CaCO_3$ to form the primitive skeleton (spicules), which is completed following gastrulation.

Glycoproteins and Gastrulation.

We have previously shown that the onset of gastrulation is preceded by an increase in the synthesis of N-linked glycoproteins (1); this finding has been confirmed in recent studies that have also shown an apparent transient burst in glycoprotein synthesis during the early cleavage stages (2). Inhibition of glycosylation by addition of the specific inhibitor, tunicamycin, prevents gastrulation without having known toxic effects on the cells of the embryo (3, 4). This observation suggests synthesis of new glycoproteins is required for gastrulation, perhaps because they serve as cell recognition signals for the cellular rearrangements that occur during this process. In this context, it is of interest that several gastrula stage specific antigens found to be localized to specific cell types are believed to be glycoproteins (5).

To better understand the factors that regulate the synthesis of glycoproteins at gastrulation, we utilized an in vitro translation/glycosylation system to examine mRNAs coding for glycoproteins (6). These studies showed that the increase in the glycosylation of proteins at gastrulation is accompanied by an increase in translatable messages that are associated with membrane-bound polysomes and encode for glycoproteins. Some of these mRNAs appear to be newly transcribed (or processed to a translationally active form) at gastrulation because they are present in the gastrula stage embryo but absent from the egg. In contrast, other glycoprotein mRNAs may be made during oogenesis and stored in silent form until activation at gastrulation because they are detected in the egg as well as in the gastrula stage embryo.

Although much work has been done on the developmentally regulated expression of genes in sea urchin embryos, there are only a few cases in which the precise function of the gene products is known. Given the apparent importance of the newly synthesized N-linked glycoproteins in the gastrulation process, we have extended these studies in several new directions by preparing monoclonal and polyclonal antibody probes to gastrula stage specific glycoproteins (7). In addition, we plan to prepare cDNA probes for studying the regulation of the mRNAs encoding for these proteins. Thus far, a library of eight monoclonal antibodies to cell surface polymannose glycoproteins has been prepared. Western blot analysis and immunofluorescence microscopy have been used to study the developmental expression of these proteins. The results indicate that these glycoprotein antigens are first synthesized at the onset of gastrulation and are found on the surface of the embryonic cells. Hopefully, microinjection techniques will allow us to determine the role of these glycoproteins in the gastrulation process.

Dolichol Synthesis and Gastrulation.

The hypothesis that N-linked glycoprotein synthesis is required for gastrulation is also supported by two findings. First, normal gastrulation does not occur when de novo synthesis of dolichyl phosphate is inhibited by the drug compactin (8, 9). Second, dolichol and dolichol phosphate synthesis is developmentally regulated, being virtually absent in the egg and increasing rapidly prior to the onset of gastrulation (9). Because β-hydroxy-β-methylglutarylCoA (HMGCoA) reductase is known to play a key regulatory role in the synthesis of dolichol, cholesterol and coenzyme Q in mammals, the activity of this enzyme has been examined in the developing sea urchin embryo. Enzyme activity was found to increase at least 200-fold during development from the unfertilized egg to the gastrula stage embryo (10). The HMGCoA reductase in the sea urchin embryo exhibits properties very different from that found in mammals; only a small fraction of the activity can be solubilized from microsomes by a variety of detergents and the solubilized material does not bind to concanavalin A agarose. In addition, soluble reductase activity cannot be released from microsomes by mild trypsinization.

In order to determine if the dramatic increase observed in HMGCoA reductase over the course of development is controlled at the level of enzyme activation, translation or transcription, efforts to obtain a cDNA probe for HMGCoA reductase are underway. A full-length rodent cDNA to the reductase has been used to screen a sea urchin genomic library. Five positive clones have been identified after three rounds of screening under low stringency. We are presently subcloning restriction fragments of one of these clones in preparation for sequencing in order to demonstrate that this clone encodes for the sea urchin HMGCoA reductase. This clone will be utilized to study the developmental regulation of this enzyme and thereby better understand the control of dolichol biosynthesis.

Cell Surface Glycoproteins and Skeleton Formation.

Late in the gastrulation process, the primary mesenchyme cells that have undergone fusion via filopodia initiate deposition of $CaCO_3$, which culminates in formation of the primitive skeleton (spicule) of the embryo. During the process of screening the monoclonal antibodies discussed above, we isolated an antibody (1223) that specifically bound to primary mesenchyme cells and retarded spicule growth. To study the effect of this antibody and the process of spicule formation in more detail, a simple method to isolate and culture primary mesenchyne cells was developed (11). Cells isolated by this method exhibit many of the properties characteristic of primary mesenchyme cells in the embryo and form spicules in vitro; this process is inhibited by antibody 1223. Studies to elucidate the pathway whereby Ca^{2+} is deposited in the spicule reveal that the spicule is dissolved from living cells in the presence of a chelator of divalent cations at pH 8.0, whereas smaller intracellular deposits having optical properties similar to that of the spicule remain unchanged. Use of rapid freezing to preserve these deposits for transmission electron microscopy and x-ray microprobe analysis indicates that electron dense deposits, found in several intracellular organelles, contain Ca^{2+}. Examination of the spicule cavity after demineralization reveals the presence of openings to the external environment, as well as numerous coated pits within the limiting membrane and the presence of residual organic matrix that stains with ruthenium red. Exogenous concanavalin A-gold conjugates were found to penetrate the spicule cavity and to specifically bind to both membrane and matrix components. On the basis of these observations, we conclude that the spicule is assembled in an extracellular cavity, and that the Ca^{2+} may be routed to the site of deposition via a secretory pathway (12).

The antigen for the 1223 antibody has been found to be a cell surface (glyco)protein; the predominant species is 130 kDa, although both a higher and lower molecular weight form are detectable (11). Studies in which the processes of Ca^{2+} uptake and Ca^{2+} deposition into $CaCO_3$ were partially separated, suggest that the 1223 antigen is involved in the uptake process. Recently, the developmental expression of this cell surface (glyco)-protein has been studied in more detail. In S. purpuratus, this protein is present in the egg and in all cell types of the early embryo. After gastrulation, its synthesis and cell surface expression are restricted to the skeleton-forming primary mesenchyme cells. In contrast, in L. pictus the 1223 protein cannot be detected in eggs or in embryos prior to the mesenchyme gastrula stage. Hybrid embryos of these two sea urchin species exhibit a pattern of expression indistinguishable from that of the species contributing the maternal genome, which suggests that early expression of the protein in S. purpuratus embryos is due to utilization of maternal transcripts from the egg. Later expression of this protein in primary mesenchyme cells appears to be the result of cell type-specific synthesis, likely encoded by embryonic transcripts. This cell type-specific expression in primary mesenchyme cells correlates temporally with Ca^{2+} accumulation during skeleton formation in the embryo. This observation is consistent with the postulated role of this antigen in Ca^{2+} uptake. Hopefully, further studies utilizing the isolated 1223 protein will better define its function at the molecular level.

Summary

Cell surface proteins have been found to play important roles in two morphogenetic events in development of the sea urchin embryos, gastrulation and skeleton formation. In the case of glycoproteins and gastrulation, a great deal has been learned about the factors that control the synthesis of these glycoproteins, but little is yet known about their possible function in cellular recognition events that are involved in this complex process. With respect to skeleton formation, it is clear that a mesenchyme cell specific protein, tentatively identified as a glycoprotein,

participates in the biomineralization process. Further studies should enable one to determine if it is directly involved in Ca^{2+} transport or in some other phase of spiculogenesis.

References

(1) Lennarz, W. J., In: Critical Reviews of Biochemistry. G. Fasman (ed.), CRC Press, Inc., Boca Raton, Vol. 14, No. 4, pp. 257-272 (1983).
(2) Grant, S. R. and Lennarz, W. J., manuscript in preparation.
(3) Schneider, E. G., Nguyen, H. and Lennarz, W. J., J. Biol. Chem. 253, 2348-2355 (1978).
(4) Heifetz, A. and Lennarz, W. J., J. Biol. Chem. 254, 6119-6127 (1979).
(5) Wessel, G. M. and McClay, D. R., Dev. Biol. 111, 451-463 (1985).
(6) Lau, J. T. Y. and Lennarz, W. J., Proc. Natl. Acad. Sci. U.S.A. 80, 1028-1032 (1983).
(7) Grant, S. R. and Lennarz, W. J., manuscript in preparation.
(8) Carson, D. D. and Lennarz, W. J., Proc. Natl. Acad. Sci. U.S.A. 76, 5709-5713 (1979).
(9) Carson, D. D. and Lennarz, W. J., J. Biol. Chem. 256, 4679-4686 (1981).
(10) Woodward, H. and Lennarz, W. J., manuscript in preparation.
(11) Carson, D. D., Farach, M. C., Earles, D. S., Decker, G. L. and Lennarz, W. J., Cell 41, 639-648 (1985).
(12) Decker, G. L., Morrill, J. A. and Lennarz, W. J., manuscript in preparation.
(13) Farach, M. C., Valdizan, M., Park, H. R., Decker, G. L., and Lennarz, W. J., Develop. Biol., submitted for publication.

Acknowledgements

It is a pleasure to acknowledge the technical assistance of J. Allen, R. Chap, R. Cruz, H. Park, F. Tan and M. Valdizan. This work is supported by NIH grants HD 18600 and HD 12718. W. J. Lennarz, who is a Robert A Welch Professor of Chemistry, acknowledges with gratitude the R. A. Welch Foundation.

DIFFERENTIAL EXPREESION OF CADHERIN CELL ADHESION MOLECULES ASSOCIATED WITH ANIMAL MORPHOGENESIS

M. Takeichi, K. Hatta, A. Nose, A. Nagafuchi and M. Hatta

Department of Biophysics, Faculty of Science, Kyoto University, Kyoto 606, Japan

Ca^{2+}-dependent cell-cell adhesion molecules, termed cadherins, are detected in most kinds of cell types forming solid tissues in vertebrates. Antibodies against cadherins tend to actively disrupt cell-cell adhesion when added to cell cultures, suggesting that these adhesion molecules are most essential for connections of cells (1, 2). Studies using monoclonal antibodies showed that cadherins are divided into subclasses which are distinct in molecular weight, immunological specificities and tissue distribution. Table 1 summarizes properties of cadherin subclasses thus far identified.

An interesting property of cadherin subclasses is their specificities in cell-cell binding. Cells tend to preferentially adhere to cells with an identical subclass of cadherins when heterotypic cells are mixed (3). This strongly suggests that cadherins play an important role in selective cell adhesions, which are regarded as a critical process for animal morphogenesis. Here, we report the spatial and temporal pattern of expression of cadherins during development of mouse and chick embryos, and show correlations of the expression patterns of cadherins with a variety of morphogenetic events. We also show some preliminary results on the biochemical relations of cadherin subclasses.

(I) EXPRESSION OF CADHERINS ASSOCIATED WITH HETEROTYPIC CELL-CELL CONNECTIONS.

Immunohistochemical studies on the pattern of expression of cadherins in developing embryos suggested that cadherins are essential not only in homotypic cell-cell connection but also in heterotypic connection.

Preimplantation mouse embryos express only E-cadherin. At the egg cylinder stage, all cells in the embryonic and extraembryonic layers, except the parietal endoderm, continue to express E-cadherin. Expression of P-cadherin begins at this stage only in the extraembryonic layers, particularly strongly at the ectoplacental cone (4).

Table 1. Cadherin subclasses identified

subclasses	MW[1]	tissue distribution in matured embryos[2]
E-cadherin[3]	124K	epithelial cells derived from the three germ layers
P-cadherin	118K	placenta, epidermis, pigmented retina, cornea endothelium, mesothelium
N-cadherin	127K	nervous system cardiac muscle, lens, primordial germ cell

[1] MW slightly changes with tissues.
[2] Only examples are shown.
[3] E-cadherin is also called uvomorulin (5), cell-CAM 120/80 (6) and L-CAM (7).

Simultaneously, the decidua, the maternal uterine tissue with which embryos make connection for placentation, begins to express P-cadherin but not E-cadherin. Thus, expression of P-cadherin at this stage appears to serve for the embryo-maternal connection.

Expression of N-cadherin also appears to be associated with connection of heterotypic cells. In the following combinations of cells, both cell types express N-cadherin: Motor neurons and myotome, genital ridge and primordial germ cells, and ectodermal visceral groove and endodermal visceral pouch. These tissues in each pair are connected with each other during development.

(II) EXPRESSION OF CADHERINS ASSOCIATED WITH SEGREGATION OF CELLS.

As described above, all embryonic cells initially express E-cadherin. However, when the mesoderm segregates from the ectoderm upon gastrulation, the former loses E-cadherin and instead begins to express N-cadherin. Such transition in expression of cadherins from E- to N-type is observed in many other morphogenetic processes in which segregation of cell populations is essential. For example, the E to N transition occurs during separation of the neural tube or the lens vesicle from the ectoderm (8).

We observed another types of pattern of cadherin expression associated with segregation of tissues, in such morphogenetic events as somite (9) and epidermis (4) differentiation which involve cell movement. In these processes, a subset of cells moving out from their original positions lose one type of cadherins.

(III) BIOCHEMICAL RELATIONS BETWEEM CADHERIN SUBCLASSES.

E- and N-cadherins can be cleaved with trypsin into smaller peptides with MW 84,000 and 87,000, respectively. We purified the 84kd fragments of mouse E-cadherin and the 87kd fragments of chicken N-cadherin, and determined amino acid sequences of their amino terminus (10). The results showed presence of identical sequences in these two cadherins, strongly suggesting that they have a common genetic origin. Thus, cadherins seem a family of molecules with diverse binding specificities.

REFERENCES

(1) Yoshida-Noro, C., Suzuki, N., and Takeichi, M. (1984) Dev. Biol. 101, 19-27.
(2) Hatta, K., Okada, T.S., and Takeichi, M. (1985) Proc. Natl. Acad. Sci. USA 82, 2789-2793.
(3) Takeichi, M., Hatta, K., and Nagafuchi, A. (1985) in Molecular Determinants of Animal Form (Edelman, G.M., ed.) pp. 223-233, Alan R. Liss, Inc., New York.
(4) Nose, A., and Takeichi, M. (1987) J. Cell Biol. in press.
(5) Hyafil, F., Morello, D., Babinet, C., and Jacob, F. (1980) Cell 21, 927-934.
(6) Damsky, C.H., Richa, J., Solter, D., Knudsen, K., and Buck, C.A. (1983) Cell 34, 455-466.
(7) Gallin, W.J., Edelman, G.M., and Cunningham, B.A. (1983) Proc. Natl. Acad. Sci. USA 80, 1038-1042.
(8) Hatta, K., and Takeichi, M. (1986) Nature 320, 447-449.
(9) Hatta, K., Takagi, S., Fujisawa, H., and Takeichi, M. (1987) Dev. Biol. in press.
(10) Shirayoshi, Y., Hatta, K., Hosoda, M., Tsunasawa, S., Sakiyama, F., and Takeichi, M. (1986) EMBO J. 5, in press.

DEVELOPMENTAL BIOLOGY OF A NEURAL CELL ADHESION MOLECULE

Urs Rutishauser

Neuroscience Program, Department of Developmental Genetics, Case Western Reserve University School of Medicine, Cleveland, Ohio 44106

The studies of Holtfreter (1) on the contribution of cell affinities to formation of tissue patterns spurred several decades of work on the nature of cell-cell interactions. One of the most important concepts to emerge from this period was the ability of cells to order themselves according to quantitative differences in adhesiveness (2). However, until recently too little has been known about the molecular nature of cell-cell adhesion to determine if hierarchies of adhesion actually exist in tissues during development.

Studies on the biochemical properties of the neural cell adhesion molecule NCAM and its function _in vivo_ (for review see 3,4) have provided several examples of how adhesive preferences can contribute to formation of some of the most intricate structures in neural tissue. In this report, this mechanism is illustrated primarily in terms of the the ability of neurons and growth cones to select targets or pathways according to the adhesive properties of their environment.

The Biochemistry of NCAM.

NCAM is an integral membrane protein with a single polypeptide chain and a large and unusual carbohydrate moiety. For the purposes of this discussion, the key properties of NCAM (Figure 1) are 1) that the molecule appears to be a ligand in the formation of cell-cell bonds, 2) that adhesion involves the participation of NCAM molecules on both adhering membranes and therefore is an example of homophilic binding, and 3) that heterogeneity in NCAM's carbohydrate moiety can alter the binding properties of the molecule. If NCAM behaves as a homophilic ligand, then cells that express this molecule and come into contact during development have the potential of forming adhesions. The duration and consequences of each adhesion would then depend on a variety of parameters, two of the most obvious being the concentration and binding properties of NCAM.

Fig. 1. Basic molecular properties of NCAM in a cell-cell bond. Homophilic binding is illustrated in its most simple form, with direct interaction between the N-terminal region of two NCAMs. The stippled area indicates the location of the polysialic acid moieties, which modulate the binding properties of the molecule.

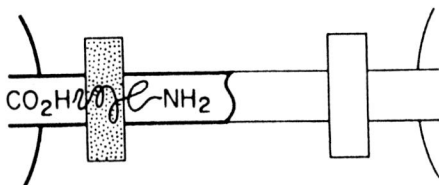

Expression of NCAM.

During embryogenesis, NCAM is expressed in many tissues including primitive neuroepithelia, a number of transient structures associated with early morphogenesis such as the notochord, placodes, and somites, and on the three primary cell types found in a differentiated nervous system, that is, neurons, glia, and muscle cells. The examples provided here will focus on adhesion of neurons to muscle, to glial precursors, and to other neurons.

NCAM and Adhesive Preferences.

At first glance, NCAM seems an unlikely candidate for mediating cell-cell recognition events that lead to the formation of specific tissue structures, in that the molecule appears to have a single binding specificity and a broad cell and tissue distribution. Recent work, however, has demonstrated that variations in the expression and form of NCAM occur on individual cells as well as across tissues, in a manner that provides critical opportunities for the choice and timing of appropriate cell-cell interactions. At the level of the cell, these interactions appear to reflect a simple preference of the cell membrane for its most adhesive environment.

Temporal Regulation of NCAM.

Two examples (Figure 2) have been found in which changes in the amount of NCAM expressed by cells at different stages of development lead to important rearrangements of tissue structure. The first concerns the migration of the neural crest cells destined to become spinal ganglia (5). Prior to their migration, these cells express NCAM and adhere strongly to the dorsal edge of the neural tube. During migration, they appear to no longer have NCAM on their surface, and instead use adhesion to a extracellular matrix-rich pathway to reach their appropriate site. Upon arrival, they once again express NCAM and are observed to reaggregate to form a compact ganglion.

Fig. 2. Examples of temporal variations in NCAM expression that contribute to tissue development.

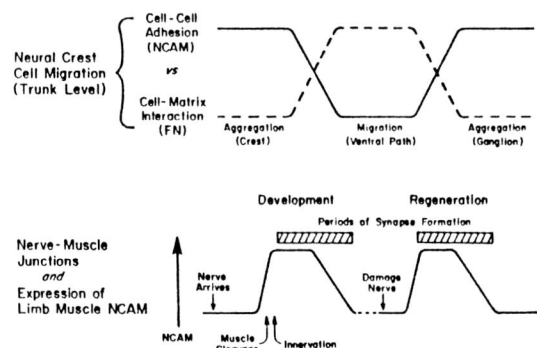

The second example involves the initial innervation of skeletal muscle by spinal cord neurons. In this case, the NCAM-rich nerve appears to wait at the periphery of the muscle until the latter, by its own internal program, also produces large amounts of the molecule (6).

At that time, the nerve-muscle association becomes more intimate, with extensive ramification of the axon into the muscle, and ultimately results in the formation of electrically active synapses. With the onset of activity, NCAM expression by NCAM is again suppressed and remains low unless the nerve's function is impaired, for example by axotomy (7).

Polarity in the Expression of NCAM Across a Cell.

When the axons of retinal ganglion cells exit the eye, they follow a stereotyped route along the outer margin of the neuroepithelium, with their growth cones in close apposition to the endfeet of radial glial-like cells. Studies on the expression of NCAM in the neuroepithelium have revealed that the molecule is not only produced by these radial cells, but that its initial distribution, prior or in response to the arrival of axons, is heavily concentrated at their endfeet (8). This remarkable pattern of expression suggested that the preference of growth cones for endfeet might reflect NCAM-mediated adhesion, a hypothesis that is supported by studies showing that antibodies against NCAM alter the route, but not the growth rate of retinal ganglion cell axons. Thus, the growth cones have sufficient adhesion in the absence of NCAM function to promote axonal growth, but it is their preferential adhesivity to NCAM on the endfeet that helps to locate them at the neuroepithelial margin.

Expression of NCAM in Different Regions of a Tissue.

Optic axons not only grow along the brain margin, but also select a zone of growth that carries them along the roof of the brain to their ultimate target, the optic tectum. In the same studies cited above (8), it was found that the neuroepithelium restricts NCAM expression to a population of endfeet that constitutes a "pathway" that conforms to this growth zone. Again, the pattern of NCAM was present prior or in immediate response to the arrival of optic axon growth cones. Thus it would appear that the regulation of NCAM, in this case within a sheet of cells, creates a situation in which a relative preference of growth cones for NCAM-mediated adhesion contributes to the specification of an axonal pathway.

While the examples given emphasize the attraction of NCAM-producing cells for axons, it is also appears that a localized absence of this molecule can help to create a specific barricade (9). In the forebrain of vertebrates, nerve fibers of the olfactory and visual systems do not intermix, even though they come into close proximity within a continuous epithelium. The separation between the two fiber systems occurs at the junction between the telencephalon and diencephalon, and the neuroepithelial cells which occupy this boundary are remarkable in their absence of both well-formed glial endfeet and of NCAM. Thus both anatomical and chemical factors appear to combine to create a zone of tissue that is refractory to the ingrowth of axons.

A similiar situation may also exist in certain muscles, where the regulation of NCAM expression can exhibit a spatial component as well as the temporal program described above. In this case, the NCAM-negative zone lies along the presumptive cleavage plane of the primitive muscle mass, and may help to prevent axons destined for one muscle from crossing over into a region destined to become a different muscle (6).

Variations in the Glycosylation of NCAM.

There are a variety of transcriptional and post-translational events that give rise to NCAM molecules with different structures (see 3). Of these, the variations in sialic acid content are of particular interest, in that a decrease in the amount of this sugar is associated with an enhancement of NCAM-mediated cell-cell adhesion. Such variations can occur both as a function of age and tissue source, and therefore represent a potential mechanism for altering adhesive preferences during development. For example, at the time when the visual system is formed, the NCAM on optic axons within the eye is of the less-sialylated (more adhesive) form, whereas outside the eye it is of the heavily-sialylated form (10). From these obsevations, has been suggested that the more adhesive form in the eye helps to keep the optic fibers together as they collect into the optic nerve, whereas the less adhesive form allows the axons to begin rearranging themselves and ultimately to respond to positional cues on the tectal surface.

References

(1) Holtfreter, J. (1939) in Foundations in Experimental Embryology B Willier and J Oppenheimer, eds. Englewood Cliffs, NJ, Prentice-Hall (1964), pp 186-225.

(2) Steinberg, M.S. (1970) J. Exp. Zool. 173:395-434.

(3) Rutishauser, U. and Goridis, C. (1986) Trends in Genet. 2:72-76.

(4) Rutishauser, U. (1984) Nature 310:549-554.

(5) Thiery, J.-P., Duband, J.-L., Rutishauser, U., and Edelman, G.M. (1982) Proc. Natl. Acad. Sci. 79:6737-6741.

(6) Tosney, K., Watanabe, M., Landmesser, L., and Rutishauser, U. (1986) Dev. Biol. 114:437-452.

(7) Covault, J. and Sanes, J. (1985) Proc. Natl. Acad. Sci. USA 82:4544-4548.

(8) Silver, J. and Rutishauser, U. (1984) Dev. Biol. 106:485-499.

(9) Silver, J., Posten, M., and Rutishauser, U. (submitted).

(10) Schlosshauer, B., Schwartz, U., and Rutishauser, U. (1984) Nature 310:141-143.

CHARACTERIZATION OF THE HEPARIN-BINDING DOMAIN OF THE NEURAL CELL ADHESION MOLECULE N-CAM

G.J. Cole and L. Glaser

Dept. of Biochemistry, University of Miami School of Medicine, P.O. Box 016129, Miami, FL. 33101

The neural cell adhesion molecule N-CAM is a well characterized adhesive protein that has been shown to play an important role in a variety of neuronal cell-cell interactions (1,2). These proposed functions include neurite fasciculation, histogenesis in the developing retina, neuron-glial interactions, and neuron-muscle adhesion. Previous studies in our laboratory have described the presence of a heparin-binding domain in the N-CAM molecule (3,4). This domain appears to be an integral component of N-CAM-mediated cell interactions. Accordingly, heparan sulfate or its structural analog heparin inhibit cell-cell and cell-substratum adhesion; a monoclonal antibody (B_1A_3) that recognizes the heparin-binding domain of N-CAM also inhibits these processes (5). In the present studies we were interested in furthering our understanding of the relationship between the structure of the N-CAM molecule and its function. We have also initiated studies designed to examine the properties of the heparin molecules that interact with N-CAM.

TOPOGRAPHIC LOCALIZATION OF HEPARIN-BINDING DOMAIN.

Studies by Cunningham et al (6) described the localization of the cell- or homophilic- binding domain of N-CAM on a linear map of the molecule. These studies indicated that a 65 kD amino-terminal fragment of N-CAM (named Fr1) contained the cell-binding domain. In the present study we wanted to determine the topographical location of a second functional domain in the N-CAM molecule: the heparin-binding domain. Preliminary experiments indicated that the B_1A_3 monoclonal antibody also reacted with a 65 kD proteolytic fragment of N-CAM, which raised the possibility that the heparin-binding site was also associated with the amino-terminal region of the polypeptide (7). Proteolytic digestion of N-CAM with S. aureus V8, which produces the Fr1 fragment, demonstrated that the B_1A_3 monoclonal antibody binds Fr1. A second monoclonal antibody, C_1H_3, which also inhibits cell-cell adhesion did not bind to Fr1 and therefore demonstrates that two physically distinct functional domains exist on N-CAM. To determine whether the 25 kD heparin-binding domain is located at the amino-terminal region of Fr1, we isolated the 25 kD polypeptide fragment and

performed amino-terminal amino acid sequence analysis. Table 1 shows the amino acid sequence obtained for the heparin-binding domain, and the previously reported sequence for the amino-terminus of N-CAM (8). Our data indicate that the heparin-binding domain of N-CAM is localized to the amino-terminus, and raise the possibility that the previously described cell-binding region is the heparin-binding domain.

Table 1. **Amino acid sequence of amino-terminus of the heparin-binding domain of N-CAM.**

Amino acid sequence	
25 kD fragment:	Leu-Gln-Val-Asp-Ile-Val-Pro-Ser-Gln-Gly
Amino-terminus, N-CAM	Leu-Gln-Val-Asp-Ile-Val-Pro-Ser-Gln-Gly-Glu-Ile-Ser-Val-Gly-Glu-Ser

We have constructed two possible linear models of the N-CAM molecule to take into account the binding of various monoclonal antibodies to N-CAM (Fig.1); these antibodies inhibit NCAM-mediated cell interactions and therefore identify potential functional domains. In A, we propose that the cell-binding domain is not located on Fr1, since the C_1H_3 MAb does not recognize Fr1 but does inhibit cell-cell and cell-substratum adhesion. In B, we show that Fr1 contains both the cell- and heparin-binding regions of N-CAM. This model incorporates data from studies in our laboratory and from Cunningham et al (6). It should be noted, however, that these models assume that only two sites on N-CAM are required for function. It is possible that other regions, which may be overlapping, are also involved in N-CAM function; these sites may therefore be affected by the C_1H_3 MAb or other MAbs.

Characterization of heparin-binding to N-CAM.

Our laboratory has also been interested in determining the structure of the heparin molecules that interact with N-CAM. Studies in other laboratories have demonstrated that specific heparin sequences are required for binding to antithrombin III (9) and for inhibiting smooth muscle cell proliferation (10). Our experiments (Cole, G.J., Maimone, M.M., Tollefsen, D., Loewy, A., and Glaser, L., in preparation) also suggest that a specific heparin structure may be required for binding to N-CAM, although the structure has not been elucidated. Our results do indicate, however, that heparin fragments

bind with lower affinity to N-CAM than do
intact heparin molecules. In order for
[3H]heparin fragments to bind to N-CAM
with high affinity, they must be at least
a decasaccharide; octasaccharide
fragments of heparin also interact with
N-CAM, although at an apparently lower
affinity. Using intact heparin molecules
we have also demonstrated that molecules
that either pass through or are retained
on an antithrombin III column can
interact with N-CAM. It thus appears
that the heparin structure necessary for
the binding of heparin to N-CAM is not
identical to that which binds to
antithrombin III.

These heparin fragments have also
been employed in cell adhesion assays to
determine whether a specific heparin
structure is required for N-CAM function.
Our data indicate that fragments smaller
than an octasaccharide do not inhibit N-
CAM-mediated cell-substratum adhesion.
These results therefore suggest that a
specific heparin structure is required
for N-CAM function, although this
structure remains to be identified.

Figure 1. Linear models depicting
proposed locations of functional domains
of N-CAM. In a, the heparin-binding
domain is aligned at the amino-terminus
of N-CAM, and the cell (homophilic)-
binding region is aligned at least 40 kD
from the heparin-binding domain. In b,
both the heparin- and cell-binding
domains are proposed to be located in the
Fr1 fragment region of N-CAM. See text
for a more detailed description.

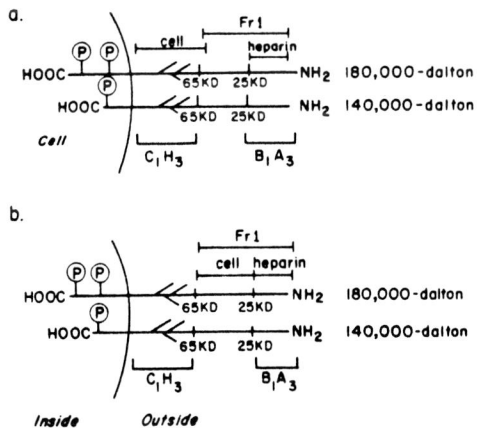

References

1. Edelman, G.M. (1983) Science 219, 450-457.
2. Rutishauser, U. (1984) Nature 310, 549-554.
3. Cole, G.J., Schubert, D. and Glaser, L. (1985) J. Cell Biol. 100, 1192-1199.
4. Cole, G.J. and Glaser, L. (1986) J. Cell Biol. 102, 403-412.
5. Cole, G.J., Loewy, A. and Glaser, L. (1986) Nature 320, 445-447.
6. Cunningham, B.A., Hoffman, S., Rutishauser, U., Hemperly, J.J., and Edelman, G.M. (1983) Proc. Natl. Acad. Sci. USA 80, 3116-3120.
7. Cole, G.J., Loewy, A., Cross, N.V., Akeson, R., and Glaser, L. (1986) J. Cell Biol., in press.
8. Rougon, G. and Marshak, D.R. (1986) J. Biol. Chem. 261, 3396-3401.
9. Thunberg, L., Backstrom, G., Grundberg, H., Riesenfeld, J., and Lindahl, U. (1980) FEBS Lett. 117, 203-206.
10. Castellot, J.J., Beeler, D.L., Rosenberg, R.D., and Karnovsky, M. (1984) J. Cell. Physiol. 120, 315-320.

COMPLETE NUCLEOTIDE SEQUENCE AND BIOLOGICAL EXPRESSION OF HUMAN 4β-GALACTOSYLTRANSFERASE

B. Bunnell[1], M.G. Humphreys-Beher[1] and V.J. Kidd[1,2]

Departments of Microbiology[1] and Cell Biology and Anatomy[2], University of Alabama-Birmingham, Birmingham, Alabama 35294

4β-Galactosyltransferase is the most common of the galactosyltransferases in glycoprotein biosynthesis. This enzyme is involved in the addition of Galβ1→4 to GlcNac in glycolipids, proteoglycans and N- and O-linked oligosaccharides. Most glycosyltransferases are membrane associated, although soluble forms of galactosyltransferase have been found in milk, serum and saliva. Membrane associated galactosyltransferases have been localized to the Golgi, but a portion of 4-galactosyltransferase activity has been found at the cell-surface (1). Numerous reports have shown that alteration of cell-surface 4-galactosyltransferase activity might be involved in cell adhesion (2), recognition (3), differentiation (4) and embryogenesis (5). To further define the potential role of this enzyme in these events we have cloned a human cDNA corresponding to 4-galactosyltransferase (6). In this report we present experimental evidence for the cloning, sequencing and expression of human 4-galactosyltransferase.

A full-length human cDNA clone has been isolated from a human liver λgt-11 cDNA expression library. A putative leader peptide of 20 amino acids as well as N-terminal sequences that correspond to previously determined bovine and human (7) 4β-galactosyltransferase proteins have been established. We have established the entire sequence of the mature unglycosylated protein, which in humans corresponds to 44,000 daltons. This clone also contains a 3' untranslated region of approximately 180 bp and a normal polyadenylation signal.

We have utilized this cDNA clone to express active recombinant human 4β-galactosyltransferase in transient and stable transformation experiments. Using either a vector or vector plus 4β-galactosyltransferase in a negative transcriptional orientation, we see only background levels of 4β-galactosyltransferase activity due to endogenous cellular and serum protein (Fig. 1, panels A and B). Conversely, when the vector contains the 4β-galactosyltransferase cDNA in a positive transcriptional orientation we observe a three-fold increase in levels of the enzyme found in media (Fig. 1, panel A) and a two-fold increase in the enzyme found in cell lysates (Fig. 1, panel B). To test whether or not this enzyme activity was due to the 4β-galactosyltransferase, and not another galactosyltransferase, we took advantage of the observation that α-lactalbumin will alter the K_m of this enzyme to utilize glucose (8). Expression experiments and high voltage paper electrophoresis indicate that the 4β-galactosyltransferase enzyme activity found in cells transfected with the vector plus the cDNA in a positive transcriptional orientation does, indeed, utilize glucose only in the presence of added α-lactalbumin (Fig. 1, panel C). This conclusively demonstrates the ability of this human cDNA clone to produce recombinant 4β-galactosyltransferase.

Further experiments are now being done to demonstrate the possible role of 4β-galactosyltransferase in various cellular functions at the molecular level. These include possible involvement in cell growth and development as well as its function as a posttranslational-modification enzyme.

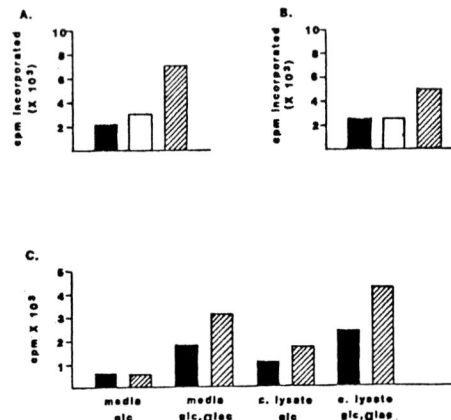

Figure 1. Bar graph representations of 4β-galactosyltransferase activity in COS M-6 Cells. Activity of the enzyme was determined by incorporation of [^{14}C]-galactose using ovalbumin as acceptor. The solid bars represent cells transfected with the vector only. Stipled bars represent the vector plus GT in a positive transcriptional orientation. Open bars represent the vector plus GT in a negative transcriptional orientation.

REFERENCES

(1) Roth, S., McGuire, E.J. and Roseman, S. (1971) J. Cell Biol. 51, 536-542.
(2) Roseman, S. (1970) Chem. Phys. Lipids 5, 270-297.
(3) Pierce, M., Turley, E.A. and Roth, S. (1980) in International Review of Cytology, pp. 2-44, Academic Press, New York.
(4) Weiser, M.M. (1973) J. Biol. Chem. 248, 2536-2549.
(5) Shur, B.D. (1982) in The Glycoconjugates Vol. 3, pp. 146-185, Academic Press, New York.
(6) Humphreys-Beher, M.G., Bunnell, B., VanTienan, P., Ledbetter, D. and Kidd, V.J. Proc. Natl. Acad. Sci., U.S.A., in press.
(7) Appert, H.E., Rutherford, T.J., Tarr, G.E., Thomford, N.R. and McCorquodale, D.J. (1986) Biochem. Biophys. Res. Comm. 138, 224-229.
(8) Brew, K., Vanaman, T.C. and Hill, R.L. (1968) Proc. Natl. Acad. Sci., U.S.A. 59, 491-497.

FLUORESCENT MICROSPHERES FOR DIRECT LOCALIZATION OF CELL SURFACE ANTIGENS AND SINGLE GENE SEQUENCES

S.W. Cheung, J.P. Crane, V. Hauptfeld*and F. Sweet

Department of Obstetrics and Gynecology and *Depart-ment of Genetics, Washington University School of Medicine, St. Louis MO 63110

During the past decade, fluorochrome labels have been used in methods for localizing cell surface antigens and detection of specific nucleic acid sequences (1-3). We report a technique for direct localization of cell surface antigens and single gene sequences with fluorochrome-labeled, 300 Å latex microspheres. An intense fluorescent signal produced by individual microspheres allows precise topological resolution of antigenic sites with minimum background fluorescence. Conjugation of DNA to the fluorescent microspheres is achieved with an avidin-biotin complex as a bridge. Rapid, high resolution mapping of single gene sequences may soon be accomplished with these new, non-autoradio-graphic probes.

Latex microspheres(300 Å) were synthesized for the present study by an established method(1). An intense fluorescent signal was produced by re-acting the methyl and hydroxyethyl ester groups on the microsphere surface with dansylcadaverine (80%) and diaminododecane(20%) overnight at room temper-ature (pH 12). The microspheres form covalent bonds by a transacylation reaction between the amine moieties and ester groups. Extent of conjuga-tion was measured by incorporation of radioactive ethanolamine. Sphere diameter and the number of spheres per unit volume was measured by electron microscopy. Approximately 6.4×10^6 molecules of amine/um^2 were covalently attached per fluorescent microsphere, well above the minimum of 1,000 mole-cules per um^2 needed for visualizing the micro-spheres by fluorescence microscopy (3). Therefore, a distinct signal is produced (Figure 1a). Either a monoclonal antibody or avidin was covalently attached to the fluorescent microspheres for local-ization of cell surface antigens or single gene sequences.

The fluorescent microspheres were tested for visualizing a specific cell surface antigens by attaching them to a monoclonal antibody. In the present work, the well established anti-H-2K puri-fied monoclonal antibody was used for localization of the major histocompatibility(MHC) class I anti-gen on the cell surface of mouse splenocytes. Flu-orescent microspheres were covalently conjugated with monoclonal antibody following a two step glu-taraldehyde reaction (1). Unconjugated antibody was removed by ultracentrifugation(Spinco SW 50.1 rotor 35K rpm 1 hr) with a sucrose density gradient (60% w/w overlayered with 10-20%). The antibody-microsphere conjugates were recovered at the 20-60% interface and dialyzed overnight against phosphate buffered saline (PBS), pH 7.4. Binding specificity of the antibody-conjugated microspheres was deter-mined by incubation with splenocytes carrying the H-2K antigen (group 1), splenocytes without rele-vant H-2K antigen (group 2), and splenocytes with H-2K antigen but preincubation with free monoclonal antibody (group 3). Group 2 and 3 served as con-trols. Up to 90% of splenocytes in Group 1 were intensely fluorescent while only 10% weakly fluore-scent splenocytes were noted in the control groups.

A second potential application of the fluore-scent microsphere technology is localization of single gene sequences with a biotinylated DNA probe.

Fluorescent microspheres were covalently attached to avidin by a glutaraldehyde reaction similar to that used for conjugating monoclonal antibody. Avi-din-conjugated fluorescent microspheres were incub-ated with a biotinylated plasmid DNA encoding the heatshock protein gene sequences. An example of DNA biotin-avidin-microsphere complex is shown in Figure 1b.

Advantages of fluorescent microsphere technology for visualizing antigenic sites and DNA sequences can be summerized as follows 1) Any fluorochrome with a primary amino group can be readily attached to 300 Å spheres, 2)the number of dansyl-cadaverine molecules per sphere is well above the resolution limits of fluorescence microscopy,thereby producing a distinct signal, 3) they are chemically stable in single suspension for at least one year, 4) cell surface antigens can be directly localized without a secondary antibody, and 5) conjugation of DNA to fluorescent microspheres is possible by taking adv-antage of the strong non-covalent interaction be-tween avidin and biotin. Langer-Safer have estim-ated that 50 biotin molecules can be incorporated per 1 kb of DNA using nick translation (2). Our preliminary data suggest that more than 175 biotin molecules per 1 kb can be incorporated using oligo-labeling (4). In situ hybridization using biotiny-lated DNA probes followed by incubation with avidin-fluorescent microsphere complexes could therefore potentially allow high resolution mapping of single gene sequences.

The presently reported, fluorescent microsphere techniques provide versatile tools with wide appli-cations for the detection of minute quantities of biological substances. The methodology is rapid, simple, safe and highly sensitive.

REFERENCES

1. Rembaum, Dreyer WJ (1980) Science 208:364-368.
2. Langer-Saver PR, Levine M, Ward DC (1982)Pro Nat Acad Sci (USA) 79:4381-4385.
3. Landegent JE, Jansen in de Wal N, Van Ommen GJB, Baas F, de Vijlder JJM, Van Duijn P, Van der Pleog M (1985) Nature 317:175-177
4. Feinberg AP,Vogelstein B (1983) Anal Biochem 132:6-13.

Figure 1a Photomicrograph shows a microsphere emitting its fluorescent light (arrow). Compare its size to a human chromosome 1 (2500 x)

Figure 1b Electron photomicrograph of DNA-biotin-avidin-microsphere complexes (26,800 X).

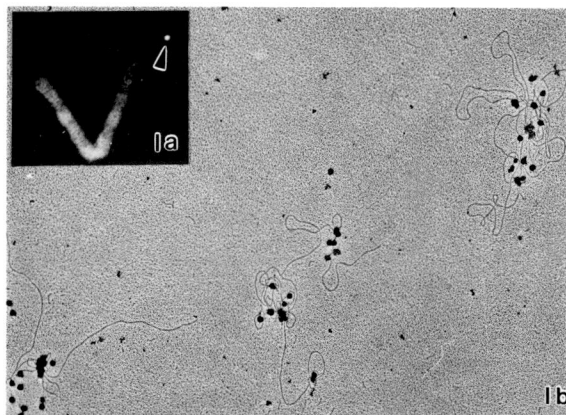

SURFACE BOUND REGULATORS OF ANTIGEN SYNTHESIS IN PARAMECIUM

I. Finger, R. Flaumenhaft, S. Vorenberg, and R. Min.

Biology Department, Haverford College, Haverford PA. 19041

Individual paramecia can make more than a dozen surface antigens (molecular weights of about 300 Kdaltons), but usually express only one at a time. A model for this mutual exclusion suggests that an antigen binds to antigen-specific inhibitor acting at the transcriptional level, thus controlling its own synthesis (1). We have shown that purified antigen indeed can specifically direct transformation (2), and that a cell fraction containing several kinds of about 70 KD molecules also appears to induce transformation (3). A comparison of several characteristics of these two kinds of mediators suggests that although they are both surface proteins they differ in several significant ways.

(I) LOCATION OF INHIBITORS. Rabbit antisera against conditioned media devoid of surface antigens can have a pronounced effect on antigen expression (as do the media themselves with different media containing different levels of individual inhibitors). All media that are effective in transforming cells have 70 KD molecules (3). Anti-media sera and homologous anti-surface antigen sera when analyzed by the ELISA test (enzyme-linked immunosorbent assay) react differentially with media eliciting transformation to different serotypes. Although neither kind of antisera by themselves immobilize heterologous cells, adding a second antibody (sheep anti-rabbit immunoglobulin) can retard or immobilize cells of a specific serotype. Which serotype (G, C or X) is affected is strongly correlated with the ELISA readings observed between these same sera and media with different inhibitor content (Figure 1). Antisera prepared against 70 KD protein eluted after electrophoresis of partially purified antigen also shows this effect, i.e. immobilization of specific serotypes following sequential treatment with 70 KD antiserum and secondary antibody.

From these data we draw two tentative conclusions:
1. Surface antigen purified by standard salting out procedures (4) contains heterologous inhibitor which copurifies with the antigen since antisera against the antigen appears to have antibody directed against inhibitor.
2. Inhibitor may be located on the cilia surface as antisera reacting with inhibitor cause the cessation of ciliary movement of specific serotypes if antibodies added are crosslinked by a non-specific anti-rabbit immunoglobulin serum.

(II) EVIDENCE FOR BINDING OF SURFACE ANTIGEN WITH INHIBITOR. The fact that inhibitor is purified together with surface antigen suggests that antigen and inhibitor complex with each other. Purified antigen when treated with mercaptoethanol and sodium dodecyl sulfate yields greatly increased amounts of the 70 KD fraction, which when eluted from an acrylamide gel transforms. Ammonium sulfate purified antigen passed through Sephadex P100 yields one major peak and one or two minor ones. The predominant fraction consists of 300 KD antigen and 70 KD inhibitor (Figure 2), consistent with the view that the two molecules bind to each other.

Inhibitor can be freed of antigen by a mercaptoethanol activated protease present in purified antigen preparations (5), which degrades the antigen and other proteins but leaves the inhibitor largely unaffected.

Thus, the synthesis of surface antigens appears to be under the dual control of two surface proteins, the antigen itself and an inhibitor. These proteins differ in molecular weight, susceptibility to a protease and, according to preliminary experiments, mechanism of regulation (6). There are also observations that indicate that the antigen binds non-covalently to the inhibitor.

Fig. 1. Comparison of anti-inhibitor titers using ELISA and secondary immobilization tests.

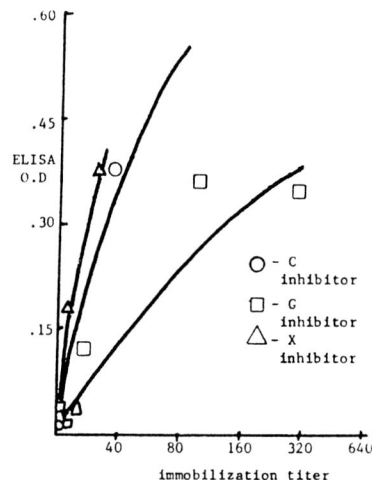

Each symbol represents an individual antiserum against heterologous antigen.

Fig. 2. Separation of surface antigen on Sephadex into 300 KD and 70 KD fractions.

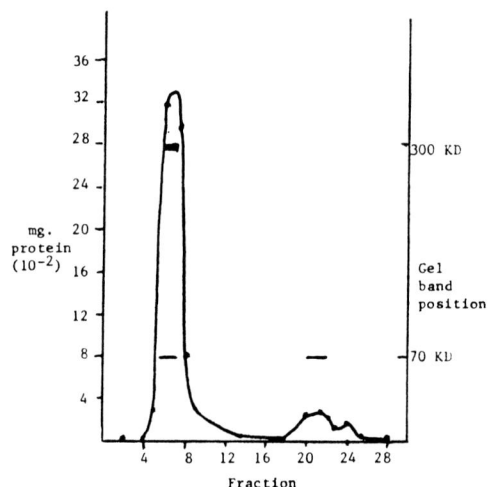

REFERENCES

(1) Finger, I. (1967) in The Control of Nuclear Activity (Goldstein, L., ed.) pp. 377–411. Prentice-Hall, Englewood Cliffs, N. J.
(2) Finger, I. (1984) Genetics, 107:S32
(3) Finger, I. and Audi, D. (1985) J. Cell Biology 101:284.
(4) Preer, J. R., Jr. (1956) J. Immunol. 77:52-60.
(5) Hansma, H. G. (1975) J. Protozool 22:257-259.
(6) Finger, I. and Birnbaum, M. (unpublished).

A ROLE FOR THE MAJOR HISTOCOMPATIBILITY COMPLEX IN NORMAL DIFFERENTIATION OF NON-LYMPHOID TISSUES

Lois A. Lampson and James P. Whelan. Children's Cancer Research Center, Children's Hospital of Philadelphia, Philadelphia, PA. 19104

Products of the major histocompatibility complex (MHC) mediate cellular interactions in the immune response. Many investigators have suggested that this is an example of a more general cell recognition function. Yet direct evidence for this hypothesis has been lacking.

We have examined the distribution of β2-microglobulin in well-fixed mouse embryos. β2-m is the invariant light chain of class I MHC products (H-2K,D,L in the mouse and HLA-A,B,C in man). It thus serves as a broad probe for the class I family.

Formalin-fixed, paraffin-embedded mouse embryos were examined in serial sections of embryos at 7 to 14 days of gestation. For the early days, the entire uterus was dissected; at later times, the embryo was dissected free of extraembryonic tissues.

DEVELOPING MUSCLE

Extraembryonic staining was seen at day 7, with the strongest stain in regions closest to the embryo proper. Stain in the embryo proper was first seen at day 8, in developing cardiac muscle. In subsequent days, transient β2-m expression was seen in different regions of the developing heart muscle, with the stain progressing from myocardium to pericardium (Table I). Transient expression was also seen developing skeletal muscle, but was never seen in smooth muscle. Transient expression was also seen in developing chondroblasts. These cell types are of interest because none of them expresses β2-m or other MHC products in the normal adult.

β2-m staining was also seen in cells that are positive in the normal adult, including fibroblasts, epithelial cells, and hepatocytes. All of this expression occurred before the development of lymphoid tissue.

TABLE I. β2-m Expression in Developing Cardiac Muscle.

Day of Gestation	Myocardium	Endocardium	Perdicardium
8	+	-	
9	++	±	
11	+	±	-
12	++	±	+
13	+++	±	+
14	±	+	+
Adult homologue	-	-	ND

NEURAL TISSUE

The expression of β2-m in developing neural tissue was of particular interest. MHC products are not expressed by the majority of normal adult neurons or glia, yet the molecules are under regulatory control.[1-3] It has been of interest to know whether this modulation could serve a non-immunological function, in addition to or instead of the established immunological role.[4]

In practice, we did not find evidence of β2-m expression in developing neurons or glial cells of the neural tube, neural crests or developing brain. In the neural tube, stain was confined to surrounding connective tissue. Consistent with these results, we have not observed β2-m expression in any cell or layer of the olfactory epithelium, which is a neuroepithelium that turns over even in the adult.[5] Nor is β2-m expression required for any stage of division or morphological form, or the establishment of processes in human neuronal cell lines in culture.[6]

CONCLUSION

The transient expression of β2-m in developing muscle and other tissues before a functioning immune system is present suggests that β2-m, and perhaps class I MHC products, may play a non-immunological cell recognition role in development. No evidence for such a role in developing neurons or glia was found. Rather, the MHC modulation that has been observed in adult neural tissue seems likely to serve a more conventional immunological role.

REFERENCES

1. Lampson, L.A. and Hickey, W.F. (1986) J. Immunol. 136, 4054-4062.
2. Lampson, L.A. and Fisher, C.A. (1984) Proc. Natl. Acad. Sci., 81, 6476-6480.
3. Wong, G.H., Bartlett, P.F., Clark-Lewis, I., Battye, F., and Schrader, J.W. (1985) J. Neuroimmunol. 7, 225-227.
4. Lampson, L.A. (1984) in Monoclonal Antibodies and Functional Cell Lines. R.H. Kennett, K.B. Bechtol, and T.J. McKearn, eds. Plenum Press, New York, pp. 153-189.
5. Whelan, J.P., Wysocki, C.J. and Lampson, L.A. (1986) J. Immunol., in press.
6. Lampson, L.A. Distribution of β2-microglobulin and class I molecules in a neuronal cell line of complex morphology. Submitted.

EXPRESSION OF THE MURINE MOX 6.3 HOMEO BOX PROTEIN IS REGULATED BY CELL-CELL CONTACT

W. Odenwald, C. Taylor, F. Palmer-Hill, M. Tani, and R. Lazzarini

Laboratory of Molecular Genetics, NINCDS, NIH, Bethesda, Maryland 20892

As an initial step toward understanding the function of mammalian homeo box genes, we have focused our efforts on the molecular biology of the murine Mox 6.3 gene. From genomic and cDNA sequences, we have determined the primary structure of the Mox 6.3 transcriptional unit and studied its expression in embryonic and adult tissues by Northern analysis. Studies with antibodies generated against synthetic peptides have revealed that the Mox 6.3 protein is present in the nuclei of primary cultured cells which possess few or no cell-cell contacts and absent from the nuclei of cells that have many cell-cell contacts.

CHARACTERIZATION OF THE MOX 6.3 TRANSCRIPTIONAL UNIT. We have cloned and sequenced a 1.6 Kb cDNA from an 18 day old whole mouse brain library utilizing the Antp homeo box as a probe. This cDNA was employed to isolate and characterize five overlapping lambda genomic clones which under reduced stringency contain additional homeo box homologies. The genomic clones were demonstrated to contain part of the Mox 6 locus located on chromosome 6 (1, 2, 3) by the following criteria: positions of homeo box homologies, restriction site homologies and overlapping homology with the lambda M-6, Mox 6 genomic clone (1, 2). Genomic sequence analysis of 2.6 Kb surrounding the Mox 6.3 homeo box demonstrated that the cDNA was derived from the Mox 6.3 transcriptional unit. The Mox 6.3 "TATA" box is 51 bp upstream from the first initiator ATG codon and this translation reading frame predicts a protein 269 amino acids in length. The protein is rich in serine (13%), glycine (10%) and proline (9%) residues. The homeo domain is positioned 15 amino acids from the carboxy terminus. The next downstream initiator codon predicts a 229 amino acid homeo domain containing protein. The transcriptional unit contains two exons which are separated by a 960 bp intron positioned 21 bp upstream from the homeo box. The 3' untranslated sequence is 792 bp long with the polyadenylation signal located 12 bp upstream of the poly A tail. In addition to the homeo box homology, another region of partial conservation has been identified. The core sequence TACCCCTGGATG (coding for the amino acids, Tyr Pro Trp Met) which is located 21 bp upstream from the 3' end of the first exon is also found in a similar position in several other homeo box genes. The hexanucleotide repeat $\frac{ACAAAA}{TGTTTT}$ present in the Krüppel and fushi tarazu genes of Drosophila was observed in both orientations 13 times with 9 of these repeats in the 3' trailer. No Opa (M) repeats were found. Comparisons to other homeo box genes revealed a striking homology to the Mox 11.4 gene located on chromosome 11 (4, 5, 6). Both the Mox 6.3 and Mox 11.4 proteins (7) are predicted to be 269 amino acids in length with their homeo domains ending 15 amino acids from the carboxy terminus. Their homeo boxes are 83.6% homologous at the nucleotide level while the amino acid sequence of the two domains are identical. Eleven of the 15 amino acids at the carboxy terminus are also identical. Southern analysis of a genomic clone containing the Mox 11.4 gene (4) with probes from either the first exon of the Mox 6.3 gene or its 3' untranslated trailer revealed that the first exons of these two genes share weak homology while no homology was detected between the 3' untranslated trailers.

EXPRESSION. Northern analysis of mRNA from 9 day gestation whole embryos with Mox 6.3 probes lacking the homeo box revealed two major transcripts of approximately 1.8 and 1.9 Kb. Three additional less abundant transcripts (approx. 4, 8, and 9 Kb) were also detected. These transcripts were also found in mRNA isolated from head, spinal column and the body of 17 day embryos. The highest levels of the major transcripts were observed in the 17 day spinal columns. mRNA isolated from adult liver, kidney, ovary, testis, spinal column and brain demonstrated that the 1.8 and 1.9 Kb transcripts were present in these tissues, but are approximately 10-50 fold less abundant when compared to the embryonic tissue. The spinal column contained the highest level of these major transcripts compared to the other adult tissues. The less abundant 4, 8 and 9 Kb transcripts were also observed in the adult liver, kidney, ovary and spinal cord mRNAs.

LOCALIZATION OF THE MOX 6.3 PROTEIN IN CULTURED CELLS. Two synthetic peptides (20 mers) were employed to immunize rabbits against the Mox 6.3 protein. Both peptides correspond to amino acid sequences derived from the first exon and not the homeo domain. Immunofluorescence studies, using Balb/c 16 day gestation embryonic fibroblasts grown under standard conditions on glass slides, yielded the following results: the nuclei and perinuclear cytoplasmic regions of non-confluent cells (cells with few or no cell-cell contacts) stained positively. In areas of the same culture where cells were confluent (numerous cell-cell contacts), the nuclei did not stain and the intensity of the perinuclear staining was reduced. In regions of transition between high and low cell density, both fluorescent patterns were observed. Cells with negative nuclei frequently had intense perinuclear staining in those areas of intermediate cell density. Cells with weak nuclear fluorescence were also found in these regions. The intensity of fluorescence appeared to be related to the degree of cell-cell contacts. Independent of the intensity of nuclear fluorescence, the nucleoli and/or heterochromatic regions did not stain. Monolayer wounding experiments revealed a similar localization profile. Cells present at the edge of the wound and those which migrated into the clearing away from the monolayer had positive nuclei while those in the monolayer were negative. Again, there was a gradient of fluorescence which reflected the degree of cell-cell contacts. Antisera to both synthetic peptides produced the same pattern of fluorescence. Peptide preabsorption experiments demonstrated that the fluorescent patterns observed were specific for the epitopes present on the peptides.

CONCLUSIONS. The detection of Mox 6.3 transcripts in a variety of embryonic and adult tissues indicates that its expression in the mouse may not be restricted in a developmental stage-specific manner. The localization of the Mox 6.3 protein in the nuclei of non-confluent cultured fibroblasts and its absence in the nuclei of cells which possessed numerous cell-cell contacts (contact inhibited) suggest that cell-cell interactions play a role in the regulation of this homeo box gene.

REFERENCES

1. Colberg-Poley, A. M. et al. Nature **314**, 713-718 (1985).
2. Colberg-Poley, A. M. et al. Cell **43**, 39-45 (1985).
3. Duboule, D. et al. EMBO J. **5**, 1973-1980 (1986).
4. Hart, C. P. et al. Cell **43**, 9-18 (1985).
5. Hauser, C. A. et al. Cell **43**, 19-28 (1985).
6. Jackson, I. J. et al. Nature **317**, 745-748 (1984).
7. Krumlauf, R. et al. EMBO Dev. Biol. Workshop (1986).

COMPLEX CELL SURFACE CARBOHYDRATES EXPRESSED BY NEURAL CREST CELLS DIFFERENTIATING IN CULTURE

Maya Sieber-Blum and Janell Duwell

Department of Anatomy and Cellular Biology, Medical College of Wisconsin, 8701 Watertown Plank Road, Milwaukee, Wisconsin 53226.

INTRODUCTION. One current issue in neural crest research is the question of the origin of the sensory and autonomic neuronal cell lineages. Is there a common precursor cell in the migrating population, or do the two lineages segregate before dissemination of the crest cells from the neural tube? To approach this question by in vitro clonal analysis, we developed culture conditions that support differentiation of neural crest cells into sensory neuroblasts. Neurons observed in crest cell cultures have so far been limited to the autonomic type. We have recently observed unipolar rounded neuroblasts that resembled sensory neurons. Some had substance P-like, vasoactive intestinal polypeptide-like, or 68kD neurofilament polypeptide immunoreactivity, some stained for carbonic anhydrase (1,2). Complex cell surface carbohydrates have been found to be specific for certain subsets of sensory neurons in the rat (3,4) and in the chick (T. Jessell, pers. comm.). In the present study we used monoclonal antibodies directed against cell surface carbohydrates (Table I) to identify subpopulations of sensory neuroblasts developing in primary culture.

Table I

Mab	Carbohydrate Epitope
1B2 (=A5)	Galβ1–4GlcNAc–R
AC4 (= anti-SSEA1)	Galβ1–4(Fucα1–3)GlcNAc–R
anti-SSEA4	NeuAcα2–3Galβ1–3GalNAcβ1–R

MATERIALS AND METHODS Primary neural crest cell cultures were prepared as described (5). Briefly, the neural tubes corresponding in length to the last six segments were isolated from 48 hr quail embryos. They were placed into collagen- (150 ug/plate), fibronectin- (25 ug/plate) and laminin- (30 ug/plate) coated culture dishes. The neural crest cells emigrated from the neural tubes onto the substratum. Twenty-four hrs after explantation the neural tubes were removed, leaving the crest cells in the dish. The culture medium consisted of 75% Alpha MEM, 15% horse serum, 10% chick embryo extract, antibiotics, and was supplemented with 66 ng/ml of nerve growth factor. For antibody staining, 3 week old formaldehyde-fixed neural crest cell cultures were exposed to the primary antibody over night at 4°C in the presence of 0.1% Triton X-100 and incubated thereafter with the secondary antibody for 1 hr at room temperature.

RESULTS. Mab 1B2 bound to the surface of rounded cells and their processes. The cells occurred in clusters throughout the culture and were only observed when the neural crest cells were co-cultured with day 8 quail back skin (Fig.1,2). Antibody against SSEA1 stained the cytoplasm and cell surface of cells and processes (Fig.3). They were located at the periphery of the culture and were observed in the absence of co-cultured skin. They were separated from each other and seemed to be in the process of a morphological transition from stellate neural crest cell to rounded unipolar neuroblast. In addition, faint staining on cells resembling nerve supporting cells was observed. Anti-SSEA4 immunoreactivity was present on the

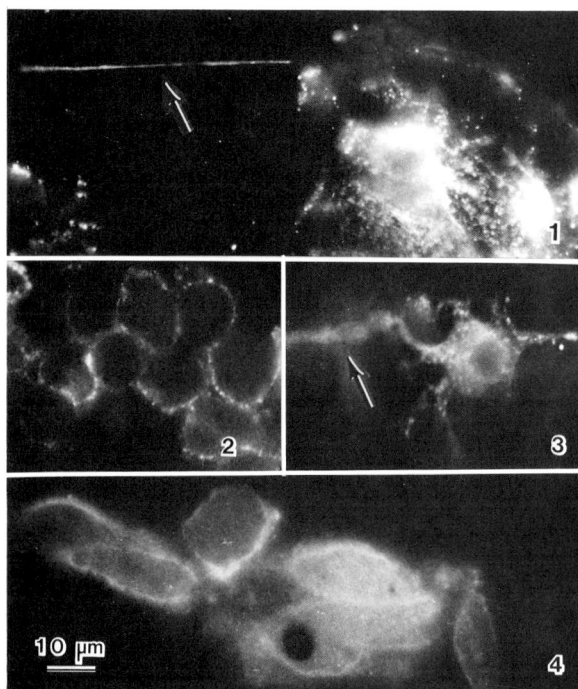

Figures: (1,2) 1B2-immunoreactivity on surface of rounded cells and their processes. (3) Intracellular staining with anti-SSEA1 antibodies. (4) Cell surface staining with anti-SSEA4 antibodies. Arrows, neuronal processes.

surface of rounded or elongated cells that were somewhat larger than 1B2- and anti-SSEA1-positive cells and were found in small aggregates throughout the culture. Fluorescence was evenly distributed on the cell surface and seemed to be absent from the cytoplasm (Fig.4). Fluorescent processes were not observed. None of the three antibodies bound to the numerous autonomic-type neuroblasts which differentiated in the same cultures.

DISCUSSION. The observation that complex cell surface carbohydrates are expressed in some neural crest cells in culture corroborates our earlier finding that neural crest cells can differentiate in culture into sensory-like neuroblasts. The results also suggest that there are subsets of sensory neuroblasts present in the cultures and that the expression of 1B2 is target-dependent.

ACKNOWLEDGEMENTS. This study was supported by NIH grant HD21423 and a grant from the Dysautonomia Foundation. We thank Dr. T. Jessell for a generous gift of carbohydrate-specific antibodies and Dr. T. Borg for providing us with laminin.

REFERENCES

(1) Sieber-Blum, M. and Patel, S.R. (1986) In: Progress in Developmental Biology, Part B, pp. 243-248 (H.C. Slavkin, ed.), Alan R. Liss, Inc., New York.
(2) Sieber-Blum, M., Patel, S.R., and Riley, D.A. (1986), submitted.
(3) Dodd, J. and Jessell, T.M. (1985) J. Neurosci. 5, 3278-3294.
(4) Jessell, T.M. and Dodd, J. (1986) In: Neuropeptides in Neurologic and Psychiatric Disease (Martin, J.B. and Barchas, J.D., eds.) Raven Press, New York.
(5) Sieber-Blum, M. and Cohen, A.M. (1980) Develop. Biol. 80, 96-106.

MORPHOGENESIS

CELLULAR INTERACTION IN VERTEBRATE LIMB PATTERNING

S.V. Bryant, K. Muneoka[*] and D.M. Gardiner

Developmental Biology Center, University of California, Irvine, California 92717
[*]Present address: Department of Biology, Tulane University, New Orleans, LA 70118

The amphibian limb has long been a model for studying the formation of complex, biological patterns. The view that has emerged during the past decade is that limb cells possess positional information, and that they can respond by cell division to positional disparities arising during development, or as a result of wound healing or grafting. This view was formalized in the polar coordinate model in which a set of rules was proposed to govern the nature of such position-specific growth stimulation, referred to as "intercalary growth" (1,2). According to this model, growth during both development and regeneration is a consequence of intercalation which generates new cells with appropriate, intervening positional values whenever positional disparities exist. Results of studies since the polar coordinate model was proposed have consistently strengthened the view that intercalation can account for the initial development of the limb, it's reformation after amputation, and the formation of extra or supernumerary limbs following grafting to create positional disparities (3). Most recently, we have been using heritable cell markers to investigate the underlying cellular mechanisms of limb outgrowth, and this report highlights three major findings:
- the demonstration of intercalary growth in the formation of supernumerary limbs in developing and regenerating limbs.
- the identification of the dermis as a major source of cells for the regeneration blastema.
- the finding that dermal cell migration is one of the earliest events in blastema formation.

SUPERNUMERARY LIMBS. The clearest evidence for the role of intercalation in limb formation comes from experiments in which limb buds or limb blastemas are grafted contralaterally so as to appose anterior and posterior cells at the graft/host junction. Supernumerary limbs develop at the site of such positional value disparities, and their orientation and handedness are consistent with their having arisen by intercalation (1). Our more recent studies of the cellular contribution to supernumerary limbs have considerably strengthened this conclusion (4). If either the host or the graft cells are marked so that they can be distinguished from one another, the pattern of contribution from each side of an experimentally created disparity can be analyzed. We have performed such analyses to study the cellular contribution to supernumerary limbs in developing axolotl limbs, regenerating axolotl limbs, and developing Xenopus limbs (4,5). In axolotls, graft and host cells differed in ploidy (diploid versus triploid), and therefore in their number of nucleoli, which were visualized using a bismuth staining technique (6). In Xenopus, graft and host cells were distinguished by a species-specific difference (X. laevis versus X. borealis) in the quinacrine staining pattern of nuclei (7). In developing and regenerating axolotl limbs, and in the limbs of Xenopus larvae, supernumerary limbs which develop following contralateral limb bud or blastema grafting to confront anterior and posterior tissues are composed of about equal numbers of cells of graft and host origin. Results for the axolotl are summarized in Fig. 1.

Fig. 1. Cellular contribution to supernumerary limbs in axolotls.

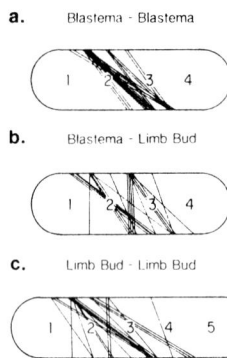

Positions of the boundaries between diploid and triploid tissues in supernumerary limbs. Each oval outline is a stylized outline of a supernumerary limb seen from the distal end. Numbers refer to the positions of digits. Superimposed lines show contribution boundaries from individual supernumerary limbs. Data from grafts in regenerating limbs (a), between regenerating and developing limbs (b), and from grafts between developing limbs (c) are shown. From (4).

The overall finding that supernumerary limbs are produced by essentially equal participation of graft and host cells implies that the interacting cells both respond to and stimulate a response from each other. In other words the interacting cells undergo intercalation. Although the boundary between graft and host derived cells falls approximately in the middle of the supernumerary limb, its position is not constant from limb to limb (Fig.1). This implies that the contribution boundary arises as a result of dynamic interactions between cells rather than as a consequence of any developmental restriction, such as a compartment boundary. The fact that the results are identical whether the grafts are performed in developing or regenerating limbs further suggests that both types of limb outgrowth are governed by similar mechanisms. Direct evidence for the similarity in mechanism between developing and regenerating limbs comes from experiments in which grafts were made between limb buds and blastemas (8,9). The supernumerary limbs which developed from such grafting experiments were identical to those described earlier - they were composed of about half developing and half regenerating cells. We conclude from these results that limb outgrowth, whether it be initial outgrowth in the embryo, or regrowth during regeneration, is driven by intercalation.

ORIGIN OF THE BLASTEMA. Despite the fact that similar mechanisms appear to direct limb outgrowth in development and regeneration, a major difference between the two is the requirement in mature limbs to first form a blastema. It has been known for many years that the blastema arises from mesodermal cells close to the amputation plane, but more detailed information about the composition of the blastema has been lacking, partly due to the absence of a suitable cell marker. We have used the diploid/triploid cell marker to analyze the contribution of dermal and skeletal tissues to the axolotl blastema (10). These two tissues were chosen because they were known from grafting experiments to have disparate effects on limb patterning. The first, dermis, has a major influence on patterning, and even very small grafts implanted so as to create positional disparities can stimulate the formation of supernumerary structures (11,12). The second, skeletal tissue, has little or no effect on patterning, even when large implants are added to limbs (13). In the experiment, either whole skin or the skeletal element was removed from an axolotl limb and replaced by the equivalent element from a similar sized, sibling triploid animal. After a period for healing, the chimeric limbs were amputated and when they formed a medium bud blastema, they were removed and analyzed to determine the fraction of blastema cells derived from the triploid graft. The results from this experiment clearly demonstrate that the differential effects that dermis and skeletal tissue have on limb patterning are reflected in their absolute and relative contributions to the blastema. Dermal cells constitute about 19% of all available mesodermal cells at the plane of amputation, whereas in the blastema, cells derived from dermis make up between 19-78% (x=43%) of the cells. Conversely, cells of skeletal tissues make up about 6% of the stump cells but contribute only 0-3% (x=2%) of the blastemal cells. These results have led to several conclusions about blastema formation. First, cells of dermal origin overcontribute to the blastema relative to their availability in the stump. Second, the fact that a major portion of the blastema is derived from dermal cells is consistent with the major effect that dermal tissue exerts on limb patterning, and suggests that this effect is mediated by means of intercalation. Third, since the cells of the dermis represent about half of all the loose connective tissue cells of the limb stump (10, 14), and contribute nearly half of the blastema cells, it is probable that the majority of the regenerate (with the exception of muscle) is derived from fibroblasts, that is, about half coming from dermal fibroblasts and about half from other connective tissue fibroblasts. By this reasoning, we presently consider fibroblasts of the limb to be the cells that most likely possess positional information and which show position-dependent growth stimulation. Thus, fibroblasts represent the most promising target to date on which to focus efforts to characterize the nature of positional information in the limb.

THE DYNAMICS OF BLASTEMA FORMATION. An explicit prediction of the polar coordinate model is that cell migration and rearrangement will be involved in blastema formation(2). Subepidermal cells at the wound edges are expected to acquire new neighbors as a result of migration, and in so doing, create the positional disparities that stimulate intercalation and hence the growth of the new limb. Since the results described in the section above suggest that cells of dermal origin play a major role in blastema formation, we have investigated their behavior during the early phases of regeneration (15). To do this, chimeric limbs consisting of triploid skin on diploid limbs were made and amputated. Whole mount preparations of the wound region were prepared at different times after amputation. Epidermal wound healing is complete by 12 hours after amputation, whereas by 5 days, cells of dermal origin are absent from the

----

Fig. 2 Cell migration in early regeneration. Computer-assisted plots of the distribution of dermal cells (+) under the wound epidermis at 5 days after amputation (a) and at 10 days (b). The centrally located cells in (a) are cells of stump origin. The outlined area in (b) is cell free. From (15).

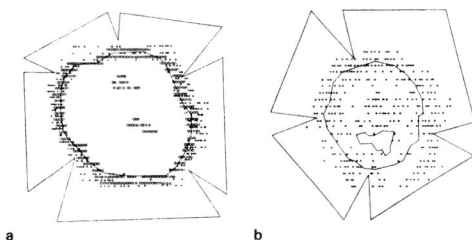

a   b

wound area. However, by 10 days dermal cells have moved across the amputation plane to populate the subepidermal region (Fig. 2). Hence, cells of dermal origin, by their migration, acquire new neighbors both of dermal origin, and of non-dermal origin, thereby establishing the positional value disparities essential for limb outgrowth. Furthermore, independent studies of the onset of cell division in amputated limbs have shown a rise in mitotic index at 4 to 6 days after amputation (16,17,18), coinciding with the onset of dermal cell migration and the acquisition of new cellular neighbors. This correlation is consistent with the view that the initiation of growth leading to regeneration is stimulated by positional disparities.

SUMMARY. Recent cell marker analyses of supernumerary limbs in regenerating and developing amphibian limbs have provided the first direct evidence for intercalation in a vertebrate organism, and have shown that the mechanisms of limb outgrowth during development are the same as those used in regeneration. Second, we have identified the dermis as a major source of cells for regeneration, and fibroblasts of the limb as the most promising cell type on which to focus efforts directed at understanding cellular and molecular mechanisms of positional information. Finally, we have shown that the positional disparities required by the polar coordinate model to bring about the regeneration of an amputated limb are generated by the centripetal migration of cells from the limb periphery early in the regeneration process.

REFERENCES

(1) French, V., Bryant, P.J. and Bryant, S.V. (1976) Science 193, 969-981.
(2) Bryant, S.V., French, V. and Bryant, P.J. (1981) Science 212, 993-1002.
(3) Muneoka, K. and Bryant, S.V. (1986) Trends in Genetics 2, 153-159.
(4) Muneoka, K. and Bryant, S.V. (1984) Devel. Biol. 105, 166-178.
(5) Muneoka, K. and Murad, E.H.B. (1986) J. Embryol. exp. Morph. (submitted).
(6) Muneoka, K., Wise, L.D., Fox, W.F. and Bryant, S.V. (1984) Devel. Biol. 105, 240-245.
(7) Thiebaud, Ch.H. (1983) Devel. Biol. 98, 245-249.
(8) Muneoka, K. and Bryant, S.V. (1982) Nature 298, 369-371.
(9) Muneoka, K. and Bryant, S.V. (1984) Devel. Biol. 105, 179-187.
(10) Muneoka, K., Fox, W.F. and Bryant, S.V. (1986) Devel. Biol. 116, 256-260.
(11) Rollman-Dinsmore, C. and Bryant, S.V. (1982) J. Exp. Zool. 223, 51-56.
(12) Tank, P.W. (1981) Amer. J. Anat. 162, 315-326.
(13) Goss, R.J. (1956) J. Morphol. 98, 89-123.
(14) Tank, P.W. and Holder, N. (1979) J. Exp. Zool. 209, 435-442.
(15) Gardiner, D.M., Muneoka, K. and Bryant, S.V. (1986) Devel. Biol. (in press).
(16) Kelly, D.J. and Tassava, R.A. (1973) J. Exp. Zool. 185, 45-53.
(17) Tassava, R.A., Bennett, L.L. and Zitnik, G.D. (1974) J. Exp. Zool. 190, 111-116.
(18) Maden, M. (1978) J. Embryol. exp. Morphol. 48, 169-175.

ACKNOWLEDGMENTS Research supported by PHS grant HD06082 and a gift from the Monsanto Company.

Figs. 1 and 2 reproduced by permission of Academic Press, Inc.

GENE EXPRESSION AND THE OPPOSING PROGRAMS OF DIFFERENTIATION AND DEDIFFERENTIATION IN DICTYOSTELIUM DISCOIDEUM

David R. Soll

Department of Biology, University of Iowa, Iowa City, Iowa 52242

One major objective in the field of Developmental Biology is to explain why things happen when they do in a developing system. The morphogenetic program of <u>Dictyostelium discoideum</u> is unusually well suited for investigating timing regulation primarily because of a number of timing characteristics (1). When amebae of this cellular slime mold are starved on the proper substratum, they aggregate after a preparatory, or interphase period, and then as a multicellular unit progress through a highly ordered sequence of morphological stages, culminating in the genesis of a fruiting body composed of a slender stalk supporting a spore cap. Because of the synchrony and temporal reproducibility of this morphogenetic program (2), it has been possible to investigate the minimum number, complexity and characteristics of that subset of essential processes which are rate-limiting for the consecutive stages of the developmental program, and which have been referred to as "developmental timers". In addition, because aggregates of D. discoideum are multicellular, they can be disaggregated at any time during morphogenesis and challenged to recapitulate development. Disaggregated cells will recapitulate the stages they had progressed through prior to disaggregation, and will do so in roughly one-tenth the original time (3,4). This capacity to rapidly recapitulate morphogenesis represents a form of developmental memory, and allows the investigator to monitor temporal progress through the program under a variety of experimental conditions. Not only is <u>D. discoideum</u> amenable to timing analyses during the forward program of development, but it is also uniquely suited for investigating the reverse program of dedifferentiation (5,6,7). When developing cultures are disaggregated and resuspended in nutrient medium, they lose the capacity to rapidly recapitulate morphogenesis in a synchronous, single step after 90 minutes, and this complete and rapid loss of developmental memory is referred to as the "erasure event" (6). The erasure event sets in motion a program of dedifferentiation during which cells lose developmentally acquired functions at different times during a period of roughly 400 minutes (7). Finally, <u>D. discoideum</u> amebae can be tricked into simultaneously progressing through the opposing programs of differentiation and dedifferentiation with apparently no communication between the opposing programs, a truly unique situation in developing systems (8).

In this presentation, I will review 1) the timing characteristics of the forward program of development, 2) the capacity to rapidly recapitulate morphogenesis, and 3) the characteristics of the reverse program of erasure and dedifferentiation. My intent will not be to describe in detail one aspect of this system, but rather to introduce the entire system as it relates to timing regulation.

The Forward Program of Development, Timing Regulation and the Timing Mutant FM-1.

During <u>Dictyostelium</u> morphogenesis, starved amebae begin to aggregate after an initial preaggregative period of 7 hours, then progress through an ordered sequence of morphogenetic stages with extraordinarily reproducible timing under controlled laboratory conditions (2). Employing the reciprocal shift experiment between long (18°C) and short (24°C) conditions (9), it has been demonstrated that the "timer" for the onset of aggregation is composed of two sequential components with a clear transition point between them and that at the completion of the second component and concomitant with the onset of aggregation, three or more parallel timers for later stages are initiated (10). Focusing upon the two components of the preaggregative period, we have demonstrated that the first rate-limiting component, which includes the first 4.5 hours, will progress in the absence of <u>de novo</u> protein synthesis and in the absence of cell-cell contact, and is reversed when cells are shifted from low to high temperature or from low to high ionic strength, and that progress through the second rate-limiting component is continuously dependent upon both <u>de novo</u> protein synthesis and cell contact (11). Associated with onset of each rate-limiting component is the cessation of synthesis of a family of "growth-specific" polypeptides and the onset of synthesis of a family of "development-specific" polypeptides (12).

A number of timing mutants have been isolated by a simple screening process which depends upon changes in the colony morphology of <u>Dictyostelium</u> clones grown in bacterial lawns (1). These mutants are unique in that each mutation affects the timing of one or more select stages, but has no significant effect upon growth or, more importantly, the actual sequence of morphogenesis. The first of this class of mutants analyzed in detail was FM-1 (fast mutant number one) (13). This mutant exhibits a decrease in the preaggregative period. By the reciprocal shift experiment, it has been demonstrated that FM-1 is completely missing the first rate-limiting component of the preaggregative period. It is relatively normal for the remaining stages of morphogenesis and is completely normal in all aspects of growth, in spore formation, in the capacity to rapidly recapitulate morphogenesis and in the erasure event and subsequent program of dedifferentiation. The FM-1 mutation is heritable and behaves as a single mutation mapping to linkage group II. However, the most surprising characteristic of the FM-1 variant is its capacity to switch heritably and reversibly at relatively high frequency ($\sim 10^{-2}$) to several other timing phenotypes with longer preaggregative periods which in turn switch at high frequency back to FM-1. Virtually every type of motility and morphogenetic variant imaginable is generated at a slightly lower frequency by FM-1 and its slower, common offspring FM-1(S1). Many of these mutants are stable. Therefore, the FM-1 system provides a high frequency, spontaneous mutation system and we would not be surprised if we discover that this system is based upon high frequency transposition. Indeed, if timing mutations were the result of a high frequency switching system based on transposition, it would have far reaching evolutionary implications, especially in regard to the theory of heterochrony.

Rapid Recapitulation. If cells are allowed to develop to the tight aggregate stage and then are disaggregated and replated for development, they will recapitulate the stages progressed through during the initial developmental program, only the second time through, they will regenerate these stages in a fraction of the original time (2,3). For instance, during the initial developmental program, the time to the ripple, loose aggregate and tight aggregate stages are 7, 9, and 11 hours, respectively. However, the second time through, the times to these stages are 20, 40 and 70 minutes. Cells can rapidly recapitulate morphogenesis roughly 5 times on average before "burning out". Recapitulating cells do not progress through the initial rate-limiting components, and they appear to reutilize the developmental machinery acquired during initial development (6). The time necessary for rapid recapitulation therefore appears to involve mechanical time for the actual morphogenetic events. It also provides the experimenter with a method for assessing temporal progress through the developmental program.

Erasure and the Subsequent Program of Dedifferentiation. If developing cultures are disaggregated and the cells suspended in either buffered dextrose solution (for cultures prior to the tight aggregate stage) or full nutrient medium, they retain the capacity to rapidly recapitulate morphogenesis for 80 minutes, then in less than a 20 minute period, they synchronously and completely lose this capacity, reverting to the slow timing of naive log phase cells. The time at which cells lose this capacity is referred to as the "erasure event", and this sets in motion a program of dedifferentiation during which developmentally acquired functions and membrane components are lost over a 400 minute period (7). When cells are initially disaggregated, the intracellular concentration of cAMP immediately drops from the very high developmental level to the low vegetative level (7). The erasure event and subsequent loss of developmentally acquired functions are inhibited by the addition of 10^{-4}M cAMP to the erasure medium (14). Surprisingly, synthesis of most growth-specific polypeptides begins prior to or concomitantly with the erasure event and synthesis of most development-specific polypeptides ceases prior to the erasure event (15). The majority of growth-specific polypeptides begin to be resynthesized even if the erasure event is inhibited by cAMP. In contrast, roughly a third of the development-specific polypeptides continue to be synthesized if the erasure event is inhibited by cAMP. The levels of two development-specific RNA's, 16G1 and 10C3, have also been monitored during erasure (15). The loss of both RNA's are inhibited by the addition of cAMP to erasure medium. Results will be presented which indicate that at the erasure event, the mechanism for cAMP mediated elongation of mRNA halflife is dismantled, but the stimulation of mRNA synthesis by cAMP may be retained after the erasure event.

A mutant, HI4, has been isolated which is selectively defective in the dedifferentiation process (16). HI4 grows normally, progresses normally through morphogenesis, and rapidly recapitulates. It progresses through the erasure event, but it does not lose its developmentally acquired cohesive properties, including EDTA-resistant cohesion and the cohesion molecule gp80 at the prescribed times in the dedifferentiation process. In fact, after the erasure event, HI4 still rapidly reaggregates, but it does so in a non-chemotactic fashion in which cells randomly collide and cohere, thus forming aggregates. HI4 and other dedifferentiation-defective mutants will be useful in dissecting the complexity of the dedifferentiation program.

Initiating Parallel Programs of Dedifferentiation and Redifferentiation in the Same Cell. Immediately after the erasure event,

cells can be stimulated to reenter the developmental program, even though they still possess most of the morphogenetic machinery acquired during the initial developmental program. One would expect a clever cell in this situation to retain these molecular components for redifferentiation. They don't. In fact, when redifferentiation is initiated immediately after the erasure event, cells lose developmentally acquired components (e.g., gp80) at the prescribed times according to the program of dedifferentiation while reacquiring these very same components according to the forward program of differentiation (8). Employing selective timer mutants, one can actually reverse the signals for removal and reacquisition. These results demonstrate that the opposing programs of differentiation and dedifferentiation can function simultaneously and independently in the same cell.

Concluding Remarks. The identification and characterization of the diverse timing characteristic of Dictyostelium discoideum, the recent isolation of timing and dedifferentiation mutants, and the discovery of a high frequency switching system have generated a new set of questions related to developmental time. Hopefully, both the manipulations which have been developed, the selective mutants and the switching system will be exploited by other developmental biologists interested in a diverse set of developmental problems, including gene regulation. More importantly, they should serve as a basis for a deeper understanding of why things happen when they do in a complex developmental program.

REFERENCES
(1) Soll, D.R. (1986) in Methods in Cell Biology (Spudich, J., ed.), in press.
(2) Soll, D.R. (1979) Science 203, 841-849.
(3) Loomis, W.F. and Sussman, M. (1966) J. Mol. Biol. 22, 401-404.
(4) Soll, D.R. and Waddell, D.R. (1975) Dev. Biol. 47, 292-302.
(5) Waddell, D.R. and Soll, D.R. (1977) Dev. Biol. 60, 83-92.
(6) Soll, D.R. and Finney, R. (1986) in The Genetic Regulation of Development (Loomis, W.F., ed.) Forty-fifth Symposium of the Society of Developmental Biology, in press.
(7) Finney, R., Varnum, B. and Soll, D.R. (1979) Dev. Biol. 73, 290-303.
(8) Finney, R., Mitchell, L., Soll, D.R., Murray, B. and Loomis, W. (1983) Dev. Biol. 98, 502-509.
(9) Soll, D.R. (1983) Dev. Biol. 95, 73-91.
(10) Varnum, B., Mitchell, L. and Soll, D.R. (1983) Dev. Biol. 95, 92-107.
(11) Finney, R., Langtimm, C. and Soll, D.R. (1985) Dev. Biol. 110, 157-170.
(12) Finney, R., Langtimm, C. and Soll, D.R. (1985) Dev. Biol. 110, 171-191.
(13) Soll, D.R., Mitchell, L., Kraft, B., Alexander, S., Finney, R. and Varnum-Finney, B. (1986) Dev. Biol., in press.
(14) Finney, R., Slutsky, B. and Soll, D.R. (1981) Dev. Biol. 84, 313-321.
(15) Finney, R., Ellis, M., Langtimm, C., Rosen, E., Firtel, R. and Soll, D.R. (1986) Submitted.
(16) Soll, D.R., Mitchell, L.H., Hedberg, C. and Varnum, B. (1984) Dev. Genet. 4, 167-184.

ACKNOWLEDGMENTS
This research was supported by N.I.H. grants GM25832 and HD18577.

DYNAMICS OF HEAD PATTERNING IN HYDRA

T. Awad, P. Bode, *O. Koizumi, and H. Bode

Developmental Biology Center, University of California, Irvine, Calif., and *Fukuoka Women's University, Fukuoka, Japan.

Patterning in hydra has focused on the phenomenon of polarity of regeneration. Excision of a piece of tissue anywhere along the length of the column always results in head formation at the original apical end of the piece, and foot formation at the basal end. Extensive transplantation experiments have demonstrated the existence of a pair of developmental gradients, the head activation gradient and the head inhibition gradient, which act to locate the head at the apical end (1-3, see 3 for review). In the intact animal head activation is a stable tissue property which is very high in the head and decreases in a graded manner down the column. Upon decapitation the head activation level of the apical tip, which is at a level much lower than the head, rapidly rises to the level of the head within 6-12 hours. At this time the tissue is committed to forming a head (2,3).

These results describe the patterning process locating the head at the apical end, but they do not explain the formation of the structure of the head. The head consists of two parts: an apical cone, the hypostome, and below that, a ring of tentacles. A further step towards understanding the patterning of the head is derived from the following three sets of results which suggest how the relative positions of the hypostome and the ring of tentacles are determined.

DETAILS OF THE MORPHOLOGY OF HEAD REGENERATION. Although the process of head regeneration is understood in general, additional details of the process were recently observed (4). 4-6 hours after decapitation, the cut edges of the two tissue layers have moved over the open apical end of the gastric cavity and formed a dome. Within a day after decapitation small initial tentacle protrusions appear near the apex of the dome (Fig 1A). A day later the dome is considerably larger and the tentacle protrusions, which have grown into short tentacles, appear further down the side of the dome close to the location of the tentacle ring. New tentacle protrusions also appear in the ring. On the third day the dome has increased further in size, and the tentacles have become longer. The new observations focus on the formation of tentacles first near the apex and later in the ring, as well as the growth of the dome. The formation of the first protrusions near the apex of the dome is emphasized by the fact that in some regenerates they are not displaced and remain as a single or a pair of tentacles well above and separate from the ring of tentacles.

DYNAMICS OF THE APPEARANCE OF A TENTACLE-SPECIFIC ANTIGEN DURING HEAD REGENERATION. Instead of using the final morphology as an indicator of the patterning process, it would be useful have a tentacle-specific marker that appears at an earlier time during the process. The monoclonal antibody, TS-19, provides such a marker. It binds to an antigen on the surface of the ectodermal epithelial cells that is present on the tentacles, but not the hypostome of the head or the upper body column in the adult animal (5,6). Further, it is an early tentacle-specific antigen as it appears before any morphological sign of a tentacle during head regeneration (6).

After decapitation, TS-19 binding was initially observed at the apex of the dome. Thereafter it spread down the sides. As the tentacle protrusions appeared within the stained area, they became more intensely stained than the dome. Subsequently, as the dome grew in size the stained area retreated toward the apex and diminished in intensity. Still later this area was reduced to a cross-like structure with the legs of the cross running across the apex of the dome from one tentacle to another. Eventually all staining of the dome vanished. At the same time the tentacles continued to grow and the staining increased in intensity resulting in the adult pattern.

The pattern of retreat of the TS-19+ area from the base of the tentacle ring back towards the apex of the head also suggested a mechanism for the growth of the dome and the tentacles during regeneration. It is unlikely that this growth is due to cell proliferation as regeneration can take place in the absence of cell division (7,8). More likely tissue is displaced from the body column into the developing head. To determine if this was actually the case, animals were marked with vital dyes somewhat below the head, decapitated, and the movement of the dyes followed. As the dome grew, the marked cells were displaced either onto the tentacles, or between tentacles up onto the dome (6).

PATTERN OF NERVE CELL APPEARANCE DURING HYPOSTOME REGENERATION. As a marker for the patterning of the hypostome, we have taken advantage of the differences in spatial location of two subsets of neurons in the hypostome. Epidermal sensory cells, which are restricted to the head, are found only near the apex in the hypostome (9). Ganglion cells found throughout the animal (10) are located along the sides and the base of the dome, but less frequently at the apex (9).

An antiserum to the peptide RFamide, which stains the sensory neurons and ganglion cells of the hypostome and tentacles, but no nerve cells of the body column below the head, provided a means for monitoring the pattern of appearance of these two subsets of neurons during head regeneration (6). During the initial stages, only ganglion cells exhibiting RFamide-like immunoreactivity (RLI) were observed. They were found at the apex as well as on the sides and base of the dome. At a later stage, but still before the appearance of tentacles in the ring, the apex was devoid of stained neurons, but the sides and base exhibited RLI+ ganglion cells. The first RLI+ sensory nerves were observed only after tentacles appeared in the tentacle ring. These and subsequent sensory cells were all located at the apex of the hypostome.

MODEL FOR THE TWO-PART HEAD PATTERN. A simple qualitative model that describes a patterning process which sets up the locations of the two parts of the head can be derived from these results and the known behavior of head activation during regeneration. Upon decapitation the level of head activation rises rapidly in the apical end, and only the apical end, of the regenerate reaching a hypostome level in 6-12 hours (2). The behavior of head activation in our simple model is derived from formal models described by Wolpert (11), and later more quantitative reaction-diffusion models of Gierer and Meinhardt (12,13). For simplicity head activation is assumed to be a single substance. A high concentration is necessary to specify, or render tissue competent, for hypostome formation.

A lower concentration would result in the tissue becoming competent to form tentacles. In the intact animal, head activation is graded with a maximum in the hypostome. Then, upon removal of the head, the apical end of the decapitated animal would have the highest level of head activation which would be well below that necessary for tentacle expression. After decapitation, head activation proceeds to rise reaching the tentacle concentration at the apex first (see Figure 1), which renders the tissue competent to form tentacles. Since the initial head

---

Fig. 1. The spread if the two-part head pattern.
A. The positions of the tentacles as they evaginate. B. The development of the underlying head pattern suggested by the presented results. From (4).

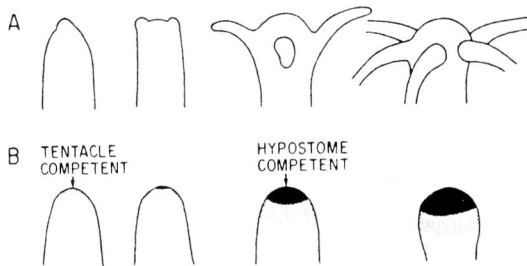

---

activation level decreases with increasing distance from the apex, the further tissue is from the apex, the later the tentacle level is reached. Hence, tentacle level of head activator concentration spreads from the apex in a radial fashion out and down the sides of the dome. As this spread occurs, head activation continues to rise at the apex so that the apical tissue is no longer specified to form tentacles, but reaches a level competent to form a hypostome. Hypostome competence would also spread radially as head activation rises in neighboring tissue. The result would be an apical hemispherical area specified to form hypostome surrounded by a ring of tentacle competent tissue. Such a mechanism would account for the relative locations of the two parts of the head.

Much of the data is consistent with this model. The tentacle-specific antigen identified by TS-19 spread from the apex in a radial fashion. Later it vanished as would be expected if the dome was no longer specified to form tentacles (6). The appearance of tentacle protrusions first near the apex of the dome indicated this tissue was competent to form tentacles (4). The later appearance of new tentacle protrusions near the base, but not near the apex, indicated that the basal dome tissue had become competent for these structures and that the apex had lost that capacity. Hence, the data fits the aspect of the model that suggests tentacle specification starts at the apex and moves radially in a ring down to the base of the head. Another common observation is also consistent with this view. Occasionally a regenerate forms only a single, or a pair, of tentacles at the apex of the dome instead of a hypostome and ring of tentacles. If the patterning process were halted shortly after the apex became tentacle-competent the observed results would have been expected.

The model predicts that apical tissue of the dome is specified to form the hypostome after it is first specified to form tentacles. The

sequence of the appearance of nerve cells is consistent with this prediction (6). Ganglion cells, typical of the base of the hypostome near the tentacles first appeared at the apex and later were confined to the sides and base following the wave of tentacle-competence. Only still later do sensory cells, typical of the hypostome apex, appear.

Thus, to a first approximation head patterning can be explained economically. The same process that locates the head at the apical end of the tissue can account for the location of the hypostome and the ring of tentacles.

This model is too simple to account for all the observations. It does not deal with the displacement of tissue from the body column into the developing head (6). This can be done by simply superimposing this displacement in an apical direction on the radial and basal spread of the rise in head activation. Of greater consequence, the model does not explain why head activation does not continue to rise further down the column rendering areas other than the apical end tentacle- or hypostome-competent. By adding the known head inhibition principle and gradient to the model as is done in the reaction-diffusion model developed by Gierer and Meinhardt (12,13) and refined by MacWilliams (14) head activation rise is confined to the apical end and defines the presumptive head region. Also, not all of the tentacle region is converted into tentacles. To account for the observed periodic pattern of tentacles, a subsequent patterning process, probably involving a spacing mechanism, must occur.

REFERENCES

(1) MacWilliams, H.K. (1983) Dev. Biol. 96, 217-238.
(2) MacWilliams, H.K. (1983) Dev. Biol. 96, 239-257.
(3) Bode, P.M. and Bode, H.R. (1984) In Pattern Formation. A Primer in Developmental Biology. (G. Malacinski and S. Bryant, eds) pp. 213-241. Macmillan, NY.
(4) Bode, P.M. and Bode, H.R. (in press) Development.
(5) Bode, H., Dunne, J., Heimfeld, S., Huang, L., Javois, L., Koizumi, O., Westerfield, J., and Yaross, M. (1986) Curr Topics in Dev. Biol. 20, 257-280.
(6) Bode, P.M., Awad, T., Koizumi, O., and Bode, H.R. (in prep.).
(7) Hicklin, J. and Wolpert, L. (1973) J. Embryol. Exp. Morph. 30, 741-752.
(8) Cummings, S.G. and Bode, H.R. (1984) Roux's Arch. Dev. Biol. 194, 79-86.
(9) Westfall, J.A. and Kinnamon, J.C. (1978) J. Neurocytol. 7, 365-389
(10) Westfall, J.A. (1973) J. Ultrastruct. Res. 42, 268-282.
(11) Wolpert, L., Hicklin, J. and Hornbruch, A. (1971) Symp. Soc. Exp. Biol. 25, 391-415.
(12) Meinhardt, H. and Gierer, A. (1974) J. Cell Sci. 15, 321-346.
(13) Meinhardt, H. (1982) Models of Biological Pattern Formation, Academic Press.
(14) MacWilliams, H.K. (1982) J. Theor. Biol. 99, 132-179.

ACKNOWLEDGEMENTS The research was supported by a grant from the National Institutes of Health (GM29130).

EMBRYONIC ANGIOGENESIS: MORPHOGENESIS OF THE EARLY HEART AND FORMATION OF THE AORTIC ARCHES

J.D. Coffin and T.J. Poole

Department of Anatomy and Cell Biology, SUNY Health Science Center at Syracuse, Syracuse, N.Y. 13210

Heart and blood vessel development have previously been studied by the use of dye injections (1) and by electron microscopy (2). We have utilized immunofluorescence in whole mount embryos to study early vascular development. Our results indicate that heart and aorta formation occur earlier and by a different mechanism than was previously believed.

IMMUNOCYTOCHEMISTRY. Whole mounts (3) of permeablized (4) Japanese quail were used. The embryos were removed from the yolk sac, rinsed in PBS, and fixed in 4% formalin/PBS overnight at 4°C. The fixative was washed away with PBS, then the embryo permeablized with successive changes of: methanol (MeOH)30 min- toluene 5min- MeOH 60min- MeOH 30min, all at 4°C with constant agitation. Rehydration in an ethanol series, a PBS rinse, and incubation in 3% BSA (Sigma) at 4°C overnight followed. Next, a QH-1 monoclonal antibody (courtesy of Dr. Paul Kitos, U.Kansas) that labels endothelium, diluted 1:200 in 3% BSA, was applied at 4°C for 6-12 hrs. Unbound antibody was rinsed away with PBS washes, and the embryo immersed in 1:200 goat-antimouse FITC conjugated IgG in 3% BSA, overnight. Again PBS washes were used to remove unbound antibody. Embryos were then dehydrated in an ethanol series, incubated 3min in toluene, and mounted in Entellan (VWR) for examination and photography.

VASCULAR MORPHOLOGY. Presumptive endothelial cells (PECs) are first seen in precardiac areas at the appearance of the first somite (S) (Fig Ia). Some cells appear singly, but most are grouped into angioblastic islets at the lateral body folds. No PECs were seen anterior to the head process. At the 2S stage, the PECs begin to move toward the anterior intestinal portal (AIP) midline between the body folds (Fig Ib), but the islets have still not formed distinctive capillaries. This occurs at 3S (Fig Ic) and 4S (Fig IIa) as the islets aggregate into bilateral capillary plexes. They then complete migration to the midline where they fuse at the future site of the sinus venosus, ventral to the AIP (Fig IIa). Heart development progresses to the 5S stage, where the ventral aorta (VA) arises as a single sprout at the point of fusion of the two lateral primordia (Fig IIb). The presumptive VA continues to grow cranially, eventually fusing with the dorsal aorta (DA) (Fig IIc). The DAs arise from 1S to 3S, as PECs and islets appear in-situ in bilateral cranio-caudal lines (Fig I). In the 4S and 5S stages, the aortic primordia enlarge and elongate, continuing their growth cranially in apposition to the head process, and caudally near the somites (Fig II). Then with the appearance of the sixth somite, the VA and DAs fuse bilaterally in the head process to form the first aortic arches (Fig II). This occurs either by a split in the VA and sprouting to meet the DAs, or by cellular bridging between the VA and DAs.

EMBRYONIC ANGIOGENESIS. Heart development has been thought to occur by the fusion of blood vessels on the ventral surface of the embryo at about 6S (1). This study indicates that a ventral surface fusion of heart primordia does indeed take place, but much earlier in embryogenesis and between capillary plexes, not patent vessels. The vessels observed previously then develop from these capillary plexes. Furthermore, the earliest report of dorsal aorta formation has been at 6S (2). Here, in-situ development of the dorsal aorta was observed at the appearance of the first two somites. Moreover, the development of the first aortic arch, by fusion of the dorsal and ventral aortae, occurs at 6S. However, this study determined the positions of cells at certain points in development, not the origin of the cells or the path taken by them in morphogenesis. Future studies will address these questions.

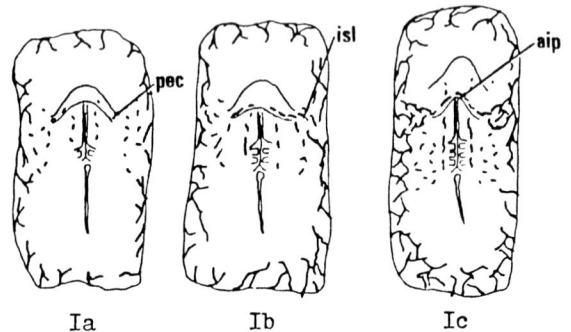

Fig. 1. Quail Whole mounts. Sketch of heart and DA formation from 1S to 3S. Ia-presumptive endothelial cells (pec) appear in the body folds as heart primordia and midline near the notochord as primative DA. The PECs aggregate into angioblastic islets (isl) in Ib, and appear closer to the midline anterior intestinal portal (aip) in Ic.

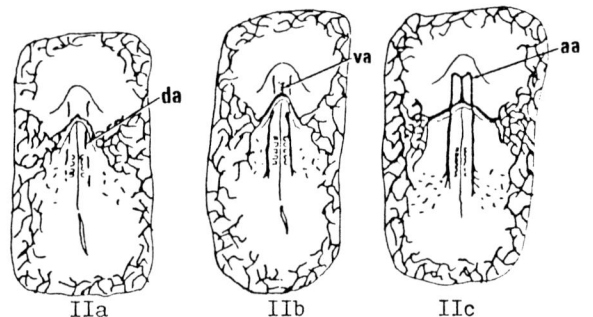

Fig. 2. Whole Mount Quail. Sketch of ventral aorta (va), dorsal aorta (da), and aortic arch (aa) formation at four to six somites. IIa- the da appears as heavy broken lines bilaterally, as the va sprouts cranially from the fusion of the heart primordia at the AIP in IIb. IIc- das are complete caudally, and fused to the va cranially to form the aortic arches.

REFERENCES

(1) Evans, H.M. (1909) Anat. Rec. 3, 498-518.
(2) Hirakow, R. and Hiruma, T. (1981) Anat. Embryol. 163, 299-306.
(3) Buck et al. (1985) J. Cell Biol. 101, 112a.
(4) Linask, K.K. and Lash, J.W. (1986) Devel. Biol. 114, 87-106.

EFFECT OF HEAD AND FOOT FACTORS ON DIFFERENTIATION IN HYDRA

S. Hoffmeister, and H.C. Schaller
Zentrum für Molekulare Biologie, University of Heidelberg, Im Neuenheimer Feld 282, 6900 Heidelberg

Head and foot factors are signals for nerve cell differentiation.

In hydra under steady state conditions and averaging over the whole animal about 40% of the cells of the interstitial stem cell pool leave this undifferentiated state per cell cycle, 10% give rise to nerve cells whereas about 30% differentiate to nematocytes (1). In the nerve cell differentiation pathway there exist at least two relay stations which are susceptible for controlling signals. The first one is the station at which determination of interstitial cells to nerve cell development occurs. This station was located in the early S-phase of the cell cycle. Head activator was shown to be the positive signal needed for this decision (2,3) head inhibitor acted antagonistically. The second station is the initiation of terminal differentiation. At this station head activator as well as foot activator are stimulating signals whereas head inhibitor again acts antagonistically. As to the localization of the second relay station in the cell cycle we found that the G_2-phase is the most likely time point where the signals can interfere.

Dominance of the head over the foot system.

As outlined above, foot activator triggers interstitial cells which are determined to become nerve cells to go through mitosis and differentiate to mature nerve cells. Foot activator had no effect on the determination of interstitial cells to nerve cells. Since only head activator and not foot activator stimulates the determination of interstitial cells for nerve cell development there exists a sort of dominance of the head system over the foot system. The determined interstitial nerve cell precursors are able to migrate along the body axis of the animal and depending on where they end their migration, whether they are in the head region where high concentrations of head activator exist or in the foot region with high concentrations of foot activator, they will finally differentiate to head specific or foot specific nerve cells (Fig. 1).

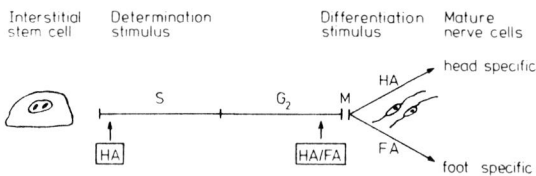

Fig. 1: Effect of head activator (HA) and foot activator (FA) on nerve cell development.

Effect of foot activator and head activator on epithelial cells

Epithelial cells of the gastric region are constantly dividing and thus represent epithelial stem cells. Epithelial cells which migrate into the head or foot region become terminally differentiated. Those in the head develop into head specific epithelial cells, those in the foot eventually differentiate to foot specific, foot mucous cells. At the bottom of the basal disk and at the end of the tentacles they finally die and are sloughed off so that they continuously have to be replaced by newly differentiated cells from the gastric region. This, however, demands proliferation of epithelial stem cells. Foot and head activator exert a stimulatory effect on the proliferative activity of epithelial cells measured either as increase in the mitotic index or as increase in the labelling index.

Foot specific differentiation of epithelial cells can be studied by measuring the peroxidase activity which is characteristically present in foot specific mucous cells (4). Due to treatment with foot activator the measured peroxidase activity shows a marked increase as compared to untreated controls. Head activator has no stimulating influence on peroxidase activity. Therefore, presence of foot activator leads to the differentiation of foot specific epithelial cells whereas head activator induces differentiation of head specific epithelial cells (Fig. 2).

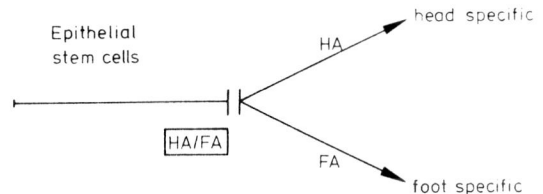

Fig. 2: Effect of head activator (HA) and foot activator (FA) on epithelial cells.

REFERENCES
(1) David, C.N. and Gierer, A. (1974). Cell cycle and development of _Hydra attenuata_. III. Nerve and nematocyte differentiation. J. Cell Sci. 16, 359-375
(2) Schaller, H.C. (1976,c). Action of the head activator on the determination of interstitial cells in hydra. Cell Diff. 5, 13-20.
(3) Holstein, Th., Schaller, H.C. and David, C.N. (1986). Nerve cell differentiation in hydra requires two signals. Develop. Biol. 114, 9-18.
(4) Hoffmeister, S. and Schaller, H.C. (1985). A new biochemical marker for foot-specific cell differentiation in hydra. Roux's Dev. Biol. 194, 453-461.

HEAT-SHOCK ACCELERATES DEVELOPMENT IN MYXOCOCCUS XANTHUS

Kevin P. Killeen and David R. Nelson

Department of Microbiology, University of Rhode Island, Kingston, R.I., 02881.

All organisms respond to elevated temperatures by altering their pattern of growth and protein synthesis (1). This is commonly termed the heat-shock response. In many organisms, heat shock genes are expressed during cellular development (2,3). The synthesis of heat shock proteins during periods of cellular differentiation suggests interrelationships between normal development and stress.

Myxococcus xanthus is a developmental bacterium. Starvation or the addition of glycerol induces vegetative rods to convert to ovoid environmentally resistant myxospores (4,5). We wished to determine the effect of heat shock on developmental M. xanthus. It is known that incubation at 40°C elicits the heat-shock response in this organism (6). We illustrate that pre-incubation of vegetative cells at 40°C for 1h accelerates the rate of myxospore formation at 28°C in both starvation-induced and glycerol-induced cells.

(I) Effect of Heat-Shock on Myxospore Formation in Starvation-Induced M.xanthus. M.xanthus cells were incubated at either 28°C or 40°C for 1h, spotted (1.5×10^9 cells/ml) onto a starvation agar medium, and incubated at 28°C. Development was interrupted after 18, 24, 30, or 36h. Cells were harvested, and vegetative cells killed by heating to 53°C for 15 min followed by sonication. The number of myxospores was determined by plating. The data presented in Fig. 1 illustrate that preincubation for 1h at 40°C accelerates myxospore formation in starvation-induced M.xanthus. After 18 h of development the heat-shocked cells had six times the number of myxospores as the control cells. This acceleration was less pronounced at later times and after 36h of development the numbers of myxospores in the heat-shocked and non heat-shocked cells were equal.

Fig. 1. **Relative rates of starvation-induced myxospore formation in heat-shocked M. xanthus cells.** Rates of myxospore formation in cells heat-shocked for 1h prior to starvation on agar medium (solid bars) were compared to control cells (striped bars). 100% values (spores/ml) are: 2.9×10^4, 18h; 3.0×10^5, 24h; 5×10^5, 30h; 6×10^5, 36h.

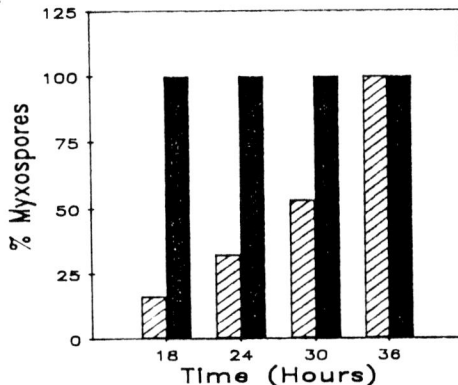

(II) Effect of Heat Shock on Myxospore Formation in Glycerol-Induced M.xanthus. M.xanthus cells grown at 28°C were divided into two aliquots. One was incubated at 28°C with the addition of 0.5M glycerol. The other was pre-incubated at 40°C for 1h, then shifted to 28°C and glycerol added to 0.5M. Development was

arrested after 1.5, 2, 3, 4, or 5h. Vegetative cells were killed by heating to 53°C for 15 min and the number of myxospores determined by plating. Figure 2 illustrates that pre-incubation at 40°C for 1h accelerates myxospore formation in glycerol-induced M.xanthus. Despite a 30-50% reduction in viable cells following heat-shock, after 90 min in glycerol the heat-shocked culture contained more than 20 times the number of myxospores as did the control culture. By 120 min, the heat-shocked culture had nearly 200 times the number of myxospores as did the control.

In E. coli, synthesis of heat-shock proteins wanes with prolonged exposure to elevated temperature. Preliminary data indicate that the same is true for M. xanthus. Further, when heat-shocked for 3h prior to glycerol-induced development the cells no longer exhibit an acceleration of myxospore formation. These data suggest that the heat-shock response is related to development in M. xanthus.

Fig. 2. **Relative rates of glycerol-induced myxospore formation in heat-shocked M. xanthus cells.** Rates of myxospore formation in cells heat-shocked for 1 hr prior to the addition of glycerol (solid bars) were compared to control cells (striped bars). 100% values (spores/ml) are: 2.7×10^2, 1.5h; 3.7×10^6, 2h; 2.4×10^5, 3h; 7.1×10^5, 4h; and 4.6×10^5, 5h.

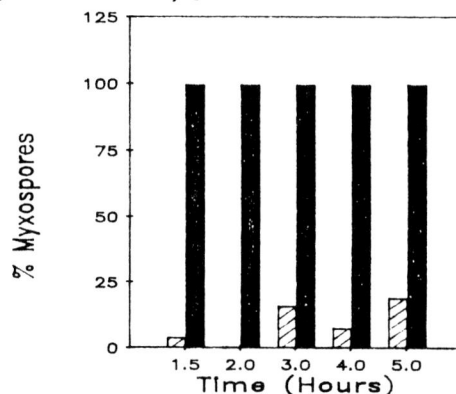

REFERENCES

(1) Neidhardt, F.C., R.A. Van Bogelen, and V.Vaughn.(1984) The genetics and regulation of heat shock proteins. Annu. Rev. Genet. 18: 295-329.

(2) Kurtz,S., J.Rossi, L.Petko, S. Lindquist. (1986) An Ancient Developmental Induction: Heat-Shock Proteins Induced in Sporulation and Oogenesis. Science.231: 1154-1157.

(3) Sirotkin,K.and N. Davidson. (1982) Developmentally Regulated Transcription from Drosophila melanogaster Chromosomal Site 67B. Dev. Biol. 89: 196-210.

(4) Dworkin,M. and S.M. Gibson. (1964) A system for studying microbial morphogenesis: Rapid formation of microcysts in Myxococcus xanthus. Science. 146: 243-244.

(5) McVittie, A., F. Messik, and S. Zahler. (1962) Developmental Biology of Myxococcus. J. Bacteriol. 84: 546-551.

(6) Nelson, D.R. and K.P. Killeen. (1986). Heat-Shock Proteins of Vegetative and Fruiting Myxococcus xanthus. J. Bacteriol. In press.

CHARACTERIZATION OF THE ROLE OF THE SEGMENT POLARITY GENES IN PATTERN FORMATION IN DROSOPHILA

T. Orenic, B. Schafer, J. Chidsey, and R. Holmgren

Department of Biochemistry, Molecular Biology, and Cell Biology Northwestern University, Evanston, Ill. 60201

Drosophila larvae are constructed from a series of repeated units called segments. The cuticular pattern within each segment is altered by mutations in the segment polarity genes (1,2). These mutations cause the deletion of a portion of each segment and its replacement by a mirror image duplication of the remaining structures. We are interested in the role of two segment polarity genes, cubitus interruptus dominant (ci^D) and Cell (Ce^2), in the generation of the pattern within segments. Both of these genes map proximally on the fourth chromosome. Animals homozygous for ci^D have the posterior half of each segment deleted and replaced with a mirror image duplication of the anterior half. The Ce^2 segmentation defect is more extreme. In Ce^2 animals the deletion includes the posterior part of each segment as well as the anterior region of the adjacent segment. To further investigate the role of the ci^D and Ce^2 genes in pattern formation, we have done studies on the maternal contribution and on the autonomy of these genes. We have also used X-ray induced mutations to better understand the genetic organization of the Ce^2-ci^D region.

Recently we have become interested in the role of the segment polarity genes in pattern formation within the nervous system, and are using antibodies directed against specific neurons (3) to observe pattern alterations in the nervous systems of segment polarity mutants.

MATERNAL EFFECT

Both ci^D and Ce^2 are recessive lethal mutations. Thus, in order to determine whether these genes have a maternal effect, it was necessary to construct germ line mosaics. This was done by transplanting pole cells from an embryo homozygous for Ce^2 or ci^D into an Ovo^{D1} embryo (4). Pole cells are germline precursor cells that also give rise to the nurse cells which surround the egg and contribute maternal gene products to it. Ovo^{D1} embryos have no germline but are otherwise wildtype. This experiment revealed that the ci^D and Ce^2 wild type gene products are not contributed to the embryo, and are not required for oogenesis.

CELL AUTONOMY

This experiment is done by using chromosome loss to generate mosaic animals which have regions that are mutant for ci^D and are marked with yellow. If ci^D is cell autonomous, only the regions marked with yellow should exhibit the ci^D segmentation defect. From analysis of both larvae and adults, it appears that ci^D function is cell autonomous.

MAPPING ci^D AND Ce^2

Both the ci^D and Ce^2 mutations cause dominant defects in adults which are the result of a novel function induced by the mutation. The dominant ci^D mutation alters the wing vein pattern. The dominant Ce^2 mutation alters the wing vein pattern and removes the ocelli in heterozygous animals. We used the elimination of these novel functions to isolate X-ray induced lesions within the Ce^2-ci^D region. 34 revertants of ci^D were isolated. All but one of these lines have become mutant for Ce^2. In several of these lines there has been a reciprocal translocation of material between the fourth and another chromosome. These results indicate that the ci^D and Ce^2 genes are intimately associated with one another.

NERVOUS SYSTEM

We have looked at the pattern defects in the nervous system of all the segment polarity mutants using flourescently labeled antibodies directed against specific neurons. ci^D, armadillo, and fused mutants show no defects in the nervous system. patched mutants have pattern defects in the nervous system that are homologous to the cuticular pattern defects. Ce^2, hedgehog, wingless, and gooseberry cause pattern defects which are still being analyzed.

DISCUSSION

We have begun a characterization of two segment polarity genes on the fourth chromosome, ci^D and Ce^2. Mutations in ci^D cause the deletion of pattern elements in the posterior half of each segment which are replaced by a mirror image duplication of pattern elements in the anterior half of each segment. Mutations in Ce^2 cause the deletion of the posterior half of each segment as well as part of the anterior region of the adjacent segment with a mirror image duplication of the remaining structures. The Ce^2 mutation causes cell death and results in animals substantially smaller than wild type. The ci^D mutation, however, does not cause cell death (5), and the animals are the same size as wild type. This indicates that the mirror image duplication does not result from regeneration of the remaining structures, but, rather, from an alteration of the fate of posterior cells. Our studies have shown that the ci^D function is strictly zygotic, and that it is cell autonomous. These results indicate that ci^D may function specifically in cells that will make up the posterior half of each segment. Further support of this will require a molecular characterization of the gene.

We believe that Ce^2 and ci^D form part of a gene complex for several reasons. They are both segment polarity genes which map proximally on the fourth chromosome. They both exhibit a similar segmentation defect, and neither has a maternal effect or is required for oogenesis. All but one of the revertants of ci^D generated by X-ray mutagenesis have acquired the Ce^2 segmentation defect. In several of the revertants there has been a reciprocal translocation of genetic material between the fourth chromosome and another chromosome. These results indicate that ci^D and Ce^2 are very tightly linked or that the ci^D gene maps within the Ce^2 gene. Complementation analysis between various alleles of the ci^D and Ce^2 genes suggest that there are probably two independent functions within the complex.

REFERENCES

(1) Nusslein-Volhard, C. and Wieschaus, E. (1980) Nature 287, 795-801.
(2) Wieschaus, E. and Nusslein-Volhard, C. (1984) Wilhem Roux's Archives of Devel. Biol. 193, 296-307.
(3) Goodman, C.S. personal communication.
(4) Lawrence, P.A., Johnston, P., and Struhl, G. (1983). Cell 35, 27-34.
(5) Martinez-Arias, A. personal communication.

GENE THERAPY

GENE TRANSFER: A POTENTIAL NEW THERAPY FOR THE TREATMENT OF INHERITED DISEASE

Richard C. Mulligan

Whitehead Institute for Biomedical Research and
Department of Biology, M.I.T., Cambridge, MA 02142

For several years, our laboratory has been interested in determining the feasibility of using gene transfer methods to treat selected disorders of the hematopoietic system in man. Most of our work in this direction has focused on determining whether retrovirus vectors can be used to transfer new genetic information into hematopoietic cells of the mouse (1). Specifically, we have asked whether the totipotent hematopoietic stems present in freshly explanted bone marrow can be infected by recombinant retroviruses in vitro, and subsequently used to permanently and completely engraft lethally irradiated recipients via bone marrow transplantation.

In order to characterize the developmental potential and general properties of the transduced hematopoietic stem cells, we have used retroviral-mediated gene transfer to clonally mark stem cells in vitro, and have tracked the fate of those cells after their introduction into lethally irradiated recipients via bone marrow transplantation (2). In order to efficiently transduce hematopoietic stem cells, a protocol was developed that involved the co-cultivation of bone marrow cells from 5-Fluorouracil (5FU) treated mice with virus producing cell lines in the presence of a source of IL-3. This protocol resulted in the transduction of 50-100% of both day 14 CFU-S and stem cells with long term reconstitution capacity, regardless of the titer of viral stocks used. To assess the contribution of individual transduced stem cell clones to hematopoiesis in completely and permanently engrafted transplant recipients, hematopoietic cells from the recipients were fractionated into pure cell populations representative of a particular lineage or anatomical location, and DNA isolated from these cell populations was analyzed in a quantitative Southern blot assay. Classes of stem cells with totipotent or apparently restricted developmental potential have been defined based on detection of specific proviral integrants in the various fractionated cell populations. These include cells whose progeny repopulate all lineages and anatomical locations as well as stem cells that appear to contribute only to certain lineages, subsets of lineages or to specific organs and anatomical locations. We further describe a large class of stem cells which although totipotent displays quantitative differences in degree of contribution to various cell populations. The predominance of this class of cells suggests that apparent developmentally restricted stem cells may represent extreme examples of unequal lineage or organ repopulation by a totipotent cell.

The Southern blot analyses also indicated that surprisingly few (one or two) clones of transduced stem cells often account for the majority of mature hematopoietic cells in a recipient. Coupled with the finding that a substantial number of transduced cells that do not actively contribute to hematopoiesis in a primary recipient, can be activated to do so by retransplantation, the results suggest that a mechanism exists in vivo for the control of stem cell utilization. Studies involving the periodic sampling of hematopoietic tissue from a primary transplant recipient suggest that significant changes in the utilization of different stem cell clones can naturally occur.

These studies underscore the dynamic nature of hematopoiesis, and may suggest that normal hematopoiesis results from the sequential activation of different stem cell clones, rather than from an averaged contribution of the entire stem cell pool.

We have also undertaken more practical experiments to determine the feasibility of 'genetic therapies' for two specific genetic diseases: severe combined immunodeficiency (due to ADA deficiency) and β-thalassemia. A major focus of these latter studies has been to examine the expression of various constructs carrying the ADA or β-globin gene. For expression of the ADA gene, a number of recombinant genomes have been constructed (In collaboration with Dr. Stuart Orkin's laboratory at Harvard) which utilize various viral and cellular promoters to drive expression of a human ADA cDNA.

Experiments with the β-globin gene have involved the generation and characterization of recombinant genomes encoding a complete intact human β-globin gene. We have already characterized in detail the expression of human β globin sequences after their introduction into murine erythroleukemia (MEL) cells. These studies indicated that the gene was expressed in a tissue-specific manner, yet at levels approximately 1-2% of the endogenous mouse β globin levels. Interestingly, we were unable to detect any interactions between the inserted β globin transcriptional regulatory sequences and the proviral transcriptional signals (3). To further investigate the properties of the β globin-retroviral constructs, we have introduced these sequences into pluripotent, hematopoietic stem cells present in bone marrow, and subsequently introduced the transduced stem cells into mice via bone marrow transplantation. In spite of the fact that the viral titers of the β globin recombinants were significantly lower than many of the viral recombinants previously used in the bone marrow system, we have been able to generate over a half a dozen bone marrow transplant recipients containing hematopoietic tissue harboring the β globin proviral sequences. More importantly, we were able to generate recipients that were permanently and completely engrafted with transduced cells. Our studies to date have focused in detail on only three of these recipients, engrafted for over four months. While the copy number of the proviral sequences in hematopoietic tissue derived from these mice has not been accurately determined, at least one of the mice appears to harbor single copy proviral sequences while the other two most likely contain only a tenth copy. Two of the mice contain cells infected with the ZipneoSV(X)-β globin recombinant, while one contains cells infected with the viral enhancer deficient derivative of ZIPneoSV(X). In all three mice, significant levels of human β globin mRNA was detected in blood. In the most efficiently transduced mouse, the level of human β globin RNA expressed was approximately 2-4% the level of mouse β globin RNA. To further assess the absolute levels of expression of the β globin gene and its tissue specificity of expression, pure population of specific hematopoietic cell types (e.g. B, T, macrophages) were prepared from each mouse and RNA was isolated. In addition, RNA was prepared from several hematopoietic organs, including spleen, bone marrow and peripheral blood. So far, we have analyzed only two of the three mice in this latter way. In the case of the mouse transplanted into the enhancer plus

construct, significant levels of β globin RNA was
found in blood, bone marrow and spleen. Very
much lower, yet detectable levels of human β
globin RNA were found in the lymphoid and macrophage cell populations. To our surprise, however,
significant levels of LTR promoted viral RNA
could be detected in the macrophage and lymphoid
cell populations. No viral RNA could be detected
in peripheral blood cells. These results contrast
sharply with the experience of our laboratory and
others using LTR-based expression vectors, in that
little or no viral transcription has ever been
detected in the past after long term reconstitution
with cells bearing LTR based constructs. In the
case of the mouse transplanted with cells infected
with the enhancer deficient construct, an entirely
different pattern of transcription was detected.
As in the case of the enhancer plus mice described
above, human β globin RNA was detected predominantly in the blood, spleen, and bone marrow.
However, in the case of the enhancer minus mouse,
the degree of tissue specificity of human β globin
gene expression was extreme, as no detectable
human β globin RNA could be detected in non
erythroid cells, even after long exposure. Most
interestingly, viral RNA expression was restricted
in this mouse to the blood, bone marrow, and
spleen, with no RNA detectable in the lymphoid
and macrophage cell populations.

A number of important conclusions can be drawn
from the above studies. First, the work represents
one of the first demonstrations of the expression
of a foreign gene introduced into hematopoietic
cells via stem cell infection. Second, the
activation of viral transcription by the insertion
of β globin sequences into the pZIPSV(X) vectors
suggest the existence of sequences with the human
β globin gene that can overcome the irreversible
block to expression of LTR based constructs in
hematopoietic cells (derived from transduced stem
cells) seen by our group and others (4, unpublished
results). The nature of such sequences and their
possible normal role in β globin gene expression
is currently being investigated. Finally, the
dramatic differences in viral gene expression and
β globin gene expression obtained with the
enhancer plus and minus versions of the ZIP vector
indicates that interactions between the vector and
insert do seem to occur in the stem cell gene
transfer system and may be exploited to help identify a number of interesting transcriptional
regulatory sequences with the β-globin sequences.

REFERENCES

144

RETROVIRAL MEDIATED TRANSDUCTION INTO HEPATOCYTES IN VITRO

J. Wolff, J. Moores, J.-K. Yee, J. Respess, H. Skelly, H. Leffert, and T. Friedmann

Departments of Pediatrics and Medicine, University of California, San Diego, La Jolla, California 92093, USA

INTRODUCTION

Transducing vectors derived from murine retroviruses have proven useful for the efficient transfer of foreign genes into a variety of recipient mammalian cells in vitro (1-5). Transduced genes in the form of randomly integrated proviral sequences are expressed relatively stably in cells infected with such vectors (6,7). The fact that the efficient expression of foreign genes in genetically marked recipient cells can lead to correction of a mutant phenotype in vitro (8,9) has led to the suggestion that such retroviral-mediated gene transfer might be useful for the clinical treatment of some kinds of genetic disease in humans (10-12). Among the many important conceptual and technical problems that must be solved before such methods become feasible are the difficulties of introducing and expressing a foreign gene in appropriate cells and tissues. Organ-specific targeted delivery of transducing viruses per se is an obvious and desirable theoretical approach toward the introduction of foreign genes into whole animals. Several conceptually simpler gene therapy models are based on genetic modification of target cells in vitro followed by the introduction of genetically and phenotypically corrected cells into suitable organs of a recipient animal. Such an approach has been used to introduce genes into the mouse bone marrow (13-15) and we now report results with hepatocytes that not only may make a similar approach feasible for the liver but that also provide a new and potentially powerful model system for the study of hepatocyte development and differentiation.

MATERIALS AND METHODS

Murine retroviral-based vectors expressing a mutagenized human hypoxanthine guanine phosphoribosyltransferase (HPRT) and bacterial neomycin resistance genes were prepared as previously described (5,16). Proliferation-competent primary rat hepatocytes (17) were infected with these viruses at several times after the establishment of the cultures and gene expression was determined by isoelectric focusing gel assays for human HPRT enzymic activity (18) and by resistance to the neomycin analogue G418 (19). Hepatocytes were identified by morphological criteria (17) as well as by the synthesis of immunoprecipitable serum albumin.

RESULTS

Fig. 1 presents a gel assay of HPRT enzymic activity in lysates of hepatocytes infected on days 0, 1, 2, 5, 8, 10 and 12 of culture. Human enzyme was present only in cells infected during the proliferative de-differentiated stage of growth on days 1-5, with no evidence of infection and gene expression during the stationary, differentiated periods earlier (day 0) or later in culture. The level of human HPRT activity was approximately 10-25% of the endogenous rat activity.

Fig. 1. Isoelectric focusing gel assay for human HPRT activity. Samples of uninfected (c) hepatocytes and cells infected at 0, 1, 2, 3 and 5 days after plating. The expression of human HPRT enzymic activity (arrow) is most easily seen on days 2 and 3.

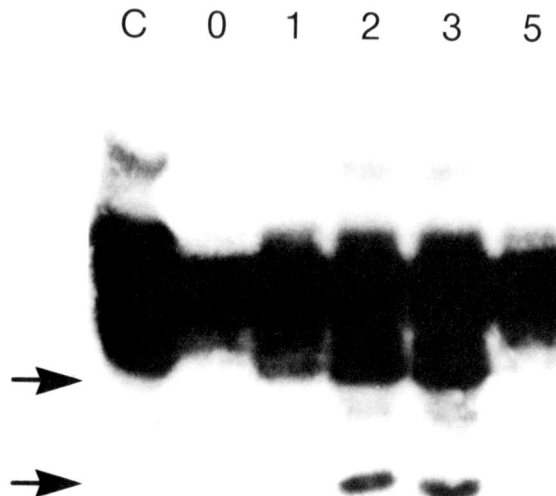

To determine whether the absence of detectable human enzymic activity in stationary cultures after day 5 was due to failure of infection or inability of the differentiated cells to express an integrated HPRT provirus stably, cells were infected on day 3 and the new synthesis of HPRT was determined on days 7 and 14 by specific immunoprecipitation of pulse-labelled HPRT antigen. An undiminished amount of labelled HPRT antigen was found on day 14 compared with day 7, suggesting that the absence of human enzymic activity in cells infected during late stationary phase of growth was due to failure of viral infection rather than shutdown of proviral gene expression.

Not only were both early and late stationary hepatocytes apparently refractory to infection, but late stationary cultures remained uninfectable even after they were stimulated to undergo cell division. This result is reminiscent of the refractoriness of regenerating rat liver to murine leukemia virus (19).

To determine if hepatocytes per se had been infected rather than only the non-parenchymal cells that constitute 5-20% of the cultured cells, cultures were infected during the susceptible phase of growth with the vector carrying the neomycin resistance gene and subsequently grown in the presence of G418 selection for neomycin resistance. The resulting cells had hepatocyte morphology (Fig. 2) and synthesized serum albumin as shown by immunoprecipitation of newly synthesized albumin.

Fig. 2. Primary culture of rat hepatocytes on day 20. The culture had been infected with N2 neomycin vector and selected with G418, and shows typical hepatocyte morphology and bile canaliculi. Immunoprecipiation of pulse-labeled albumin from uninfected cells at day 14 (a) and from day 14 cultures of N2-infected cells selected with G418. The band visible at approximately 68,000 daltons represents serum albumin.

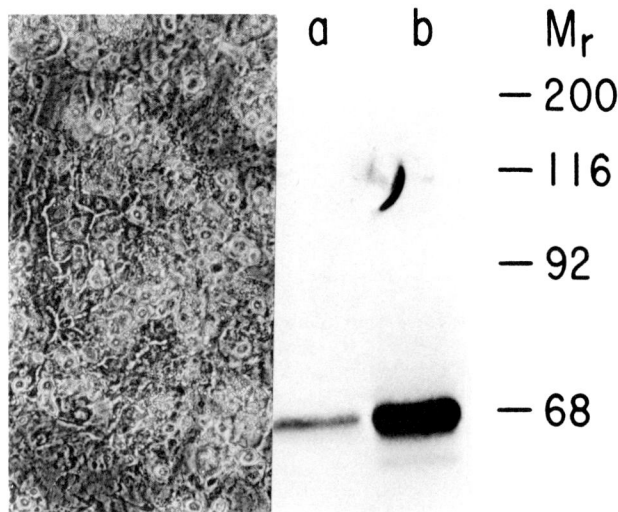

DISCUSSION

Murine leukemia viruses cannot generally infect differentiated hepatocytes in vivo even during liver regeneration, but can infect fetal liver cells. The methods reported here allow for the first time the efficient introduction of new genes into non-transformed hepatocytes. The de-differentiated state in culture may simulate the MLV infectable state of fetal hepatocytes in vivo. Furthermore, it seems likely that hepatocytes efficiently infected in vitro with retroviral vectors and expressing a foreign gene can be introduced into a variety of sites in recipient animals to provide a new function to such animals. It is very likely that such an approach will eventually be useful to complement enzymic deficiencies of hepatic function in a number of genetic diseases.

REFERENCES

(1) Wei, C., Gibson, M., Spear, P. and Scolnick, E. (1981) J. Virol. 39, 935-944.

(2) Shimotohno, K. and Temin, H. (1981) Cell 26, 67-77.

(3) Tabin, C., Hoffmann, J., Goff, S. and Wernberg, R. (1982) Mol. Cell Biol. 2, 426-436.

(4) Miller, A., Jolly, D., Friedmann, T. and Verma, I. (1983) Proc. Natl. Acad. Sci. USA 80, 4709-4713.

(5) Eglitis, M., Kantoff, P., Gilboa, E. and Anderson, W. (1985) Science 230, 1395-1398.

(6) Jolly, D., Willis, R., and Friedmann, T. (1986) Mol. Cell Biol. 6, 1141-1147.

(7) Varmus, H., Quintrell, N. and Wyke, J. (1981) Virol. 108, 25-46.

(8) Willis, R., Jolly, D., Miller, A., Plent, M., Esty, A., Anderson, P., Chang, H., Jones, O., Seegmiller, J. and Friedmann, T. (1984) J. Biol. Chem. 259, 7842-7849.

(9) Kantoff, P., Kohn, D., Mitsuya, H., Armentano, D., Sieberg, M., Zwiebel, J., Eglitis, M., McLachlin, J., Wiginton, D., Hutton, J., Horowitz, S., Gilboa, E., Blaise, R. and Anderson, W. (1986) Proc. Natl. Acad. Sci. USA 83, 6563-6567.

(10) Friedmann, T. (1983) Gene Therapy-Fact and Fiction in Biology's New Approaches to Disease, Cold Spring Harbor Laboratory Press.

(11) Anderson, W. (1984) Science 226, 401-409.

(12) Williams, D. and Orkin, S. (1986) J. Clin. Inves. 77, 1053-1056.

(13) Joyner, A., Keller, G., Phillips, R. and Bernstein, A. (1983) Nature 305, 556-558.

(14) Miller, A., Eckner, R., Jolly, D., Friedmann, T. and Verma, I. (1984) Science 225, 630-632.

(15) Williams, D., Lemischka, I., Nathan, D. and Mulligan, R. (1984) Nature 310, 476-480.

(16) Miller, A., Law, M. and Verma I. (1985) Mol. Cell. Biol. 5, 431-437.

(17) Koch, K. and Leffert, H. (1980) Ann. N.Y. Acad. Sci. 349, 111-127.

(18) Johnson, G., Eisenberg, L and Migeon, B. (1979) Science 203, 174-176.

(19) Jaenisch, R. (1980) Cell 19, 181-188.

TARGETING OF GENES TO SPECIFIC SITES IN THE MAMMALIAN GENOME

M.R. Capecchi and K.R. Thomas

Department of Biology, University of Utah, Salt Lake City, UT 84112

Gene-targeting, homologous recombination between DNA sequences residing in the chromosome and newly introduced DNA sequences, provide the means for systematically altering the mammalian genome (1). A desired alteration would first be introduced into a cloned DNA sequence and gene-targeting would transfer the alteration into the genome. Gene-targeting should be equally effective for correcting or mutating the desired chromosomal locus.

We have initiated our analysis of gene-targeting in cultured mammalian cells, by studying recombination between a defective gene residing in the chromosome and newly introduced plasmid DNA carrying a different mutation in the same gene. For these experiments, we first established cell lines containing a mutant neomycin resistance gene (Neo) integrated into the genome of mouse L cells and then sought to specifically restore the gene via homologous recombination by injecting DNA carrying a different mutation in the Neo gene. Productive gene-targeting events were identified by selecting for cells resistant to the drug G418. We characterized the corrected genes by Southern transfer analysis, by rescuing them from the genome and by sequencing them.

In the course of the experiments we have observed three classes of events: 1) the correction of the chromosomal gene by gene conversion or double reciprocal recombination with the newly introduced homologous sequence. This resulted in sequences in the genome being replaced by sequences in the introduced DNA; 2) the correction of the incoming DNA by gene conversion with the chromosomal sequence. This resulted in sequences in the incoming DNA being replaced by chromosomal sequences; and 3) induction of mutations in the chromosomal gene. In the latter case the mutations appear to result from incorrect repair of a heteroduplex formed between the newly introduced DNA and the homologous chromosomal sequence. We term this phenomena 'heteroduplex induced mutagenesis'.

RESULTS

Recombinant Plasmids and Recipient Cell Lines

The recombinant plasmids used for these experiments are designated pRH4-14/TK and pRH140ΔNae/TK (1). These plasmids were derived from the parental plasmid pRH140, which contains sequences from the bacterial plasmid pBR322 and the Neo^r gene coded for by the bacterial Tn5 transposon. The Neo^r gene was engineered to be functional in both bacteria and in mammalian cells. In bacteria, the Neo^r gene confers kanamycin resistance; in mammalian cells the Neo^r gene confers resistance to the drug G418. The Herpes simplex virus thymidine kinase gene, HSV-tk, was introduced into the above plasmid at the unique BamHI site to generate the plasmid pRH140/TK.

pRH4-14/TK contains an amber codon near the 5' end of the Neo gene. The presence of this premature polypeptide chain termination signal renders the gene product defective in both bacteria and mammalian cells. This mutation concomitantly creates a new DdeI side which is used as a diagnostic test for the presence of the amber mutation. pRH140ΔNae/TK contains a deletion of 284 bp at the 3' end of the Neo gene which removes 52 amino acids from the carboxy-terminal end of the Neo protein rendering it nonfunctional.

We have used three recipient cell lines designated LM1, LM4 and LM5. They differ in the number of integrated plasmids that they contain, the arrangement of the plasmids within the host chromosome and whether they contain pRH4-14/TK or pRH140ΔNae/TK. Each was derived by transforming LMtk$^-$ cells to tk$^+$ by injecting either pRH4-14/TK or pRH140ΔNae/TK DNA linearized with HindIII and selecting for HAT^R cells.

LM1 contains a single copy of the pRH4-14/TK plasmid integrated into chromosomal DNA by its HindIII ends. The cell line LM4 contains four copies of the pRH4-14/TK plasmid integrated by their HindIII ends at four different sites in the mammalian genome. LM5 contains five copies of the pRH140ΔNae/TK plasmid integrated at a single site in the mouse chromosome as a head-to-tail concatemer.

Class I Events: Correction of a Deletion Mutation Residing in the Chromosome

To test for gene-targeting, we injected 5-10 copies of pRH4-14/TK into nuclei of LM5 and selected for resistance to G418. G418^r cell lines arose at a frequency of approximately 1 per 1000 cells receiving an injection. Three of the G418^r cell lines were analyzed in detail. Digestion of genomic DNA with the restriction endonuclease DdeI yields a series of fragments that allow us to identify the presence of the wild-type Neo^r gene and each of the mutant alleles of Neo since the 4-14 amber mutation creates a new DdeI site and the ΔNae deletion removes a DdeI site.

By Southern transfer analyses, in the parental cell line LM5 we observe only the DdeI fragment produced by the ΔNae deletion fragment. In addition to the ΔNae deletion fragment, each of the G418^r cell lines also contain the DdeI fragment produced by the presence of a wild-type Neo^r gene. Further Southern transfer analysis using a series of restriction enzymes (HindIII, BamHI, Bgl II, EcoRI and PvuII) demonstrated that in each of the G418^r cell lines one of the pRH140ΔNae/TK sequences residing in the chromosome of LM5 was corrected to wild-type by recombination with the incoming pRH4-14/TK sequences. None of the G418^r cell lines contained a DdeI fragment 1048 bp long. Such a fragment would indicate the presence of a 4-14-ΔNae double mutant. The presence of a double mutant would be predicted if correction of the pRH140ΔNae/TK sequence resulted from a single reciprocal homologous recombination event between the 4-14 amber mutant and the ΔNae deletion mutation. The absence of the double mutant indicates that the correction of the deletion mutation occurred via double reciprocal homologous recombination or by gene conversion.

In order to extend our analysis we rescued the Neo sequences residing in the genome of the G418^r cell lines. Analysis of the rescued sequences provided an independent check of the results obtained by Southern transfer and permitted a more detailed characterization of the recombination events. A wild-type unit length pRH140 sequence was rescued from each of the G418^r cell lines. These sequences were indistinguishable from the parental pRH140 or pRH140/TK sequences. We also sequenced the region corresponding to the ΔNae deletion and observed no differences in this region compared to the pRH140/TK sequence. Thus, in the process of correcting the pRH140ΔNae/TK sequence residing in the chromosome, the information donated by the incoming pRH4-14/TK plasmid was faithfully transferred.

Class II Events: Correction of the Incoming Plasmid by Sequences Residing in the Chromosome

Following injection of LM1 or LM4 with 5-10 copies of pRH140ΔNae molecules, we obtained G418^r cells at a frequency of approximately 1 per 1000 cells receiving an injection. The observed frequency of G418^r is five orders of magnitude greater than the spontaneous reversion frequency of LM1 or LM4 to G418^r.

We have analyzed a number of G418^r cell lines derived from LM1 or LM4 and they fall into two classes. In the first class a wild type gene was generated by homologous recombination between the introduced plasmid, pRH140ΔNae and the chromosomal sequence pRH4-14/TK. An example of such an event, as described below, is seen in the cell line LM1-3. In the second class a pseudo-wild-type gene is generated by the amber mutant acquiring a compensating mutation by 'heteroduplex induced mutagenesis'. We will describe this class later.

Southern transfer analysis of LM1 genomic DNA digested with Bgl II reveals a single hybridizing band J-1. In LM1-3 we observe the J1 band and an intensely hybridizing high molecular weight band not present in the parental LM1 cell line. This band represents a head-to-tail concatemer containing approximately four copies of the input pRH140ΔNae sequence and one copy of the wild-type pRH140 sequence. We will show that this wild-type sequence was generated by the correction of one of the incoming pRH140ΔNae plasmids by the pRH4014/TK sequence residing in the chromosome of LM1.

DdeI Southern transfer patterns of LM1 and LM1-3 show that in LM1 we observe the DdeI fragments indicating the presence of the 4-14 amber mutant. In LM1-3 we observe the DdeI fragments indicating the presence of the 4-14 amber mutant, the ΔNae deletion mutant, and the corrected wild-type Neo^r gene. Thus, the single copy of the pRH4-14/TK sequence in LM1 was not corrected since it is still present in LM1-3; instead one of the incoming pRH140ΔNae sequences was corrected. This result was confirmed by plasmid rescue. From LM1-3 we rescued the pRH4-14/TK sequence residing in LM1-3. It contains the appropriate 5' chromosomal junction DNA and is indistinguishable from the corresponding sequence isolated from LM1. From LM1-3 we rescued two additional classes of sequences, one corresponding to the unit length pRH140ΔNae and the other corresponding to the corrected wild-type pRH140 sequence.

Class III Events: 'Heteroduplex Induced Mutagenesis'

Following injection of LM1 and LM4 with pRH140ΔNae/TK we obtained a series of G418^r cell lines whose Southern transfer patterns were indistinguishable from the parental pattern. These results show that conversion to G418^r was not accomplished by acquisition of new Neo sequences or by detectable rearrangements of old ones. All of these cell lines exhibited only the DdeI fragments characteristic of the 4-14 amber mutant. These results were quite unexpected since all of the derivative cell lines were G418^r. The above results were confirmed by plasmid rescue. From each of the G418^r cell lines we rescued a sequence that conferred Kn^r phenotype when propagated in E. coli. When we examined the DdeI polymorphism present in these rescued sequences, all had the 4-14 DdeI polymorphism. Thus each of these sequences still retained the amber mutation yet conferred Kn^r on bacteria.

We sequenced approximately 300 bp at the 5' ends of the Neo^r genes isolated from the G418^r cell lines. These were then compared to the corresponding sequences isolated from the parental cell lines. The only changes found in the Neo^r genes were as follows: the insertion of a thymidine residue 11 bp downstream from the amber mutation in pLM1-8-J1 and pLM4-9-J1; the insertion of four bases, GGCT, in pLM1-1-J1, pLM1-21-J1, pLM1-24-J1, pLM4-10-J1 and pLM4-29-J4; the insertion of a guanosine residue 8 or 9 bp downstream from the amber mutation in pLM1-36-J1.

To determine how Neo^r genes containing both an amber mutation and a frameshift mutation confer Neo^r activity we determined the amino terminal amino acid sequence of the Neo^r gene product. We will refer to the gene containing both the amber mutation and an insertion mutation as the pseudo-wild-type gene. As seen below wild-type Neo^r protein gave the predicted sequence: met ile glu gln asp gly ... However, the protein from the pseudo-wild-type gene began with the sequence met asp cys thr gln ... Examination of the DNA sequence of the Neo gene indicates that the latter protein was initiated from an AUG 14 bp downstream from the normal AUG in the -1 translation reading frame.

```
                          0
wt-Neoʳ      ATGATTGAACAAGATGGATTGCACGCAGGTTCT
             met ile glu gln asp gly leu his ala .... wt protein

                          -1
pseudo wt-Neoʳ ATGATTGAACAAGATGGATTGCAGGCAGGTTCT
                met asp cys thr gln ... pseudo wt protein
```

Since the pseudo-wild-type gene is translated from an AUG in the -1 reading frame, it became apparent how the insertion mutation reverted the amber mutation. After initiation of protein synthesis in the -1 frame, the ribosome passes through the amber mutation, which is in the 0 reading frame and therefore not read, and regains proper phase when it reaches the +1 frameshift mutation (insertion of a T, G, or GGCT). Although this mechanism alters the NH$_2$-terminal amino acid sequence of the Neo protein, this portion of the protein has been shown to be dispensable.

Two models could explain how translation of the pseudo-wild-type gene begins at the -1 AUG. In model 1 we assume that the ribosome can enter at two sites, AUG in the 0 and -1 reading frame. The second model makes use of a single entry point followed by translation reinitiation. In this situation the ribosome enters the AUG in the 0 reading frame and translates the mRNA until it reaches the amber codon. There it terminates and releases the polypeptide fragment. At this point the ribosome scans back until it reaches the -1 AUG where it reinitiates protein synthesis. Genetic analysis demonstrated that the second model is correct.

DISCUSSION

The frequency of the described targeting events was much higher than we anticipated, particularly considering the complexity of the mammalian genome and the propensity of mammalian cells to incorporate exogenous DNA into their genome by nonhomologous recombination. We observed a productive targeting event in approximately 1 of every 1000 cells receiving DNA.

It is interesting that in the cell LM5, where the deletion mutation resides in the chromosome, homologous recombination with the incoming gene carrying the point mutation led to the correction of the deletion mutation in the chromosomal gene, whereas in the cell lines LM1 and LM4 where the point mutation resides in the chromosome, homologous recombination with the incoming DNA carrying the deletion mutation led to correction of the incoming plasmid DNA. This observation may reflect a preference for correction of a deletion mutation over the point mutation.

The unexpected finding that correction of the pRH4-14/TK sequences residing in the host genome of LM1 and LM4 often resulted from insertion of a few base-pairs downstream from the amber mutation raises a number of questions. How did the insertion of these bases occur? How did injection of pRH140ΔNae/TK into either LM1 or LM4 mediate these events?

On examining the DNA sequence surrounding the position of the base-pair insertions we observe the six base-pair direct repeat GGCTAT. The fact that each of the insertions, T, G or GGCT is a tandem repeat or part of this sequence, suggests that a duplication generating mechanism is active in this region.

A number of observations argue that the insertion or duplication events at this site were induced by the initiation of recombination between pRH4-14/TK sequences residing in the chromosome and the incoming pRH140ΔNae/TK sequence. First, the frequency of generating this class of G418^r cell lines was comparable to the frequency of generating cell lines by legitimate gene-replacement or gene-conversion. Second, this frequency is five orders of magnitude greater than the spontaneous reversion frequency of LM1 or LM4 to G418^r. Third, we isolated a cell line in which both a homologous recombination event and a 'het-induced mutagenic event' occurred (1). Since the homologous recombination events and the mutagenic events each occur at a frequency of approximately 1 per 1000 cells receiving an injection, the predicted frequency of both events occurring independently in the same cell is 1 per 10^6 injected cells. Having obtained such a cell line after a few thousand injections, it is tempting to postulate that the two events occurred as a result of a concerted reaction. Fourth, we did not obtain G418^r cells from LM1 or LM4 following injection of a plasmid DNA, pBR322, that does not contain homology to the Neo gene. Thus injection of DNA, per se, does not induce rampant mutagenesis. Similarly, we did not obtain G418^r cells from LM1 and LM4 following injection of pRH4-14/TK or pRH4-14ΔNae/TK. The latter

experiments speak to the specificity of the reaction. Since injection of the pRH4-14/TK vector does not induce the mutations, an important factor for triggering the mutagenic response may be the single base-pair mismatches at the amber mutation and/or the large mismatch at the ΔNae deletion. The observation that injection of the double mutant, pRH4-14ΔNae/TK, also does not induce mutations argues for the requirement of at least the single base-pair mismatch. The only mechanism that we can envision whereby single base-pair mismatches between a chromosomal sequence and an exogenous sequence can influence the mutation process requires the formation of a heteroduplex.

In the future, we will examine 'heteroduplex-induced mutagenesis' that directs the loss of gene function rather than the correction of gene function. These events should occur at a higher frequency as well as reveal a wider spectrum of changes at the DNA sequence level. The only apparent requirement for a 'hot spot' for 'het-induced mutagenesis' is small direct repeats proximal to base-pair mismatches between the newly introduced plasmid sequence and the homologous sequence in the genome. Comparable small direct repeats are encountered in the DNA coding sequence of most genes, making them susceptible to this type of mutagenesis. Permutations of this methodology should provide the means for efficiently introducing mutations into specific mammalian cellular genes.

CONCLUSIONS

In the process of studying the targeting of exogenous DNA sequences to specific loci in the mammalian genome, we have uncovered two mechanisms for altering chromosomal sequences. The first involves the transfer of information, by homologous recombination, from the newly introduced DNA into the cognate chromosomal sequence. The second involves inducing mutations in the homologous chromosomal sequence by what appears to entail incorrect repair of a heteroduplex formed between the newly introduced DNA and the cognate chromosomal sequence. Each has its own advantages. The transfer of information by homologous recombination allows the transfer of defined DNA sequences. This procedure should be equally effective for correcting or mutating, in a defined manner, the desired chromosomal locus. On the other hand, the frequency of altering chromosomal sequences by 'heteroduplex induced mutagenesis' promises to be higher than via homologous recombination. A higher frequency would simplify extension of this methodology to altering endogenous genes in cultured mammalian cells or mammalian organisms, such as the mouse.

REFERENCES

(1) Thomas, K.R., Folger, K.R. and Capecchi, M.R. (1986) Cell 44, 419-428.
(2) Thomas, K.R. and Capecchi, M.R. (1986) Nature submitted.

HEMATOPOIETIC STEM CELL DEVELOPMENT WITH AND WITHOUT FOREIGN GENES

Beatrice Mintz, Luis Covarrubias, and
Robert G. Hawley

Institute for Cancer Research, Fox Chase Cancer
Center, Fox Chase, Philadelphia, Pennsylvania
19111

Genetically defective hematopoietic stem cells of mice (W-series mutants) can be replaced, even before birth (1), with normal stem cells. During hepatic hematopoiesis of the fetal host, fetal liver or adult bone marrow cells are microinjected via a placental blood vessel. The competitive superiority of the donor cells enables irradiation of the host to be circumvented and seeding occurs in the fetal liver. Genetic markers document engraftment by self-renewing totipotent hematopoietic stem cells (THSC) whose mitotic progeny progress to the bone marrow and give rise to the definitive myeloid as well as lymphoid lineages (2). The objectives of early THSC replacement in mice are twofold: to analyze the developmental histories of stem cells and their derivatives during the normal progression in intact hematopoietic tissue environments; and, ultimately, to utilize this experimental system to introduce genes of interest into early THSC, as probes of development and of tumorigenesis and as models of gene therapy.

Here we briefly summarize the experimental observations on normal hematopoiesis made in this laboratory. And we describe our initial results on gene transfer by retroviral infection of THSC -- at first limited to adult donors and hosts as a relatively convenient means of evaluating the vectors.

MONOCLONAL ORIGIN OF MYELOID AND LYMPHOID

CELLS. The intraplacental-inoculation assay has provided several lines of evidence for the developmental totipotency of individual stem cells in fetal liver, despite the difficulty that THSC have not been recognized or isolated: (a) The frequency and the rate of seeding of W-mutant fetuses by normal donor cells decrease with decreasing severity of the endogenous stem-cell genetic defect. In mutant hosts with only a mild defect (e.g., W^f/W^f), success declines to 5-10% of cases, suggestive -- and in principle equivalent to a limiting-dilution experiment -- of engraftment by one or very few cells (2). Thus, the myeloid and lymphoid lineages that are of donor genotype can have arisen monoclonally. (b) When the mildly defective W^f/W^f fetal hosts were offered a "choice" from an injected mixture of two normal genetic strains of fetal liver cells, most of those that were seeded had retained only one or the other donor strain. From statistical analyses, true single-stem-cell seeding had probably occurred in at least 50% of the cases (3). In addition to these points, even in severely defective (W/W) hosts, very few bright-red colonies (the normal-donor type) were seen in the fetal livers a week after engraftment.

THSC CLONAL SUCCESSION.

In an engrafted host, the presence of very few (donor/host) THSC of distinguishable genotypes enables the pedigrees of individual stem cells to be followed over time. The results revealed a complementary rise and fall of the respective genetic subpopulations of cells, with a periodicity of approximately 20-30 weeks (3). Each new wave apparently represents the amplified proliferation of a single THSC clone. In intact normal hosts, the numbers of endogenous

THSC may possibly be greater than in experimentally engrafted W-mutant fetuses, and amplified clones may perhaps succeed each other more quickly. Nevertheless, the cycling behavior that comes to light in the experimental animals appears to be a physiological phenomenon of clonal succession, thus validating a theoretical model once propounded by Kay (4).

We have further proposed that the stem cell population is hierarchical (5). There appears to be a small reserve compartment, designated THSC I, and consisting of slowly dividing cells of long self-renewal potential; from this, a more actively proliferating but ephemeral THSC II compartment periodically arises. The latter would be expected to include a heterogeneous series of cells -- albeit all THSC -- due to gradually reduced renewal potential and increasing expression of genes involved in specialization.

DEVELOPMENTAL CHANGES IN THSC. Other ways in which the THSC population was found to change during development are the following: Evidence of THSC in adult bone marrow was also found by inoculation of cells into fetuses, but the bone marrow stem cells tend to have less self-renewal potential than fetal liver THSC (6), possibly partly because of the presence of relatively more THSC II than THSC I in marrow. In addition, although histocompatibility differences at the H-2 locus between donor and fetal host are of no consequence when the donor cells are from fetal liver, allogeneic combinations prevent retention of the THSC when the donor cells are from bone marrow (6). And marrow THSC that are experimentally placed in a fetal environment have lost the capacity to generate erythroid cells on which an erythrocyte-specific and fetal-specific antigen, Ft, is expressed (7).

RETROVIRAL VECTORS. The feasibility of using retroviral vectors to introduce recombinant genes into hematopoietic cells (CFU-S), without causing viremia, was first demonstrated in Mulligan's laboratory (8). With further technical advances, recombinant retroviruses have been used to infect bone marrow hematopoietic cells in vitro and the cells have given rise to differentiated myeloid and lymphoid derivatives after transfer to irradiated adult recipients (9-11). In all these studies, the sole gene of interest in the retroviruses was the bacterial neomycin resistance (neo) gene conferring resistance to the antibiotic G418 in mammalian cells. Expression of the neo gene was demonstrated in two of the studies (9, 10).

Important objectives still to be met include the capability of introducing nonselectable genes into THSC; the introduction of transcriptional regulatory sequences near the gene of interest, and removal of the viral transcriptional controlling sequences; and elimination of host irradiation with its detrimental (and possibly irreproducible) effects on the hematopoietic microenvironment.

We have designed and utilized a series of novel retroviral vectors whose major feature involves a deletion containing the enhancer and promoter sequences in the retroviral 3' long terminal repeat (LTR) (12). Because the region encompassing the deletion serves as the template for a portion of the 5' LTR during the reverse transcription of viral RNA to DNA, the result is a virus lacking controlling sequences in both LTRs. This should eliminate potential deleterious effects of those viral sequences on expression of the genes of interest inserted into the body of the vectors; and it should allow transferred genes to function only in those cell types in which any associated controlling sequences are normally active. In addition, the change should decrease the potential of activating cellular genes by insertional mutagenesis or of generating infectious viruses through recombination with endogenous viruses.

Our prototype vectors contain the neo gene transcribed from an internal promoter supplied by the herpes virus thymidine kinase (tk) gene. The tk promoter was chosen because of its ability to function in a wide variety of cell types. In our HHAM(tk-neo) vectors, discussed here, the direction of transcription of the neo gene coincides with transcription promoted by the retroviral LTR (12). (The rubric HHAM designates the genetically handicapped status of these viruses, due to their 3' and internal deletions, and their chimeric origin from Harvey, Abelson, and Moloney murine retroviruses.)

RETROVIRAL SEQUENCES IN LYMPHOID CELL LINES.
The c-myc proto-oncogene, implicated in plasma-cytomagenesis in BALB/c and NZB mice, was inserted into the HHAM(tk-neo) vector under the control of immunoglobulin kappa (κ) chain gene enhancer-promoter sequences. The enhancer was coupled with the κ promoter because of a possible synergistic action between the two; this has in fact recently been shown to occur (13). A Psi-2(ψ2) cell culture line was derived that produces helper-free HHAM(tk-neo)$E_\kappa P_\kappa$myc virus at a titer of 8 x 10^4 G418^r cfu per ml. Supernatants from this cell line were used to infect WEHI 231 B-cells and 70Z/3 pre-B-cells for in vitro tests. G418-resistant subpopulations of cells were isolated and expanded for DNA and RNA analyses. The structure of the proviruses was as expected, with the deletion originally introduced in the 3' LTR successfully transferred to the 5' LTR. Transcription of the c-myc gene from the κ promoter was observed both in the B-cells and in mitogenically stimulated pre-B-cells.

RETROVIRAL SEQUENCES IN VIVO.
The adult mice chosen as recipients for long-term bone marrow reconstitution have different genetic defects favoring competitive replacement by genetically normal cells, so that irradiation could be omitted. These included W/W^v mice, discussed above, with a stem cell defect and an associated macrocytic anemia; and C.B-17 scid mice, defective in lymphopoiesis and devoid of mature B- and T-cells (14). Bone marrow from normal syngeneic animals that had been treated 2 or 4 days previously with 150 mg/kg of 5-fluorouracil was cocultured with ψ2 cells producing helper-free HHAM(tk-neo)$E_\kappa P_\kappa$myc virus. The growth factors interleukin-3 and hemopoietin-1 were included in the culture medium as they have been shown to enhance proliferation of multipotent bone marrow cells (15). After 48 hours, the nonadherent bone marrow cells were collected and injected via the tail vein of recipients. Two months later, peripheral blood of the W/W^v animals was tested for the presence of a hemoglobin electrophoretic variant characteristic of donor-strain red blood cells. All six W/W^v hosts had greater than 90% donor-type hemoglobin. Spleen DNA, prepared from partial splenectomies three months after engraftment, was analyzed for a restriction fragment length polymorphism (RFLP) in the immunoglobulin μ heavy chain gene that is specific for DNA of the donor strain. The results established that more than 90% of the splenic

(lymphoid) cells of all six W/W^v animals were of donor origin. Thus, self-renewing donor stem cells had been seeded in these mice and had generated myeloid and lymphoid derivatives. In the C.B-17 scid mice that received comparable numbers of cells, the RFLP marker was present in approximately 50% of spleen DNA.

Subsequent hybridization of spleen DNA with labeled neo or myc probes demonstrated the presence of intact recombinant viral sequences in spleen DNA of all these W/W^v and C.B-17 scid recipients. When the DNA was digested with a restriction enzyme that could discriminate between different viral integration sites, it became apparent that multiple integrations had occurred. The animals were therefore either repopulated with multiple stem cells having dissimilar viral integrations, or -- less likely -- with very few stem cells each with several integrations. The alternatives can be examined in further experiments, including re-grafts. In either case, this result is noteworthy in its implications for therapy with retroviruses: Adults reconstituted with unselected bone marrow cells may thus come to harbor variant cell clones. It will be of interest to compare reconstitutions of adults engrafted with THSC bearing such handicapped viruses and reconstitutions of prenatal hosts, in which engraftment entails only single or few hematopoietic stem cells.

REFERENCES

(1) Fleischman, R. A. and Mintz, B. (1979) Proc. Natl. Acad. Sci. USA 76, 5736-5740.
(2) Fleischman, R. A., Custer, R. P. and Mintz, B. (1982) Cell 30, 351-359.
(3) Mintz, B., Anthony, K. and Litwin S. (1984) Proc. Natl. Acad. Sci. USA 81, 7835-7839.
(4) Kay, H. E. M. (1965) Lancet 2, 418-419.
(5) Mintz, B. (1985) in Fetal Liver Transplantation (Gale, R. P. et al., eds.) pp. 3-16, Alan R. Liss, Inc., New York.
(6) Fleischman, R. A. and Mintz, B. (1984) J. Exp. Med. 159, 731-745.
(7) Blanchet, J. P., Fleischman, R. A. and Mintz, B. (1982) Dev. Genet. 3, 197-205.
(8) Williams, D. A., Lemischka, I. R., Nathan, D. G. and Mulligan, R. C. (1984) Nature 310, 476-480.
(9) Dick, J. E., Magli, M. C., Huszar, D., Phillips, R. A. and Bernstein, A. (1985) Cell 42, 71-79.
(10) Keller, G., Paige, C., Gilboa, E. and Wagner, E. F. (1985) Nature 318, 149-154.
(11) Lemischka, I. R., Raulet, D. H. and Mulligan, R. C. (1986) Cell 45, 917-927.
(12) Mintz, B., Covarrubias, L. and Hawley, R. G. (1986) in International Workshop on Human Preleukemia (Breda, R., ed.) (In press).
(13) Garcia, J. V., Bich-Thuy, L. T., Stafford, J. and Queen, C. (1986) Nature 322, 383-385.
(14) Bosma, G. C., Custer, R. P. and Bosma, M. J. (1983) Nature 301, 527-530.
(15) Stanley, E. R., Bartocci, A., Patinkin, D., Rosendaal, M. and Bradley, T. R. (1986) Cell 45, 667-674.

ACKNOWLEDGMENT

This work was supported by grants HD-10646, CA-06927, and RR-05539 from the U.S. Public Health Service, and by an appropriation from the Commonwealth of Pennsylvania. R.G.H. was the recipient of a fellowship from the Medical Research Council of Canada.

THE GENES CODING FOR HUMAN PRO α1 (IV) AND PRO α2 (IV) COLLAGEN ARE BOTH LOCATED AT THE TERMINAL END OF THE LONG ARM OF CHROMOSOME 13

Charles D. Boyd[1], SuEllen Toth-Fejel[3], Inder K. Gadi[1], Markku Kurkinen[2], Margret Kolbe[1], I. Kathleen Hagen[1], James W. Mackenzie[1] and Ellen Magenis[3]

Departments of [1]Surgery, and [2]Medicine, UMDNJ-Robert Wood Johnson Medical School, New Brunswick, NJ 08903
[3]Department of Medical Genetics, Crippled Childrens Division, The Oregon Health Sciences University, Portland, Oregon 97201

Introduction

The genes coding for fibrillar collagens have previously been shown to be widely dispersed in the human genome. (1) Recent evidence from several groups have shown that a non-fibrillar collagen gene, coding for a basement membrane component, pro α1(IV) collagen, is located at the distal end of the long arm of chromosome 13. (2-3) This chromosomal localization is different to any previously reported for collagenous coding sequences. To investigate the possibility that other non-fibrillar collagen genes are also dispersed through-out the genome, we undertook the chromosomal localization of the gene coding for human pro α2(IV) collagen, using a recently isolated cDNA clone, containing DNA sequences coding for human proα 2(IV) collagen. This manuscript presents results demonstrating in contrast to the dispersed nature of the fibrillar collagen genes, that both genes coding for the non-fibrillar proα 1(IV) and pro α2(IV) collagens have the same chromosomal localization.

Materials and Methods

A previously constructed library of recombinant cDNAs, prepared using poly A+ RNA isolated from a fibrosarcoma cell line, (HT-1080), was screened with mouse pro α1(IV) and pro α2(IV) collagen cDNA clones. The human α1(IV) procollagen cDNA, HT-21 has been previously characterized. (4) Two additional human cDNA clones (HT-68, HT-39) have been recently identified that contain DNA sequences coding for part of the non-collagenous NCl domain of human pro α2(IV) collagen. These recombinant DNA sequences were tritium labelled by nick translation, hybridized in situ with meta phase chromosomes obtained from lymphocytes of normal donors and subject to autoradiography and chromosome staining, using previously published procedures. (5)

Results and Discussion

A significant percentage of autoradiographically positive grains were associated with the q33-q34 bands of chromosome 13. Table 1 clearly illustrates this. The hybridization of the α2(IV) cDNA to the region of chromosome 13 is considerably greater than the results achieved with the 1(IV) cDNA, using the same conditions of in situ hybridization. Although the significance of this is at present unclear, Southern blot data have demonstrated that it is not due to cross hybridization of the α2(IV) cDNA sequences to homologous sequences in the α1(IV) gene.

Table 1. Summary of in situ hybridization results

cDNA	Number of cells scored	Total number of grains	Number and % of grains associated with 13q33 - 34	
HT-21 (α1(IV))	260	447	63	14.1%
HT-68 (α2(IV))	100	203	57	28.1%

Figure 1A illustrates the grain distribution following in situ hybridization with the pro α2(IV) collagen cDNA clone. Hybridization is clearly localized to bands q33-q34 on chromosome 13, with no evident hybridization to any other chromosomal region. A similar result is obtained using the α1(IV) cDNA clone, HT-21; the hybridization pattern to the q33-q34 bands of chromosome 13 by HT-21 is shown in Figure 1B.

The proximity of chromosomal locality of two non-fibrillar collagen genes is significant both to considerations of the evolutionary development of the family of collagenous coding gene sequences and to the possible role non-fibrous collagen genes play in inherited human diseases.

Figure 1. Grain distribution from cells examined following in situ hybridization with an α2(IV) procollagen cDNA (A) and α1(IV) procollagen cDNA (B)

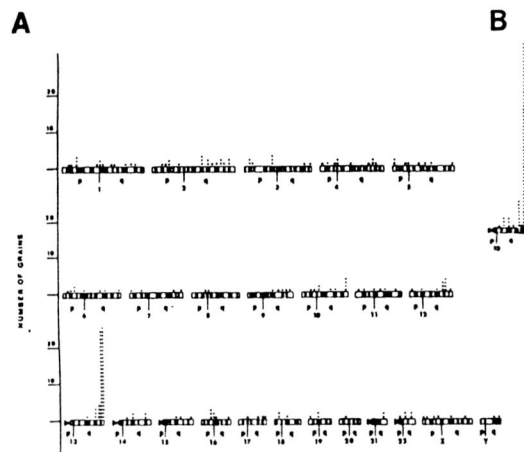

References

1. Huerre-Jeanpierre, C., Mattei, M-G., Weil, D., Gizeschik, K.H., Chu, M-L., Sangiorgi, F.O., Sobel, M.E., Ramirez, F. and Junien, C. (1986) Am. J. Hum. Genet., 38, 26-37.

2. Emanuel, B.S., Sellinger, B.T., Gudas, L.J., Myers, J.C. (1986) Am. J. Hum. Genet., 38, 38-44.

3. Boyd, C.D., Weliky, K., Deak, S.B., Christiano, A.M., Mackenzie, J.W., Sandell, L.J., Tryggvason, K. and Magenis, E. (1986) Hum. Genet. In Press.

4. Pihlajaniemi, T., Tryggvason, K., Myers, J.C., Kurkinen, M., Lebo, R., Cheung, M-C., Prockop, D.J. and Boyd, C.D. (1985) J. Biol. Chem., 260, 7681-7687.

5. Harper, M.E. and Saunders, G.F. (1981) Chromosoma, 83, 431-439.

EXPRESSION OF HUMAN TISSUE PLASMINOGEN ACTIVATOR IN TRANSGENIC MICE

Sharon J. Busby, Mason C. Bailey, Eileen R. Mulvihill, Michael L. Joseph, A. Ashok Kumar

ZymoGenetics, Inc.; 2121 North 35th Street; Seattle, WA 98103

Introduction

Tissue plasminogen activator (t-PA) functions in the regulation of fibrinolysis by converting plasminogen to plasmin, a protease that hydrolyzes fibrin and fibrinogen as well as other plasma proteins[1]. Because of its affinity for fibrin and its low enzymatic activity in the absence of fibrin, t-PA has been proposed as a fibrin-specific, therapeutic agent for thrombolytic disorders. However, there have been a number of problems with this type of therapy, among them 1) the short half-life of the t-PA administered to patients and the need to inject large doses to effect thrombolysis, and 2) the systemic activation of plasmin that can lead to bleeding complications[2].

One way of gaining a greater understanding of the effects of elevated t-PA levels is by examining animals in which a high steady state level of t-PA is maintained in plasma. To this end, we have produced transgenic mice that express human t-PA in their plasma at levels up to 30 times endogenous t-PA levels.

Results and Discussion

A cDNA containing the sequences for human t-PA was fused to the mouse metallothionein (MT1) promoter and the human growth hormone poly A addition site and then microinjected into the pronuclei of mouse eggs[3]. The injected eggs were returned to the oviducts of foster mothers. Of the 55 progeny we obtained, 14 had integrated the MT1-t-PA DNA into their genomes. Five animals were expressing human t-PA in their plasma as determined by an enzyme-linked immunosorbent assay (ELISA) specific for human t-PA. One mouse expresses human t-PA at levels of 60-80 ng/ml. His F1 and F2 progeny express at a variety of levels from 10 ng/ml to 110 ng/ml. In some progeny, addition of 25 mM $ZnSO_4$ to the drinking water resulted in a 2-3 fold increase in t-PA levels. The highest levels observed were 120 ng/ml. The transgenic t-PA was also determined to be biologically active by a fibrin plate lysis assay performed on euglobulin fractions of transgenic plasma. The levels of t-PA observed in this mouse line are approximately 2-30 times the normal levels of endogenous mouse t-PA and yet we have observed no abnormalities in blood clotting or wound healing and no other obvious phenotypic differences between the transgenic mice and their siblings.

We have also examined the molecular form of the transgenic t-PA by immuno-precipitation and Western analysis. Our results show that greater than 90% of the transgenic t-PA is circulating as the single chain form (mw of 68,000) and little, if any, as the two chain form. Our results also suggest that most of the transgenic t-PA is not complexed with the known inhibitors of t-PA in mouse plasma. We are examining more closely the interaction of the transgenic t-PA with mouse plasma proteins to understand how these animals can tolerate the elevated t-PA levels.

References

1. For review, Francis, M.D. and Marder ,V.J. (1986). Ann Rev. Med. 37: 187-204.

2. Van de Werf et al. (1984). Circulation 69: 605-610.
   Collen et al. (1984). Circulation 70, 1012-1017.
   Verstraete et al. (1985). Lancet 1: 842-847.
   The Thrombosis in Myocardial Infarction (TIMI) Trial. (1985). N. Engl. J. Med. 312: 932-936.

3. Gordon, J. and Ruddle, F. (1983). In Methods in Enzymology, eds. Wu, R., Grossman, L. and Moldave, K. (Academic, New York), vol. 101, pp. 411-433.
   For review, Palmiter, R. and Brinster, R. (1985). Cell 41: 343-345.

154

IN VIVO MODEL SYSTEMS FOR THE ANALYSIS OF EXPRESSION OF GENES TRANSFERRED USING RETROVIRAL VECTORS

M.A. Eglitis, P.W. Kantoff, A.W. Flake, J.R. McLachlin, A. Gillio, R.C. Moen, J.A. Zwiebel, D. Kohn, C. Bordignon, R.M. Blaese, E. Gilboa, R. O'Reilly, M.R. Harrison, E.D. Zanjani, and W.F. Anderson

Laboratory of Molecular Hematology, NHLBI, NIH, Bethesda, MD 20892; Department of Surgery, University of California, San Francisco 94143; Memorial-Sloan Kettering Cancer Center, New York, NY; Metabolism Branch, NCI, NIH, Bethesda, MD; Veterans Administration Medical Center, University of Minnesota, Minneapolis, MN 55417.

We have been developing animal models to analyze the efficiency of retroviral vectors for gene transfer and expression. Initial studies with mice (1,2) showed that N2, a vector derived from Moloney murine leukemia virus containing the bacterial neomycin resistance (neo^R) gene, could transfer the neo^R gene to as many as 85% of CFU-S and that 25-50% of infected progenitors expressed the transferred gene as determined by enzymatic and colony forming assays.

Recent results in sheep have shown that fetal blood progenitors may also be efficiently infected with retroviral vectors. Using the same N2 vector, 10-40% of CFUs were initially infected, using cells from 100 day fetuses. Seven days after birth (roughly 60 days after gene transfer and marrow transplant in utero (3)), up to 33% of CFUs in the marrow of the newborn lambs continued to be drug resistant (Table I). Although the proportion of resistant CFUs in the marrow subsequently declined somewhat, an apparently stable level of approximately 10% has been found in one animal 153 days post-transplant (over 3 months after birth).

TABLE I. G418 resistant hematopoietic colonies one week after birth following in utero gene transfer.

Lamb	CFU-E	BFU-E	CFU-C	CFU-Mix
2519	557/2456 (23%)	385/1339 (29%)	28/137 (20%)	19/57 (33%)
2999	N.D.	140/840 (11%)	75/905 (12%)	10/52 (23%)
3000	N.D.	25/295 (8.5%)	25/260 (9.6%)	5/27 (19%)
Untr. Ctl.	48/5550 (0.9%)	7/1863 (0.4%)	6/735 (0.8%)	0/96 (0%)

We have also been using cynomologous and rhesus monkeys to develop a model system for testing human gene therapy protocols. In these experiments, the N2-derived vector SAX, containing an SV40 promoted human adenosine deaminase (ADA) cDNA in addition to the neo^R gene, has been used. Twelve monkeys have received SAX-treated marrow by autologous transplantation. Initial infection efficiencies of progenitors, as assayed by drug resistance in colony assays, have been about 10-25%. However, in vivo proportions of infected cells have been too low for vector DNA to be detected. Nonetheless, human ADA activity has been detected in six of seven monkeys analyzed. Although only trace amounts of human ADA could be detected in four of these animals, moderate amounts were found in the other two. In the monkey with the greatest activity, human ADA levels approximately 0.5% of endogenous monkey ADA were detected 103 days after transplant. In situ hybridization of bone marrow cells from this monkey using a probe for the neo^R gene has shown that 1 in 200 cells express neo^R mRNA. Overall, neo^R gene enzymatic activity has been detected in four of the six human ADA positive monkeys.

Therefore, we conclude that, in monkeys, the proportion of vector-containing cells in vivo appears to be low. However, the assay of human ADA combined with the in situ analysis suggest that the few cells infected are expressing the transferred genes. It has also been found that the level of measurable human ADA declines in these animals to negligible levels over a period of weeks. Whether this decline in activity is due to an inactivation of the vector or the supplanting of the infected pool of progenitor cells by an uninfected one is being investigated.

These animal models provide us with the means to analyze a variety of parameters in our efforts to determine the utility of retroviral vectors for gene transfer into somatic cells in vivo. In addition to the sorts of long-term in vivo stability studies mentioned above, these systems enable us to study gene regulation of specific genes in vivo in a variety of species. It also appears that infectivity with amphotropic vectors varies between species (our results, as well as 4,5). Therefore, these animals are also being used to determine optimal infection conditions.

REFERENCES

1. Keller, G., Paige, C., Gilboa, E. and Wagner, E.F. Nature, 318: 149-154 (1985).
2. Eglitis, M.A., Kantoff, P.W., Gilboa, E. and Anderson, W.F. Science, 230: 1395-1398 (1985).
3. Flake, A.W., Harrison, M.R., Azdick, N.C. and Zanjani, E.D. Science, 233: 776-778 (1986).
4. Hock, R.A. and Miller, A.D. Nature, 320: 275-277 (1986).
5. Kwok, W.W., Schuening, F., Stead, R.B. and Miller, A.S. Proc. Natl. Acad. Sci. USA, 83: 4552-4555 (1986).

GENE TARGETING TO THE IMMUNOGLOBULIN HEAVY CHAIN LOCUS IN A MOUSE B-CELL HYBRIDOMA LINE

H.Eibel and G.Köhler
Max-Planck-Institut für Immunbiologie
Stuebeweg 51, 7800 Freiburg, F.R.G.

1. Introduction.

Gene targeting of introduced DNA to homologous chromosomal sequences is a very rare event in comparison to unspecific integration. We chose an approach by which we can directly detect the repair of a mutated chromosomal constant region of the immunoglobulin mu gene by gene targeting using the hybridoma line igm662 (1). Two plasmid vectors carrying the wildtype mu allele and the Tn5 Neomycin resistance gen (G418^R) were constructed. A successful homologous recombination event between the introduced wildtype allele and the mutated mu gene will restore the functional mu gene expression in G418^R stable transfectants and result in IgM antibody production. As the recombination between plasmid and chromosomal sequences can take place either via a single crossover event or via a gene conversion process, we constructed different plasmid vectors which would allow the detection of both recombination mechanisms.

2. Results and Discussion.

We introduced the constructs shown in Fig.1 into the igm662 hybridoma line by electroporation (2). Stable transfectants were screened for wildtype antibody production by an agglutination assay and by RIA using labelled monoclonal anti-mu antibodies. Plasmid pSZ31, linearized with BamHI prior to transfection, allows the detection of single crossovers. Cleavage of pSZ31 with XbaI results in two fragments, one of them carries the mu constant part. Cotransfection of an excess of the c-mu fragment together with pRSVneo allows the detection of gene conversion between the c-mu carrying fragment and the mu gene in G418^R stable transfectants. Plasmid pXX518 has the selectable G418^R marker gene under the control of a heavy chain promoter integrated at the EcoRV site within the large exon of the c-mu part. The V$_H$ promoter activity in B-cell hybridomas is strongly dependent on the presence of functional enhancer sequences. The transfection with this plasmid should therefore reduce the number of stable transfectants which are due to unspecific integrations. Ideally, they should only arise from a homologous recombination event at the c-mu locus, because the incoming DNA fragments carrying the G418^R gene controlled by the V$_H$ promoter would be placed in close vicinity to its genuine enhancer sequences on chromosome 12. The results of the individual experiments are shown in table 1.

They demonstrate, that homologous recombination between the transfected plasmid DNA sequences and the immunoglobulin heavy chain locus is an extremely rare event. With a transfection rate of approximately 10^{-5} we did not detect a homologous recombination event in 7.6 x 10^8 transfected igm662 cells resulting in repair of the mutated c-mu gene. Possible explanations are:

i) Mitotic recombination via single crossover or gene conversion in these cells is very infrequent (about 1/10^6) as they rarely show changes in their antibody expression pattern which could be explained by recombination events (3).

ii) Electroporation results in a low number of integrated fragments in the host genome (2), which possibly reflects the small number of DNA fragments entering a transfected cell. The recombination frequency however, might be proportional to the number of introduced DNA molecules and consequently low in our experiments. Both potential drawbacks could be circumvented by introducing a high copy number of DNA molecules over a long period of time. This is accomplished by placing the DNA sequences to be targeted to the genome on autonomously replicating plasmids, e.g. on a Polyoma based vector. Experiments which are currently under way to examine this possibility.

Fig.1

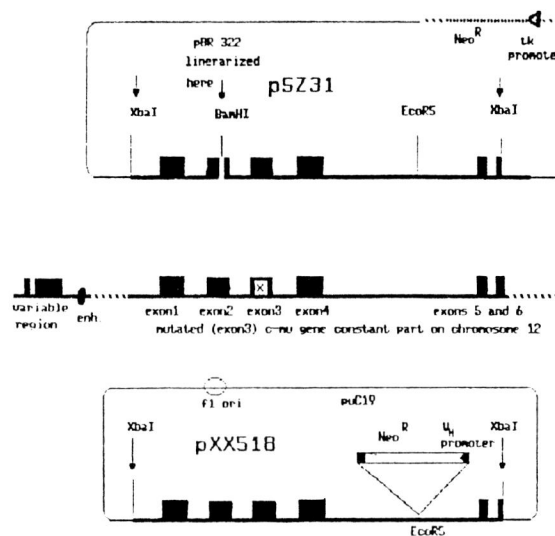

Figure: plasmid maps for pSZ31, the c-mu gene on chromosome 12, and pXX518.

Table 1. Analysis of Stable Transfectants

plasmid	cells/ exp. x 10^8	G418^R clones x10^{-5}	transfection rate	anti-TNP specific IgM$^+$ clones
pSPZ31 BamHI	1.1	>850	0.9	none
pSPZ31 XbaI	1.6	1100	1.1	none
pXX518 XbaI (dsDNA)	1.3	2	n.d.*	none
pXX518 (ssDNA)	3.6	11	n.d.*	none

*Using a different plasmid (p6.11, immunoglobulin enhancer, V$_H$ promoter, Neo^R, the transfection rate was comparable to the first two experiments.

References:
(1) B.Baumann, M.J.Potash,G.Köhler. (1985),The EMBO Journal 4, 351 - 359
(2) H.Potter, L.Weir, P.Leder. (1984), Proc. Natl. Acad.Sci. 81, 7161 - 7165
(3) U.Krawinkel et al.(1983), Proc. Natl. Acad. Sci. 80, 4997 - 5001

156

IN VITRO ENCAPSIDATION OF PARVOVIRUS DNA: DNA
SEQUENCE SPECIFICITY AND HOST CELL FACTOR REQUIRE-
MENT

E.A. Faust[1,2], R. Salvino[1,2], and K. Brudzynska[1]

Cancer Research Laboratory[1] and Dept. of Biochemistry[2], University of Western Ontario, London,
Ontario, Canada N6A 5B7

Parvoviruses are ubiquitous mammalian patho-
gens, distinctive for their linear, single-stranded
DNA genomes. The parvovirus replicative cycle
necessarily requires that one strand of a duplex
replicative intermediate be sequestered into pre-
formed empty capsids to form infectious progeny
virions (1-3). Here we describe the in vitro
encapsidation of a bacterial plasmid bearing 910 bp
from the 5'-end of the DNA genome of MVM, a murine
parvovirus (4). The encapsidation reaction is
strictly dependent on a DNA sequence within the 910
bp viral segment and requires a soluble host cell
factor(s).

(I) IN VITRO ENCAPSIDATION. A 910 bp AccI/Bam-
HI 5'-terminal fragment of MVM DNA obtained from
an infectious MVM clone (5) was transferred to
pML-2, a derivative of pBR322. The resulting plasmid, pRS5, was cleaved at its unique BamHI site and
incubated with crude extracts from Ehrlich ascites
mouse cells that had been infected with MVM.
Approximately 0.05 percent of the input plasmid DNA
became encapsidated and sedimented as a discrete
90S peak between empty capsids (70S) and infectious
virions (110S). Encapsidated pRS5 DNA was detected
in agarose gels following sequential treatment of
reaction mixtures with DNaseI (to degrade free
plasmid DNA) and NaDodSO$_4$-pronase (to deproteinize
the encapsidated plasmid DNA) (Fig. 1).

pRS5 DNA recovered from reaction mixtures was
completely digested by the single strand-specific
S$_1$ nuclease but was not cleaved by XbaI, whereas
input pRS5 DNA was quantitatively cleaved by XbaI
and was completely resistant to S$_1$ nuclease.

Encapsidation was catalyzed by crude S-10 and
P-100 fractions obtained from infected cells. An
S-100 fraction from infected cells was inactive as
was an S-10 fraction from uninfected cells.
Purified empty capsids exhibited a relatively low
level of activity that was enhanced at least 10-
fold by factors present in S-100 fractions from
infected cells and in S-10 fractions from uninfected cells, indicating a soluble host cell factor
requirement for parvovirus morphogenesis in vitro
(Table 1).

Fig. 1. Encapsidation of pRS5 DNA in vitro. Single-stranded pRS5 DNA released from capsids by
treatment with NaDodSO$_4$-pronase was analyzed by
agarose gel electrophoresis. Lanes a-c; S-10,
S-100, P-100. pRS5[i], input DNA; pRS5[e], encapsidated DNA.

(II) DEPENDENCE ON DNA SEQUENCE. The question
of sequence specificity was examined by comparing
different plasmids as substrates. pML-2 and pRS3,
a plasmid containing 1,084 bp from the 3'-end of
the MVM genome, were not encapsidated. By contrast,
pRS5, and two of its deletion derivatives, pRS5Bi
and pRS5ΔB, lacking nucleotides 4962-5002 and 4929-
5024 respectively, were each encapsidated to the
same extent (Table 1). Thus, a cis-dominant genetic element within the 5'-terminal 910 bp of MVM
DNA, excluding nucleotides 4929-5024, is required
for in vitro parvovirus encapsidation.

Table 1. Sequence specificity of in vitro encapsidation.

plasmid	S-10[i] 10	S-100[i]	P-100[i]	S-10	MVMeHA 32/160	MVMe32 S-10/100[i]
pRS5	+++	---	+++	---	-- +	+ +
pRS5Bi	+++		+++		+	
pRS5ΔB	+++		+++			
pRS3	---				-	
pML-2	---		---		-	

+++ indicates encapsidation activity; i, infection;
HA, haemaglutination; MVMe, empty MVM capsids.

REFERENCES

(1) Berns, K.I. and Hauswirth, W.W. (1984) in The
Paroviruses (K.I. Berns, ed.) P.P. 1-31 Plenum
Press, New York.
(2) Astell, C.R., Chow, M.B. and Ward, D.C. (1985)
J. Virol. 54, 171-177.
(3) Johnson, F.B. (1984) in The Parvoviruses (K.I.
Berns, ed.) p.p. 259-295 Plenum Press, New
York.
(4) Astell, C.R., Thomson, M., Merchlinsky, M. and
Ward, D.C. (1983) Nucl. Acids Res. 11, 999-
1018.
(5) Merchlinsky, M.J. Tattersall, P.J., Leary,
J.J., Cotmore, S.F., Gardiner, E.M. and Ward,
D.C. (1983) J. Virol. 47, 227-232.

ACKNOWLEDGEMENTS

We thank J. Morgan for expert technical assistance and E. Orphan for typing the report. We are
grateful to C. Astell and P. Tattersall for
providing E. coli JC8111 recBCsbcBrecF and pMM984
and to J. Hassell for pML-2. This work was funded
by the National Cancer Institute of Canada and the
Medical Research Council of Canada.

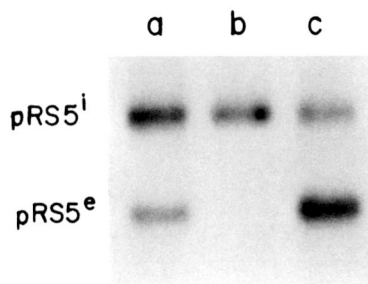

USE OF EPISOMAL REPLICONS FOR GENE EXPRESSION IN HL-60 CELLS

Christopher A. Hauer, Robert R. Getty, and Mark L. Tykocinski

Institute of Pathology, Case Western Reserve University, 2085 Adelbert Road, Cleveland, Ohio USA

The tripotential human leukemic cell line HL-60 can be induced to differentiate along neutrophilic, monocytic or eosinophilic pathways using a variety of chemical inducers (1). Transfection analysis offers a useful experimental approach for studying hematopoietic cellular differentiation programs in this and other inducible leukemic cell lines. To enable the derivation of stable HL-60 transfectants capable of high level expression of transfected genes, a feature of particular importance for anti-sense RNA expression work, we have explored the use of episomally-replicating eukaryotic shuttle vectors to drive gene expression in HL-60 cells. Here, we report the analysis of promoter function in HL-60 cells for a panel of promoter/enhancer elements inserted into an Epstein-Barr virus (EBV)-based replicon.

(I) PLASMID CONSTRUCTS. The starting plasmids were RSVCAT (2), I10CAT (3) and SV2CAT (4), which correspond to the Rous sarcoma virus, rat p3C5 gene ionophore-inducible and early SV40 promoters, respectively, linked to the prokaryotic chloramphenicol acetyltransferase (CAT) gene. We modified these three plasmids, each of which possesses a BamHI restriction enzyme site downstream of the CAT gene, by using BamHI linkers to insert an additional BamHI site upstream of the promoter/enhancer region. The promoter-CAT DNA segments were then mobilized by BamHI digestion and subcloned into the unique BamHI site of the high copy number episomal replication vector p220.2 (5). This plasmid vector contains the Epstein-Barr virus (EBV) oriP, a portion of the EBNA-1 gene and a resistance gene for the eukaryocidal antibiotic hygromycin (hyg). Constructs with the promoter-CAT inserts present in both orientations were recovered and designated A and B indicating either a proximal (A) or distal (B) positioning of the promoter relative to the EBV signals in the p220.2 vector. A representation of the basic plasmid design is illustrated in Figure 1.

Fig. 1. Schematic diagram of a promoter-CAT/A plasmid. P, promoter; CAT, CAT gene; hyg^R, hygromycin resistance gene; E, prokaryotic signals (pBR ori and amp^R); N, portion of EBNA-1 gene; O, EBV ori; B, BamHI site.

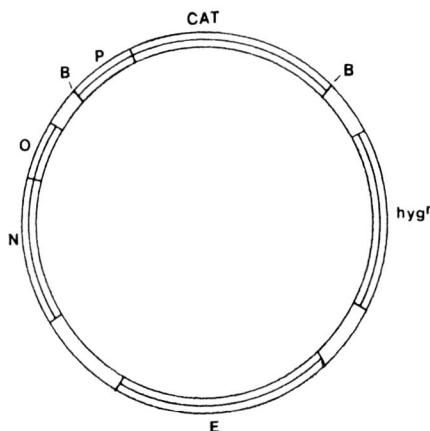

(II) PROMOTER FUNCTION IN EBV-BASED VECTOR. The promoter-CAT/220.2 plasmid constructs were transfected into HL-60 cells by protoplast fusion at a 5000:1 protoplast:target cell ratio in the presence of 50% PEG 1540 (Koch-Light, Ltd.). CAT assays (4) to evaluate ^{14}C-chloramphenicol acetylating activity were performed after 48 hours (Table I). Both RSVCAT/A and RSVCAT/B exhibited strong and comparable promoter activities. I10CAT/A demonstrated strong promoter activity as well and appeared to function maximally in the HL-60 cells even in the absence of its reported calcium-ionophore inducer A23187. In contrast to the RSV constructs, a significant promoter orientation effect was observed for this promoter in that minimal activity was observed for I10CAT/B. A similar orientation dependence was apparent for the weaker early SV40 promoter constructs, and here again the A orientation displayed greater promoter strength. The explanation for this orientation effect on promoter function remains to be determined. The relative activities seen for the RSV, I10 and SV2 promoters in p220.2 parallel our previous findings for these promoters alone in HL-60 cells (6). Transient CAT expression assays have thus established that promoter function is preserved in p220.2. Additionally, we have successfully derived stable hyg^R HL-60 transfectants using p220.2. Appropriate EBV-based eukaryotic shuttle vectors can now be assembled that will permit high level gene expression in HL-60 and perhaps other differentiating hematopoietic cells.

Table I. CAT expression in HL-60 transfectants. Estimates of CAT activities based upon densitometric scans of autoradiographs of ascending chloroform:methanol (95:5) thin layer chromatograms; calcium-ionophore inductions performed with 0.3 μM A23187 for 16 hours; + = negligible to low, ++ = moderate, +++ = high CAT activities.

plasmid	CAT activity level
RSVCAT/A	+++
RSVCAT/B	+++
I10CAT/A	+++
I10CAT/A + A23187 inducer	+++
I10CAT/B	+
I10CAT/B + A23187 inducer	+
SV2CAT/A	++
SV2CAT/B	+

REFERENCES

(1) Koeffler, H.P. (1986) Seminars in Hematology 23, 223-236.
(2) Gorman, C., Merlino, G., Willingham, M., Pastan, I. and Howard, B. (1982) PNAS 79, 6777-6781.
(3) Lin, A., Chang, S. and Lee, A. (1986) Mol. Cell. Biol. 6, 1235-1243.
(4) Gorman, C., Moffat, L. and Howard, B. (1982) Mol. Cell. Biol. 2, 1044-1051.
(5) Yates, J., Warren, N. and Sugden, B. (1985) Nature 313, 812-815.
(6) unpublished observations.

GENE TRANSFER INTO MURINE HEMATOPOIETIC STEM CELLS BY ELECTROPORATION

R. Narayanan, M.M. Jastreboff, C.F. Chiu and
J.R. Bertino

Department of Pharmacology, Yale University School
of Medicine, New Haven, CT 06510

INTRODUCTION. Gene transfer by high voltage electrical impulses (electroporation) is an efficient method for introducing genes into cells that are refractory to conventional transfection procedures (1,2). Although gene transfer by retrovirus vectors is highly efficient, the risks associated with retroviruses such as activation of the cellular protooncogene (3) or the spread of helper virus (4) may limit their application for human gene therapy. Hence we have been interested in transferring drug resistant genes into hematopoietic stem cells by alternative technologies to retrovirus mediated transfer. We show that genes can be transferred and expressed in mouse hematopoietic progenitor stem cells by electroporation.

MATERIALS AND METHODS. The apparatus used for electroporation has been recently described (2) and is similar to one used by Potter et al. (1). The plasmids used for electroporation RSVCAT and SV_2NEO have been described (5,6). Bone marrow was removed from 6-8 week old female CBA/J mice from both femurs by flushing with serum free Fischer's medium, centrifuged, resuspended in sterile cold PBS and viability determined by trypan blue. Linearised plasmid DNA (RSVCAT-Aat II or SV_2NEO-EcoRI) was added (100 ug/ml) to the bone marrow cells (2×10^7 cells/ml) in cold PBS. The DNA and cells were incubated for ten minutes at 4^oC in a 1-ml disposable cuvette (Biorad). Following electroporation (2 KV, 0.9 mA with current and wattage setting at 3 on a scale of 100) cells were diluted with 4 ml of complete medium and incubated at 37^oC for 48-72 hours and assayed for transient CAT (Chloramphenicol Acetyl Transferase; EC23.1.28) expression (5). For cotransfection of RSVCAT and SV_2NEO 1:1 molar ratio was used (6) and cells were plated onto methylcellulose with or without 2 mg/ml of G-418. Granulocyte-Macrophage (CFU GM) colonies were scored 10-14 days later with the aid of an inverted microscope. For in vivo spleen colony forming activity (CFU_S) assays, 48 hours after electroporation the cells were centrifuged, their viability determined, and $2-5 \times 10^4$ cells were injected into lethally irradiated (9 Gy) mice. The mice were sacrificed at 14 days, and tissues dissected (bone marrow, spleen, thymus, lymph nodes, and liver), and CAT activity was determined in the supernatant (5).

RESULTS AND DISCUSSION. CAT activity was present in mouse bone marrow cell extracts 72 hours after electroporation with RSVCAT. No CAT activity was detected in extracts of SV_2NEO transfected bone marrow cells. These results establish that genes can be transferred and expressed in primary bone marrow cells.

IN VITRO EXPRESSION. To examine whether genes could be introduced into hematopoietic stem cells, mouse bone marrow cells were transfected either with RSVCAT alone or cotransfected with 1:1 molar ratio of SV_2NEO and RSVCAT and then cultured in methylcellulose in the presence and absence of the neomycin analogue G-418 (2 mg/ml) for the CFU_{GM} assay. G-418 resistant CFU_{GM} colonies were pooled at 14 days and analyzed for CAT expression. CAT expression was readily detected in extracts of RSVCAT transfectants as well as in co-electroporated

bone marrow cells, whereas in the SV_2NEO transfected extracts of CFU_{GM} no CAT activity was detectable. CFU_{GM} recovery, following electroporation and plating in nonselective medium varied from 95-98%. The G-418 resistant CFU_{GM} colonies were morphologically indistinguishable from the control colonies. Hybridization of total cellular RNA from pooled G-418 resistant CFU_{GM} with NEO specific probe revealed a high level of NEO expression. No G-418 resistant CFU_{GM} were obtained from marrow cells subjected to electroporation in the presence of RSVCAT plasmid.

IN VIVO EXPRESSION. RSVCAT electroporated bone marrow cells were injected into the tail vein of lethally irradiated mice and fourteen days later mice were sacrificed and extracts from bone marrow, spleen, thymus, lymph node, and liver were assayed for CAT expression. CAT activity was detected in all of the hematopoietic tissues, but not in non hematopoietic tissues suggesting the stable transfer of this gene into hematopoietic stem cells. These results suggest the specificity of gene transfer into hematopoietic cells by electroporation.

In the present study, we chose to use the CAT gene which is absent in eucaryotic cells and can be assayed within 24-48 hours after introduction, to establish gene transfer by electroporation into primary bone marrow cells. The persistence of CAT expression in the CFU_{GM} and in the tissues of irradiated mice in the absence of drug selection implies that electroporation can be used to introduce nonselectable exogenous genes into hematopoietic cells, at least for short periods of time, and that the introduced genes are expressed in different hematopoietic lineages.

REFERENCES

(1) Potter, H., Lawrence, W. and Leder, P. (1984) Proc. Natl. Acad. Sci. USA 81, 7161-7165.
(2) Reiss, M., Jastreboff, M.M., Bertino, J.R. and Narayanan, R. (1986). Biochem. Biophy. Res. Commun. 137, 244-249.
(3) Blair, D.G., McClements, W.L., Oskarsson, M.K., Fischinger, P.J. and Vandewoude, G.F. (1980) Proc. Natl. Acad. Sci. USA 77, 3504-3508.
(4) Hock, R.A. and Miller, A.D. (1986) Nature (London) 320, 275-277.
(5) Gorman, C.M., Moffat, L.F. and Howard, B.H. (1982) Mol. Cell. Biol. 2, 1044-1051.
(6) Turkaspa, R., Teicher, L., Levine, B.J., Skoultchi, A.I. and Shafritz, D.A. (1986) Mol. Cell Biol. 6, 716-718.

REGULATION OF HUMAN ALPHA-1-ANTITRYPSIN GENE EXPRESSION IN TRANSFECTED CELLS AND TRANSGENIC MICE

R-F Shen, R.N. Sifers, J.C. Carlson, S.M. Clift, J.L. DeMayo, C.P. Hardick, F.J. DeMayo, D.W. Bullock and S.L.C. Woo

Howard Hughes Medical Institute, Dept. of Cell Biology and Institute of Molecular Genetics, Baylor College of Medicine, Houston, Texas, 77030.

Alpha-1-antitrypsin (AAT) is the major plasma serine-protease inhibitor which exhibits a wide spectrum of substrate specificity. Although the protease inhibitor is synthesized primarily in the liver, its major physiological function is to inhibit neutrophil elastase in the lung(1). Genetic deficiency of AAT predisposes affected individuals to development of chronic obstructive pulmonary emphysema. We recently cloned the normal and mutant AAT genes and demonstrated that the lesion is a point mutation in exon V of the gene (2). Using the chromosomal human AAT gene, we wish to study the cis-acting elements in the gene that specify its liver-specific expression.

(i) EXPRESSION OF THE HUMAN ALPHA-1-ANTITRYPSIN GENE IN TRANSGENIC MICE.

A 14Kb EcoRI fragment containing the entire human AAT gene plus 2.5 Kb of sequences flanking both the 5' and 3' termini was microinjected into mouse embryos to generate transgenic mice. Five F_0 mice bearing the human gene were obtained, and F_1 progenies were generated from 4 F_0 lines. Analysis of the serum proteins using an antibody preparation that is specific for the human antigen showed the transgenic mice expressed human AAT at concentrations ranging from 0.2 to 10 times that present in human plasma. S1 analysis of total RNA isolated from various organs of the transgenic mice showed that the human AAT gene was expressed in the liver of all transgenic lines, although in 2 lines the gene was also expressed in the kidney. These results indicate that the cis-regulatory elements controlling the liver-specific expression of the human AAT gene resides within the 14 Kb EcoRI fragment, and that these elements must be able to properly interact with the trans-acting factor(s) in the mouse liver in vivo.

(ii) EXPRESSION OF THE 5'AAT/CAT CHIMERIC GENE IN HUMAN HEPATOMA CELLS.

Eukaryotic gene expression is controlled by multiple cis-acting elements including the promotor, enhancer and other regulatory sequences. The interaction between specific cellular transacting factor(s) with the cis-acting elements presumably determines the expression of the gene in a tissue or cell specific manner. To investigate whether all the cis-acting elements reside in the 5' flanking sequence of the human AAT gene, a 1.2 Kb(-1200 to -30) 5' flanking sequence of the human AAT gene was inserted immediately 5' to the structural gene of chloramphenicol acetyltransferase(CAT) in the vector $pA_{10}CAT_2$. Transient expression of CAT activity was examined in transfected cells. It was found that CAT activity was expressed in two human hepatoma cell lines, HepG$_2$ and PLC/PRF/5, but not in Hela cells,

consistent with that observed in a previous report (3). The CAT activity was about 10 times greater in hepatoma cells transfected with the chimeric construct than those transfected with the pA10CAT$_2$ vector alone, suggesting the presence of trans-acting factors in the hepatoma cells that interacted with the 5' flanking sequence of the AAT gene. The enhancement of CAT activity by the 5' flanking sequence of the AAT gene was observed only when the 5' AAT sequence is positioned in the sense orientation with the CAT gene. Reversing the orientation abolishes completely the fragment's ability to enhance expression of CAT activity. To determine the minimal length of the 5' sequence necessary for cell specific expression of the CAT gene, a series of 5' deletion mutants were constructed and assayed in hepatoma cells for expression of CAT activity. As shown in fig. 1, deletion of the 5' sequence to -730 bp resulted in a slight but consistant increase of CAT activity, while the deletion to -345 bp retained most of the CAT activity. Further deletion to less than -345 bp drastically diminished the enhancing effect on the expression of the CAT gene. Again, the enhancement by the 5' sequence was orientation-dependent. The results suggest that the 5' flanking sequence of the AAT gene may contain multiple cis-acting elements within a few hundred bp 5' of the cap site of the gene.

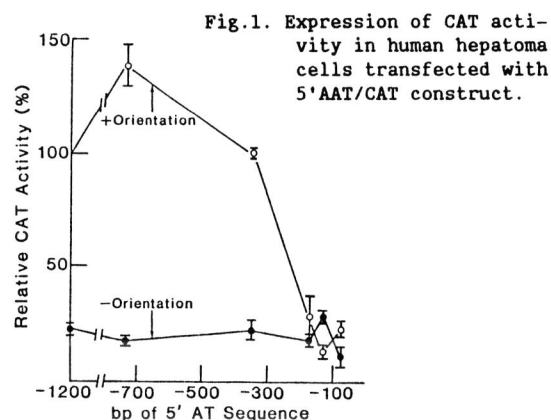

Fig.1. Expression of CAT activity in human hepatoma cells transfected with 5'AAT/CAT construct.

In order to verify whether the 5' flanking sequence of the human AAT gene is sufficient to confer specificity for liver exression in vivo, we have generated transgenic mice that harbor 700 bp of the 5' flanking sequence of the gene fused with the CAT gene. Analysis of the expression of the CAT gene in various tissues of these transgenic mice both qualitatively and quantitatively will provide a clue to the regulation of liver-specific AAT gene expression.

REFERENCES

(1) Laurell, C.B. and Erickson, S. (1963). Scand. J. Clin. Lab. Invest. 15, 132-140.

(2) Sifers, R.N. and Woo, S.L.C. (manuscript in preparation).

(3) Ciliberto, G., Dente, L. and Cortes, R. (1985). Cell 41, 531-540.

CHANGES IN UV-SENSITIVITY OF THE IN VITRO FERTILIZED MOUSE ZYGOTES DURING THE PRONUCLEAR STAGE

T. Yamada[1], H. Ohyama[2], K. Okuda[3] and N. Uke[3]

Div. [1]Biol. & [2]Radiat. Health, National Institute of Radiological Sciences, Anagawa 4-9-1, Chiba 260, Japan.
[3]Dept. Biol., Faculty of Science, Toho University

During the past decade, a number of techniques for manipulating the mammalian embryo have been developed that allow production of the transgenic mice. Culture of the fertilized egg, the zygote, in vitro is one of the most important techniques among them. Culture initiated from the zygote stage to blastocyst stage can only be achieved under the most stringent conditions, because mouse embryos at the zygote stage are highly sensitive to many environmental agents including UV-light routinely used in laboratories. The present study was therefore undertaken to quantitatively measure ultraviolet light sensitivity of the mouse zygotes, using the in vitro fertilization technique with higher efficiency we developed recently[2].

MATERIALS AND METHODS. Fertilization of the mouse eggs in vitro and subsequent culture of the zygotes were carried out as described previously[2]. Briefly, eggs were obtained from $BC3F_1$ (C57BL/C3H) strain females hormonally superovulated. Sperm had been obtaied from cauda epididymis of ICR male mouse and capacitated by incubation with the fertilization medium[2] at 37°C for 1 hr. Sperm suspension was added to the medium containing eggs for insemination. At 3 to 5 hr after insemination, the fertilized eggs, which were identified by the extrusion of the 2nd polar bodies, were washed and transferred to the culture medium[2]. Efficiency of blastocyst formation of the normal zygotes was more than 95 %.

UV irradiation of the zygotes was performed at various times after insemination. The zygotes in the culture medium without bovine serum albumin were irradiated with 1 to 12 J/m^2 UV from HITACHI GL germicidal lamp, and its 254 nm-light intensity at the irradiation spot was measured by an UVX-Radio-meter (Ultra Violet Product, Inc., CA. USA). After UV-irradiation, the zygotes were allowed to develop in vitro for about 4 days to blastocyst stage.

With development to the expanded blastocyst stage as the end point, the LD_{50}, the UV fluence causing 50 % lethality, was determined by the probit transformation of the data using a computer program.

RESULTS AND DISCUSSION. The in vitro fertilization technique allows irradiation of mature germ cells just before fertilization or zygotes at a precise time after fertilization. Using this system,

we measured the radio- or UV-sensitivity of mouse zygotes as a function of time after fertilization to the first cleavage. Fig. 1a shows the normal expanded blastocyst developed from the in vitro fertilized zygotes. When the zygotes were exposed to UV-light, the embryos became arrested at the stage earlier than blastocyst stage, degenerated and exhibited abnormal features shown in Fig. 1b.

The half lethal UV fluence (LD_{50}) values were plotted against the time of UV-irradiation after insemination (Fig. 2). The UV sensitivity expressed as LD_{50} varied markedly depending on the times of irradiation. The UV sensitivity was maximum (LD_{50} is 1.4 J/m^2) at 4 hr after insemination which corresponds to the beginning stage of pronuclear formation. Thereafter, the sensitivity decreased progressively through the completion of the pronuclear formation to minimum at 12 hr after insemination (LD_{50}; 10.7 J/m^2). The sensitivity increased again with progression of development to the first cleavage.

The results indicate that the zygotes have the highest sensitivity to UV at the beginning of pronuclear formation stage, and the lowest at the end of pronuclear stage. The pattern is exactly similar to the pattern of the X-irradiated zygotes[3], and is in contrast to that of UV-irradiated somatic cells[4].

The present study is the first successful attempt to determine the half lethal UV fluence for the zygotes at the various phases from fertilzation to the first cleavage. UV sensitivity varies markedly depending on the time of irradiation. Resistancy of the zygotes for manipulation must vary depending on the time of manipulation.

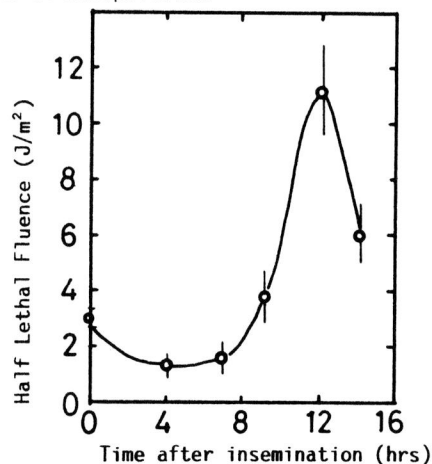

Fig. 2. Changes in UV sensitivity of the in vitro fertilized mouse zygotes during the pronuclear stage. The zygotes were irradiated with UV light at the times indicated on abcissa and then allowed to develop in vitro to blastocyst stage. The half lethal fluence (LD_{50}) values were determined as shown in the text. Bars represent 95 % confidence limits. The first chromosome formation and cleavage division begin at about 15 and 17 hr after insemination respectively.

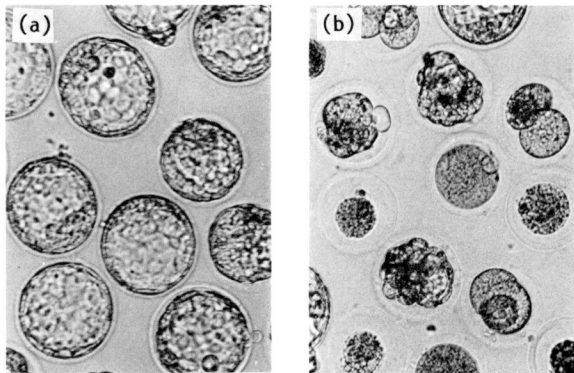

Fig. 1. Normal expanded blastocysts (a) developed from the zygotes fertilized in vitro. Abnormal development of the embryos irradiated with UV light at the zygote stage (b).

REFERENCES

(1) Pedersen, R.A. and Goldstein, L.S. (1979) Genetics 92, s141-s151.
(2) Yamada, T., Yukawa, O., Asami, K. and Nakazawa, T. (1982) Radiat. Res. 92, 359-369.
(3) Yamada, T., Yukawa, O., Matsuda, Y. and Ohkawa, A. (1982) J. Radiat. Res. 450-456.
(4) Eddington, A. (1982) in The Cell Division Cycle Temporal Organization and Control of Cellular Growth and Reproduction (Lloyd, D., Poole, R.K., and Edwards, S.W. Eds.) pp. 419-442.

AUTHOR INDEX